THE GOLD HUNTER'S HANDBOOK

Panning for gold on a north Georgia mountain stream.
(Tourist Division, Georgia Department of Industry and Trade)

THE GOLD HUNTER'S HANDBOOK

George Sullivan

STEIN AND DAY/*Publishers*/New York

First published in 1981
Copyright © 1981 by George Sullivan
All rights reserved
Designed by Louis A. Ditizio
Printed in the United States of America
STEIN AND DAY / *Publishers*
Scarborough House
Briarcliff Manor, N.Y. 10510

Library of Congress Cataloging in Publication Data

Sullivan, George, 1927-
 The gold hunter's handbook.

 Bibliography: p.
 Includes index.
 1. Gold. 2. Prospecting. I. Title.
TN271.G6S84   622′.3422   80-5718
ISBN 0-8128-2788-0

Contents

Acknowledgments

I AM GRATEFUL to the many people who helped to make this book possible. Special thanks are due Charles Garrett, Garrett Electronics: Frank Forrester, U.S. Geological Survey; Bob Erwin and Don Albrecht of Auburn, California; Jerry Keene, Keene Engineering; Guy Schroeder, Wade Hampton, A. J. DuMais, Aime LaMontagne, Bill Sullivan, and Tim Sullivan.

Others who have been particularly helpful are: Thomas J. Joiner, Alabama State Geologist; Margaret I. Erwin, U.S. Geological Survey, Anchorage, Alaska; Tom McGarvin, Arizona Department of Mineral Resources; John T. Alfors, California Division of Mines and Geology; Stephen W. Kline, Georgia Geologic Survey; Earl H. Bennett, Idaho Bureau of Mines and Geology; James C. Bradbury; Illinois Institute of Natural Resources; R. Dee Rarick, Indiana Geological Survey; Ray Anderson, Iowa Geological Survey; Carolyn A. Lepage, Geologist, Maine Department of Conservation; Karen Kuff, Geologist, Maryland Geological Survey; John Splettstoesser, Minnesota Geological Survey; Willis M. Johns, Montana Bureau of Mines and Geology; Arthur W. Hebrank, Missouri Division of Geology and Land Survey; Richard B. Jones, Nevada Bureau of Mines and Geology; Robert Davis, New Hampshire State Geologist; Robert M. North, New Mexico Bureau of Mines and Mineral Resources; P. Albert Carpenter III, Senior Geologist, North Carolina Geological Survey; John P. Bluemle, Senior Geologist, North Dakota Geological Survey; M. Civis, Oklahoma Geological Survey; Arthur A. Socolow, Pennsylvania State Geologist; Agnes Streeter, South Carolina Geological Survey; J. P. Gries, South Dakota School of Mines and Technology; James L. Moore, Geologist, Tennessee Department of Conservation; Palmer Sweet, Virginia Division of Mineral Resources; W. Dan Hausel, Geological Survey of Wyoming.

Unattributed photos are my own.

G. S.

CHAPTER 1

The Great Gold Rush

NOT LONG AFTER nine o'clock on a warm, sun-filled July morning, Tim Hurley, standing thigh deep in the north fork of central California's American River, tugs hard on the rope, and kicks the 5-hp, 4-cycle Briggs & Stratton to life. His partner, grasping the hose in both hands, checks for suction, nods, and then eases the nozzle beneath the surface to begin sucking mud and gravel from the bedrock.*

The next several hours, with only a break for lunch, the two men will spend tending their dredge, a floating vacuum cleaner that pulls up tons of river bottom and passes it over a sluice box that screens and separates the mixture, trapping what looks to be a black sludge. After panning that sand, Tim knows that he will end up with some specks of glittering metal—gold.

Tending a dredge on the north fork of the American River in central California.

---

*See Glossary
for unfamiliar terms or technical usages.

Hurley, 36 years old, a Sacramento school teacher, is not expecting to get rich. Gold hunting is a summertime recreation. "It's like fishing," he says. "You're out in the open; you're getting some exercise. And when you see a few flecks of gold in the bottom of your pan, and you almost always do, it's better than hitting a trout."

Scattered up and down the river from Tim Hurley and his partner are other miners operating dredges, minding sluices and rockers, and panning. Everyone pans.

They're part of an army of thousands of prospectors who are prowling the rivers and streams, hills and valleys, and even the desert areas of America, participants in the great gold rush of the 1980s. All kinds of people are involved. There are full-time prospectors, professionals, individuals to whom gold seeking is an adventure-filled vocation. There are families in campers and trailers, vacationers who are hoping they find enough gold merely to cover expenses. There are the weekenders who pack their gear into their vans or pickups every Friday afternoon and head up into the gold country.

How many people make up this army is anybody's guess. California's Department of Fish and Game, which issues permits to operate suction dredges, began experiencing a sharp upswing in applications in the late 1970s, with the number reaching 10,000 in 1979, and beyond that today. You don't need the official statistics, however, to realize there are many thousands of people involved. Visit the American, Bear, Yuba, or any one of several other rivers in California's Mother Lode country on a Sunday morning, and the *pucka-pucka* of gasoline engines documents the boom.

Another statistic is provided by the Bureau of Land Management of the federal Department of the Interior, which reports that there are now more than 66,000 gold claims on file in California alone. During the summer of 1980, in the city of San Francisco, it was impossible to purchase the form titled "Filing a Placer Claim." Stationery stores that supply legal forms had exhausted their supplies. Claims that do exist change hands at fees of $5,000 and more. Of course, tens of thousands of recreational miners pan for gold on a casual basis, without any claim or any thought of ever entering one.

The membership roster of the Gold Prospectors Association of America also gives evidence as to what's happening. The organization now boasts 52,000 members, an all-time high. The association says that for each member there are 30 nonmembers out on the streams and rivers, meaning the number of active prospectors in America is something over 1.5 million.

There have always been weekend gold hunters in California and other Western states as well—Colorado, Oregon, Washington, South Dakota, to mention only some of them. But the wave of gold seekers now involves a number of Eastern states too. Prospectors are active in the Green Mountains of Vermont, on half a dozen streams in

Maine, on the eastern slope of the Blue Ridge Mountains of Virginia, in the Piedmont and Mountain regions of North Carolina, once the nation's leading gold-producing state, and in north Georgia, near Dahlonega, in particular, where the gold deposits led the federal government to establish a mint in 1832.

It is a matter of fact that 32 of the 50 states have enough gold to make it worth searching for. (The Appendix of this book contains a state-by-state and, for Canada, province-by-province report on exactly where gold is found or geologically indicated.)

What's triggered all of this activity is the mushrooming price of gold. The gold that started the rush of 1848 was pegged at $12 an ounce. Most miners of that day bothered only with larger nuggets. When the rich deposits ran out, they packed up their mules and headed for a new site.

Many decades later, when the government revalued gold at $35 an ounce, mining was still not an attractive proposition. In fact, things didn't begin to change until 1974, the year the federal government went off the gold standard and allowed the price to float. At the same time, restrictions that had to do with the private ownership of gold were lifted. The results were almost immediate: Gold zoomed to over $150 an ounce.

In 1978, the price began to rise steadily. Inflation, and a general perception among investors that the Carter administration was powerless to deal with it, triggered this stage of price escalation, pushing the metal to $242 an ounce on the eve of the administration's decision in November 1978 to sell some gold to defend the dollar.

Don Albrecht displays nuggets he found in rivers and streams of California's Mother Lode country.

When the price climbed to more than $300 an ounce in 1979, professional miners started getting interested. Gold mines began reopening and tailings from old mines were reworked. "There's practically a geol-

ogist for every rock," said Robert S. Shoe-
maker, a metallurgist with a large mining corporation operating in Nevada. "Helicopters are flying around all over the place. They're staking claims by helicopter."

But $300 an ounce was really only a jumping-off point. The Shah of Iran was
deposed, OPEC kept boosting oil prices, Iranian students captured the U.S. Embassy
in Tehran and 50-odd American hostages, and the Russians invaded Afghanistan.
Gold prices shot up, breaking the $500 barrier in December 1979. They cracked
through the $600, $700, and $800 barriers the next month. That spring, when fledgling
prospectors visited their local mining equipment shops to buy their pans, shovels,
dredges, and other paraphernalia, they had to stand in line to get waited on.

Latter-day prospectors have an edge over their predecessors, and it isn't only in
terms of the price they get for the metal. "The old-time prospectors didn't know that
much," says a Downieville, California, miner. "They didn't know geology. They didn't
have any state or federal geological experts to help them, to provide them with maps
and books."

Placer miners pan out sand obtained from their gold rocker (*foreground*).

Another factor that works to the advantage of today's miner has to do with the land itself. The Forty-Niners and their immediate successors had to cope with land that was virtually untouched. Only where a major upheaval in the earth had occurred or a river cut through the terrain could they get any idea of the geological structure of an area. But today, with the never-ending construction of highways, homes, and commercial enterprises, the land is constantly being opened up and its secrets revealed. As a result, there is a greater chance of finding a productive quartz deposit or related placer concentrations than there was a century ago.

The mining equipment available today represents another advantage the modern prospector has. Modern-day power suction dredges enable him to clean out bedrock fissures with enormous efficiency. Sophisticated metal/mineral detectors aid in finding nuggets in streambeds, ore-bearing veins, and black sand deposits.

The Forty-Niners' basic tool was the gold pan, and it's still as necessary to your success as a good breakfast and a warm place to sleep. But the gold pan has gone through some startling changes in the last decade. It is as likely to be made of plastic as metal, and recessed within the inner surface of the pan sides are riffles, or "traps," that make it much easier to capture the tiny gold particles and flakes.

But despite all the geological information now at the miner's disposal and despite all equipment advances, only about one individual in a thousand makes a living at mining gold, and those who do move thousands of tons of ore to get the few ounces of gold that represent their profit.

This shouldn't discourage you, however. Gold hunting is fun; it provides a special excitement. And most of the gold-bearing regions of the country are to be found in areas of breathtaking scenic beauty. The scenery itself can be worth the effort.

Can a novice prospector find gold? I put that question to Bob Erwin, a geology teacher in the town of Auburn, California, who has been prospecting for more than two decades, as we stood together on the bank of the American River. "It's not hard to find gold," said Erwin. To demonstrate, he filled his pan with sand and gravel taken from the nooks and crannies where the river's water might have left gold particles eroded from chunks of quartz upstream. Then he submerged the pan and began plucking out the rocks.

Constantly swirling the pan and shaking it from side to side, and occasionally letting the lighter-weight materials spill over the edge, Erwin soon had nothing left in the pan but black silt. "Nothing left but the heavy stuff," he said, "the iron pyrite and the gold—if there is any."

He kept rocking the pan and deftly tilting it, allowing small portions of black sand to wash over the pan edge. Then he drained off the water and swirled the remaining sand around in the pan bottom, and there, gleaming in the bright sun, were several tiny particles of gold.

"It's not hard to find gold," Erwin repeated. "The problem is finding a lot of it."

About Placer Mining

If you're an Easterner, your image of a gold miner is likely to be that of a grizzled, bearded prospector who ekes out his existence by extracting gold-bearing ore from an excavation in the ground. Forget that. That's *lode* mining. This book has little to do with lode mining, which involves underground mining. It concerns *placer* (rhymes with "passer") mining.

Gold is found in many different types of rocks and geological environments. Whenever deposits of gold occur in hard rock, they're classified as lode gold. They're primary deposits.

Lode gold is subject to weathering and erosion. The rocks and ore-bearing minerals disintegrate into mineral grains and are washed into the rivers and streams. Being relatively heavy and resistant to corrosion, the gold particles accumulate amidst the sediments of the streams. These are secondary deposits—the placer deposits that will be our chief concern.

In other words, the natural process of the streams works in much the same way as a miner's pan or sluice box in sorting and concentrating the heavy minerals and gold particles. Prospectors thus look for gold wherever these heavy minerals have accumulated.

There's really not much chance that you, as a novice prospector, are going to discover new placer deposits of gold. The best opportunity you have of finding gold is by working those areas where placer concentrations have been found in the past. While

Typical placer mining area on California's American River.

early miners often worked the bedrock, as you'll be doing, much of it was carelessly or improperly cleaned, and some gold was usually left behind. In addition, "new" gold accumulates anytime a river floods. In recent years, quite satisfying amounts of placer gold have been recovered by individuals who carefully cleaned the bedrock in old-time placer districts.

Placer mining in the United States spans a period of more than 200 years. As early as 1792, gold was being taken from the placers of the southern Appalachian region, but these deposits, although rich, were relatively small and quickly depleted, at least from a commercial standpoint.

The earliest production in the West occurred at Old and New Placer Diggings near Golden, New Mexico, in 1828. Several other deposits were mined in the years that followed but all were overshadowed by James Marshall's discovery of gold at Sutter's mill on the south fork of the American River at Coloma, California, on the morning of January 24, 1848. Marshall placed the gold particles in the crown of his hat and hurried over to where a group of carpenters were working. "Boys, by God," said Marshall, "I believe I've found a gold mine."

"What is it?" asked one of the men.

"Gold," said Marshall.

"Oh, no, it can't be," said another man.

A third carpenter took one of the particles from Marshall and tested it by biting on it. Holding up the glittering yellow specimen, he said, "Gold, boys! Gold!"

The greatest gold rush of all time followed. Diggers by the tens of thousands descended on California's Sierra Nevada in 1848 and 1849, coming from as far away as Europe, China, and Australia. By 1851, the Sacramento Valley had yielded some 2.5 million ounces of gold. By the end of the decade, some 24 million ounces had been extracted. About half a million people immigrated to California during that period.

Early-day miners used shovels and picks to gather the sand and gravel, then washed it in sluices or rectangular wooden boxes called "cradles." "Rocking the cradle" was somewhat like panning on a large scale. Sand and gravel were shoveled into the top of the box. As water was then poured in, the cradle was moved briskly from side to side. Wooden riffles over a canvas or burlap mat that lined the bottom of the cradle captured the gold particles.

As competition grew, more efficient methods of separating the gold from the sand and gravel were devised. The sluice box, a riffle-lined trough for washing sand and gravel, evolved into the Long Tom, a series of troughs that sped up the sluicing process.

The streambeds of the Mother Lode country were sifted over and over again. After the first miners had moved on, Chinese laborers arrived upon the scene, and picked

Sutter's mill as it appeared to a nineteenth century artist. (*New York Public Library*)

Gold-rush miners of 1849. (*New York Public Library*)

Placer mining in 1858 (*New York Public Library*)

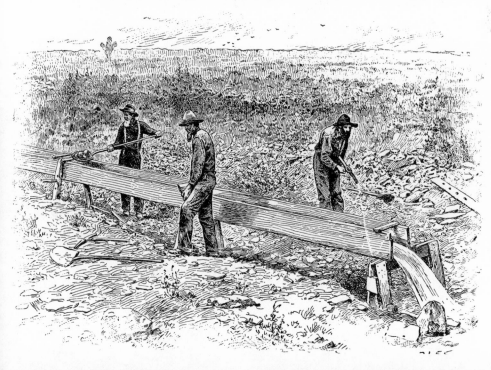

the streambeds clean of just about anything the Forty-Niners and their successors had happened to leave behind.

In 1859, in Virginia City, Nevada, two indigent miners found a gold reef, but sold their claim to Harry Comstock. The Comstock Lode was America's chief source of gold for 20 years.

During the 1870s, hydraulic mining became the order of the day. Powerful jet blasts of water were turned upon the mountains, ripping them apart, to get at the gold inside. If hydraulic mining had not been stopped by law in 1884, the Mother Lode country might be one enormous rubble pile today.

In 1874 and 1875, rumors of gold lured miners to the Black Hills of South Dakota. Out of these discoveries evolved the largest gold mine in North America, the Homestake. Still producing, it is spread over 200 miles. It yields more than 300,000 ounces of gold per year, making South Dakota the leading state in gold production.

Gold finds were reported in Alaska as early as 1848. It wasn't until 1897 and 1898, however, that the rich placer discoveries of the Klondike—in Canadian territory— were made. More than 60,000 men fought their way into the frozen Klondike wastes; fewer than 10 percent of them found gold.

South Dakota's Homestake Mine, the biggest gold mine in North America. (*Public Affairs Department; Homestake Gold Mine*)

Miners stampeded to the Nome area of Alaska in 1898 to exploit the rich discoveries there. In 1902, they were drawn to Fairbanks, where the last placer deposits of any importance were found.

The Carlin Mine, near Carlin, Nevada, opened in 1965, represents the biggest gold discovery since the 1930s in the United States. This lode deposit was found through geological research conducted by the U.S. Geological Survey.

Where base metals such as copper, lead, and zinc are deposited in veins, small amounts of gold are usually deposited with them. More than one third of the gold recovered in the United States today is obtained as by-product in the processing of other ores. Gold recovered from copper ore mined at the vast open-pit mine at Bingham, Utah, for example, amounts to almost as much as that produced by the aforementioned Homestake Mine.

The federal government keeps most of its gold—about 147 million troy ounces—in the vault of the Fort Knox Bullion Depository, located about thirty miles southwest of Louisville, Kentucky. The gold takes the form of bars about the size of ordinary building blocks. Each weighs approximately 400 troy ounces (27½ pounds).

Although Fort Knox has achieved widespread fame as a gold depository, a far greater accumulation of gold is to be found under Nassau Street in the Wall Street area of New York City, specifically, in the vaults of the Federal Reserve Bank of New York. About 366 million troy ounces—more than 25 percent of all the official monetary reserves of the non-Communist nations—is held by the Bank. About 10 million troy ounces of United States government gold is also on deposit there. The Bank's gold vault is a popular tourist stop, incidentally, attracting about 16,000 visitors a year.

In terms of total gold production, the United States ranks fourth among the nations of the world, behind South Africa, the Soviet Union, and Canada. South Africa, in fact, turns out about two thirds of the world's supply, with about four tons a day coming from the Rand Refinery in Johannesburg.

South African is used for investment. Gold mined in Russia, because it is purer, is used for jewelry. It is 999.9 parts per thousand pure as opposed to 999.5 parts per thousand pure in the case of investment gold.

Gold held in the Federal Reserve Bank of New York for foreign accounts. (*Federal Reserve Bank of New York*)

The province of Ontario ranks as North America's greatest source of gold. Most of the gold mined there is found in a large, deep quartz gold-carrying vein. British Columbia, the Northwest Territories, and Quebec are other important gold-producing provinces of Canada. Other leading gold-producing countries are Peru and Australia.

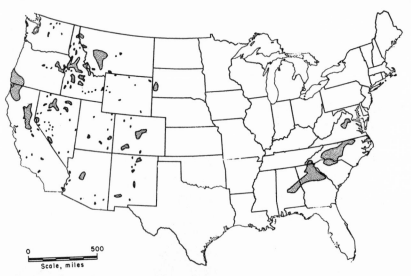

Areas of conterminous United States where gold has been commercially mined from placers. (*U.S. Bureau of Mines*)

The United States produces about 45 tons of gold annually. But little of this is placer gold in the common sense. Most of it is the placer gold that has been commercially produced, primarily as a by-product of the washing of sand and gravel for use as aggregate by the construction industry.

According to statistics compiled by the Bureau of Mines of the Department of the Interior, California and Alaska are the leading states in placer-gold production, accounting for more than two thirds of all the placer-gold output between the years 1792 and 1968. But government statistics concerning the amount of gold derived from placer operations have to be questioned. People recovering gold from rivers and streams frequently squirrel it away. It's their hedge against the dollar's decline in value and other economic upheavals. Or, as one miner told me, "I just like to look at it."

If they do sell what they find, they deal directly with a jeweler or with the owner of a mining equipment store. Dive shops on the West Coast often deal in the metal. In California, there is known to be a sizeable underground market in the buying and selling of native gold, with all transactions in cash. How much placer gold is being recovered today is anybody's guess.

Leading States in Placer-Gold Production

1. California	7. Colorado	13. North Carolina
2. Alaska	8. Georgia	14. Utah
3. Montana	9. New Mexico	15. South Carolina
4. Idaho	10. Arizona	16. Virginia
5. Oregon	11. South Dakota	17. Wyoming
6. Nevada	12. Washington	18. Alabama

Gold Facts

Placer gold is found in many different forms and sizes. The most sought-after size has a well-known name—*nugget.* Technically speaking, a nugget is a lump of native gold that is one grain in size or larger.

Anything smaller than a nugget is called a *color,* defined as any particle of gold large enough to be seen readily by the naked eye. When you pan for gold, that's what you hope to get—colors.

Other terms used in classifying particles of gold are as follows:

• Coarse gold—gold that will not pass through a 10-mesh screen (about the size of a grain of rice).

• Medium gold—gold that will pass through a 10-mesh screen, but not a 20-mesh screen (about half the size of a pinhead). Medium gold averages about 2,000 particles to a troy ounce.

• Fine gold—gold that will pass through a 20-mesh screen, but not a 40-mesh screen (about one-fourth the size of a pinhead). Fine gold averages about 12,000 particles to the troy ounce.

• Very find gold—gold that will pass through a 40-mesh screen. Very fine gold averages about 40,000 particles to the troy ounce.

• Flour gold—particles of gold that are powdery fine. Flour gold won't be trapped by the riffles in a sluice box and floats away when attempts are made to pan it.

Gold is widely measured in *troy ounces.* The *troy system of weights* is based on a *pound of 12 ounces* and an *ounce of 20 pennyweights* or *480 grains.*

A troy ounce is about 10 percent heavier than an avoirdupois ounce. To convert avoirdupois ounces to troy ounces, multiply by 0.911.

One troy ounce of pure gold has a volume equal to a cube that is 0.464 inch (11.8 mm) to a side.

A *karat* is a unit used for measuring the fineness of gold. Pure gold is 24 karats fine.

A ring marked "24 karat" is 100 percent gold; "14 karat," the American standard, is fourteen parts gold and ten parts of some other metal; "10 karat" is ten parts gold, and

Much of the gold recovered today becomes investment gold. (*St. Joe Minerals; Ed Goldstein*)

fourteen parts another metal. Anything less than 10 karat can't be called gold in the United States.

The term *karat* is believed to be a holdover from ancient times, when carob-tree seeds were used to balance scales in Oriental bazaars. The word for "carob" in Arabic is *qirat*; in Greek, *keration*; and in Italian, *acrato.*

The term *fineness* is also used to refer to the proportion of pure gold in an alloy. It is expressed in parts per thousand. For instance, a gold nugget containing 885 parts of pure gold, 100 parts silver, and 15 parts of copper would be considered 885-fine.

Placer gold invariably occurs as an alloy with silver. In fineness it usually ranges from 700- to 950-fine.

The term *gold filled* is used to describe articles of jewelry made of one base metal or another that are coated with one or more layers of gold alloy. If the gold alloy portion does not amount to at least 1/20 of the total weight of the article, it may not be marked gold filled.

Electroplated jewelry items carrying at least 7 millionths of an inch of gold on significant surfaces may be labeled *electroplate.* Plated thicknesses of less than this amount may be marked *gold flashed* or *gold washed.*

Gold's natural color is yellow. If, in manufacturing jewelry, both copper and silver are added to gold, it remains yellow. Pink or red gold, popular in Europe, is achieved by adding copper only. White gold results when copper, nickel, and zinc are added. Green gold is created when silver and small amounts of copper and zinc are added.

Chemists have given gold the chemical symbol Au, from the Latin *aurum,* meaning gold.

Gold has a specific gravity of 19.3 when pure, a density exceeded only by that of platinum. Although it is relatively soft—2.5 to 3 on the Mohs scale of hardness—gold has a high melting point, $1,063°$ C ($1,945°$ F), and a boiling point of $2,600°$ C ($4,712°$ F).

Pure gold is one of the most malleable and ductile of all metals. The ancient Egyptians knew how to hammer gold into leaf so thin that it took 367,000 leaves to make a stack one inch high. A single ounce of gold can be drawn into a hairlike wire 60 miles long.

Gold is impervious to air, heat, or moisture, and is affected by only a handful of chemicals. It can be dissolved chemically in aqua regia, an acid compound of three parts hydrochloric acid and one part concentrated nitric acid. The result is chloauric acid, from which gold chloride is derived.

Native gold is not difficult to sell.

A dilute solution of sodium or calcium cyanide can also be used to dissolve gold. This solvent, in a procedure known as the *cyanide process,* is used in extracting gold from massive deposits of low-grade ore.

Of the 297 tons of gold that were consumed in 1978, the last year for which figures are available, 128 tons went into jewelry, 105 tons were used for investment purposes, 41 tons in the electronics industry, and 23 tons in dentistry.

Physicians now use gold as a healing agent in the treatment of rheumatoid arthritis, and gold leaf is used in treating chronic ulcers. Radioactive gold is used in the treatment of cancer.

Gold has also been found to have value in conserving energy. When a thin film of gold is applied over a window, it reflects up to 92 percent of the glare of the sun and it improves the insulation quality of the glass. It thus cuts air conditioning costs. It also reduces heating costs by reflecting central heat back into rooms. Gold-coated windows are now in use in more than two hundred American offices.

One of the chief reasons that gold is so valuable is because of its relative scarcity. Being one of the most durable metals known, it lasts indefinitely, but, as centuries pass, it becomes harder and harder to find and extract from the earth. It has been estimated that all the gold mined in the world over the past 6,000 years amounts to about 80,000 metric tons, equal to a 2½-foot slab the size of a football field.

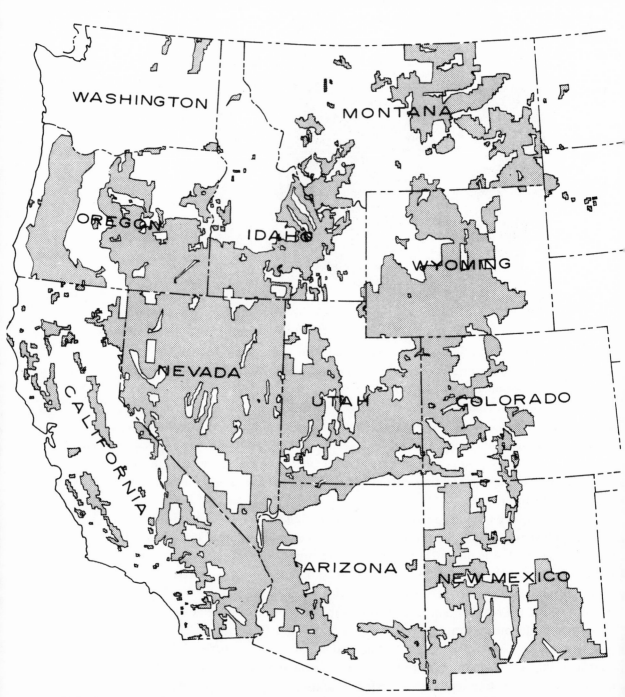

Bureau of Land Management land in the Western states. (*U.S. Department of the Interior*)

CHAPTER **2**

Finding Placer Deposits and Mining Sites

IF THERE'S A "secret" to successful gold hunting, it's prospecting. Once you know how to pinpoint sites that give an indication of the presence of native gold, the rest is fairly routine. You set up your sluice box, rocker, or suction dredge and go to work. If you're determined and work hard, you'll get the gold.

How much investigation and prospecting you do depends on the quantity of gold you're seeking. If you're merely out for a day of recreation, hoping to recover a few flakes and particles to show your friends, then you won't have to do very much research. You can simply set up your equipment where other people are already mining, that is, on the same river or stream. But if your goal is to make a good day's pay and, at the same time, to earn enough money to pay for your equipment and other expenses, you have serious work to do.

Fortunately, there are a good number of federal, state, and local agencies that can be helpful to you. Geographical maps and literature are available from a wide range of sources, but undoubtedly the best is the U.S. Geological Survey, an agency of the Department of the Interior. (In Canada, it's the Geological Survey of Canada.) One of the chief responsibilities of the Geological Survey is to prepare and distribute topographic maps that show in minute detail the natural and man-made features of the land's surface. Such maps, explained in detail later in this chapter, are an ideal starting point for any prospector.

The U.S. Geological Survey also conducts a broad program of field and laboratory research concerning the geology of the United States. Some of these investigations are carried out in cooperation with individual states, counties, and municipalities. Information derived from these studies is often available through state geological offices (listed in the Appendix). If your state has such an agency, it can be of enormous importance to you. Take, for example, the state of Arizona's Bureau of Geology and Mineral Technology. The Bureau acquires and disseminates basic geological data about Arizona and provides information of maps, circulars, bulletins, and special papers, including such publications as *Arizona Lode Gold Mines and Gold Mining* and *Gold Placers and Placering in Arizona.* If you're interested in mining in Arizona, there is no better source of information than its Bureau of Geology and Mineral Technology. The same help is available from the comparable agencies in other states.

Exactly what land is open to you depends to a great extent on where you plan to do your prospecting and mining. Naturally, to prospect on private land you need the permission of the owner. In the case of state parks, you're likely to require the permission of the park administrative officer.

Most of the National Forests in the Western states are open to prospecting and mining. However, some of the mineral-rich areas of the Forest lands have been claimed by other prospectors. To enter upon a claim already staked in order to prospect or remove mineral samples is trespassing and you become subject to the penalties involved. Before you begin prospecting, check local county records for claims. Also, search the area in which you're interested for physical evidence that a claim has been staked.

You should also be aware that within the boundaries of many National Forests there is a considerable amount of privately owned land, and this land is not open to prospecting or mining without the owner's permission. Maps available at Forest Service offices (listed in the Appendix) show the location of these privately owned tracts.

It's always wise to check with the local District Ranger before prospecting on National Forest land. (National Forests are administered by the Department of Agriculture.) He'll tell you what regulations apply for the region in which you're interested. He can also advise you of any permits that may be required by other federal agencies or state or local agencies.

Much of what's said above also applies to those vast areas of land in the Western states that are administered by the Bureau of Land Management (BLM) of the Department of the Interior (see map on page 00). These lands are open to prospecting and mining, although certain portions of them are said to be "withdrawn" for such uses. Be sure to consult the land records at a local BLM office (listed in the Appendix) for information.

National parks and national monuments are closed to prospecting and mining, as are Indian reservations, most reclamation projects, military reservations, scientific testing areas, and some wildlife-protection areas.

Recognizing Placer Deposits

To find placer gold, you don't have to have a degree in geology. But you should have an understanding of how placer deposits are formed and the different types it's possible to encounter.

In the geologic beginning, all gold was lode gold, or vein gold. Scientists believe that gold was deposited by gases or liquids rising from beneath the earth's surface. The gases and liquids traveled toward the surface through cracks in the crust (called *fractures*). What resulted were single veins that outcropped hillsides, or whole networks of veins, which, individually, held only tiny amounts of gold.

Most of the material contained within the veins is today identified as quartz or quartz and calcite. It also contains some metal-bearing compounds, such as those of copper, lead, iron, and others.

In many veins, brasslike iron disulfide—pyrite—is abundant. Pyrite and iron-stained mica are frequently mistaken for gold, and have thus earned the nickname "fool's gold."

Through millions of years, lode deposits were subjected to weathering and erosion. The ore minerals and even the rocks themselves were broken down into fragments of various sizes. Products of this disintegration—boulders, gravel, sand, and finer particles—were washed into gullies, creeks, and rivers.

During this weathering, most metal-bearing minerals went through chemical decomposition. Or some of their constituents were carried away by surface water. Not gold, however. Being extremely resistant to weathering, the gold, when freed from the host rocks, would be carried downstream, usually in the form of flattened flakes.

Running water sorts the various types of material with which it is laden into different strata according to the specific gravity of each type. Anytime the velocity of the stream diminishes, this classification process accelerates, with the heavier material filtering to the stream bottom, working its way downward until it finds a resting place in the stream bedrock or in the clay stratum supporting it.

Gold, being about seven times as dense as sand or gravel, quickly seeks the bedrock. Unusually heavy minerals, commonly referred to as black sand, accumulate with the gold. Magnetite is the most common, but other heavy minerals that may be present are platinum, cassiterite, monazite, ilmenite, columbite-tantalite, chromite, and even some gemstones.

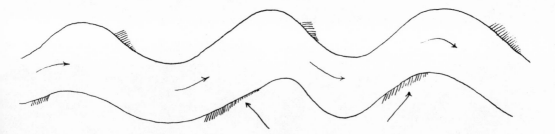

If the bedrock is smooth, the gold and associated material will not remain permanently in place, but will be coaxed slowly downstream until lodging against a ledge or settling in a pothole or natural riffle.

There are no fixed methods you can follow in pinpointing where placer deposits are apt to occur in a streambed, because the velocity of the stream is not the same at all points of its cross-section. But concerning placers of different types, there are some general rules that can prove helpful.

Stream Placers Stream placers are gravel deposits in the bed of a stream or within the stream's adjacent flood plain. You're more likely to encounter stream placers than any other type.

Naturally, the richest stream placers are those to be found in the valleys that collect the drainage from the mountains that are known to contain deposits of lode gold. The coarsest particles of gold will occur at the heads of ravines or gulches, unless, of course, the stream is particularly swift-running, in which case the gold particles and associated material will be swept downstream along the bedrock. What to look for are streams with a moderate gradient, a fall of about 30 feet to the mile. This condition offers the best chance for the sorting and grading of stream material.

Any area where the stream widens or changes direction should be worked. If the movement of water is swiftest at the center, there is a better chance for gold concentrations along the stream sides.

In a stream placer that is formed in a normal fashion, the highest concentration of gold particles is likely to be found in the inner side of the bends of the stream at or near the bedrock. Concentrations are also likely to be found wherever the streambed widens, or anywhere else the velocity of the water's flow slackens. Don't overlook natural riffles or any break in the bedrock.

Still another good place to look is at the base of any large boulder in the stream channel. Since boulders offer an obstruction to the movement of materials in the channel, heavy materials tend to lodge around them. "Big boulders mean big gold," is an axiom many miners follow.

In a narrow V-shaped valley, there may be concentrations along the entire stream bottom. But in broad, flat valleys, the pay dirt is likely to occur in a narrow stretch along one side of the stream.

▲ Shaded areas indicated likely placer sites along stream channel. (*Department of Natural Resources, state of Washington*)

◄ Quiet cover waters offer placer mining opportunities.

There's a miner here, but he's underwater, working the bedrock.

Potholes created by rapids or falls would seem to be an ideal collection point, but they're not likely to be. Such formations act as milling machines, grinding the gold to powder, or flushing it into the stream waters beyond to be deposited in the quieter water below.

Keep in mind that streams are constantly replenishing themselves. Winter storms erode the lode gold deposits in the mountains. In addition, streams never stop changing the arrangement of gold-bearing materials, sometimes exposing concentrations that have never been touched before. Some prospectors go back to the same streams year after year, and carry away quantities of gold on each trip.

Bar Placers Such deposits are formed on rivers and large streams during periods of high water, and left exposed when the water level drops. Bar placers are seldom permanent, and may either shift in location or dematerialize entirely. The gold is usually distributed throughout the bar and generally it's more fine than medium or coarse.

Bench Placers When a stream changes course and cuts a new channel, leaving the former streambed high and dry, bench placers can be formed. These can be located anywhere from a few feet to several hundred feet above the present level of the stream. The presence of well-rounded gravel, material carried and sorted by the water, is the tip-off to the existence of a bench placer.

To get to the gold, you often have to move big rocks. ▶

Beach Placer These placers form as a result of ocean-wave action that reworks sand and gravel, concentrating the heavy materials in an area where they can be conveniently worked. The sand and gravel may have been originally derived from the alluvium brought to the coastal area by streams. Or the deposits may have resulted from the erosion of the bedrock of the sea cliffs or the glacial sediments that overlie the shore bedrock. Gold particles released by these processes accumulate with the other materials in the black sand. The best time to look for beach placers is right after a storm, when the wave action has been violent, or at least vigorous.

Finding workable placer deposits doesn't necessarily mean that lode deposits are nearby. Flakes and particles can travel many hundreds of miles in a swift stream or muddy water.

Some indication of the proximity of the lode gold can be derived by examining the gold grains under a microscope. If the particles are smooth and well rounded, they've obviously been carried a long distance. But flakes or particles with sharp angles and, occasionally, slivers of quartz adhering to the metal are evidence that the original vein is not far off.

Gathering gravel for sluicing from a stream's flood plain.

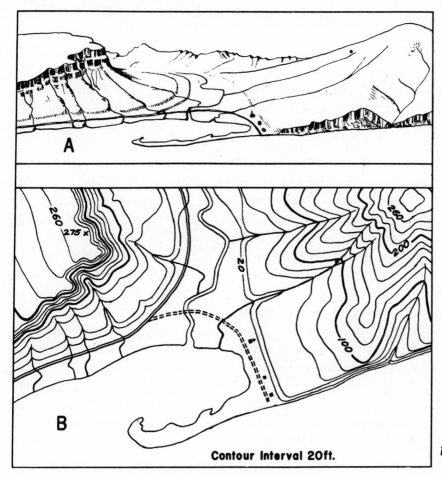

A

B

Contour Interval 20ft.

Topographic Maps and How to Use Them

For prospecting in any remote section of the country, you need detailed and accurate *topographic maps*. Maps of this type show on a flat paper surface the actual form of the land by means of curved lines called *contour lines*. With a little practice and imagination, you can use a topographic map to recognize the height of a hill, the depth of a valley, and the approximate elevation and slope of the ground at any given point.

The accompanying illustration (above), depicting a river valley and surrounding hills, shows the principal features of a topographic map. Part A depicts a river that flows into a bay partly enclosed by a hooked sandpit. Both sides of the valley are terraces laced with small streams. The hill on the right slopes gradually toward the bay. The hill on the left has a steep slope, but falls off gently to form a stretch of tableland.

Part B shows the same features as presented on a topographic map, with the elevations represented by contour lines. The vertical difference between any two adjacent contour lines is the contour interval—20 feet in this illustration.

To facilitate reading, every fourth or fifth contour line is heavier than the others. This line is known as an index contour. Notice that it is given elevation figures. Supplementary contour lines, shown as dashed or dotted lines, are added to better depict areas of little relief when the basic contours happen to be widely spaced. This system of contour representation is used on all topographic maps.

By "reading" the contour lines, you can form a mental picture of any particular piece of terrain. You can thus understand how helpful a topographic map can be when planning to penetrate any roadless area on foot.

You can use contour lines to pinpoint the location of flood plains or gravel bars that sometimes fringe mountain streams or rivers. Widely spaced contour lines at the bottom of a canyon and adjacent to a river bed are a tip-off to the existence of these formations, well known for their placer concentrations.

In addition to natural features, these maps also depict such man-made objects as roads, trails, mineshafts, tunnel and cave entrances, dams, buildings, and various boundary lines. In part B, discussed above, symbols stand for an unimproved dirt road and bridge that provide access to a church and two houses, situated across the river from a light-duty road that follows the seacoast and curves up a river valley.

An obvious way to use a topographic map is simply to pinpoint former mining sites. Some of these mines may have been of the hydraulic type, wherein huge water cannons played upon banks of gold-bearing gravel. By visiting the area, or perhaps by consulting geological reports of the mine, you may be able to establish where the mine tailings were dumped; some of them could contain gold-bearing ore.

In maps prepared and distributed by the U.S. Geological Survey, land features are printed in light brown, water features in light blue, and cultural features in black. Woodlands are shown by means of a light green overprint. Recent maps use red for principal highways and township and other land-subdivision boundaries. Contour lines are brown.

Former hydraulic mining sites can be rich in abandoned tailings.

Each of the maps issued by the U.S. Geological Survey is a quadrangle bounded by north-south lines of longitude and east-west lines of latitude. The type of map you're most likely to use will measure 7.5 minutes of latitude across the top and bottom, and 7.5 minutes of longitude along the sides. There are also quadrangles that measure 15 minutes by 15 minutes.

It's important to understand that topographic maps are prepared in several different scales. For maps issued by the U.S. Geological Survey, the scales are as follows:

Scale	One inch represents	Quadrangle size	Quadrangle area (square miles)
1/24,000	2,000 feet	7.5 x 7.5 min.	49 to 71
1/62,500	about 1 mile	15 x 15 min.	197 to 282
1/100,000	over 1.5 miles	30 min. x 1°	1,145 to 2,167
1/250,000	about 4 miles	1° x 2°	4,580 to 8,669

You'll want to select the largest-scale map available, the 1/24,000-scale map. Since one inch on this map represents something a bit less than one-half mile on the ground, there's plenty of detail. Overall paper size is 17 by 21 inches.

Unfortunately, maps for many of the less-developed areas of the United States are available only in the 1/62,500 size. Although these are a bit less detailed than the 1/24,000-scale maps and a bit more difficult to read, they are useful.

How to Obtain Maps

Geological or topographic maps published by the U.S. Geological Survey are inexpensive and easy to obtain. They range in price from $1.75 to $3 and are available by mail or by means of over-the-counter purchase at Survey offices or more than a thousand private sources.

In securing a topographic map, the first thing to do is to obtain an index map for the state in which you're interested. The index map depicts a map of the state that has been overprinted with rectangles corresponding to the map quadrangles available. You simply order the quadrangles you want. There is no charge for the index map.

The index map also gives information as to map scales and lists map agents and map reference libraries located within the area covered by the index. The date on which the area was most recently surveyed is also given.

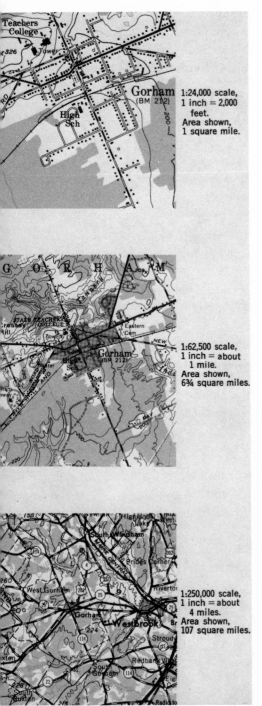

Gorham
(BM 212) 1:24,000 scale,
1 inch = 2,000
feet.
Area shown,
1 square mile.

Gorham
(BM 212) 1:62,500 scale,
1 inch = about
1 mile.
Area shown,
6¾ square miles.

1:250,000 scale,
1 inch = about
4 miles.
Area shown,
107 square miles.

U.S. Geological Survey

Index maps and the topographic maps themselves for states east of the Mississippi River, including the state of Minnesota, should be requested from

Washington Distribution Branch
U.S. Geological Survey
1200 South Eads St.
Arlington, VA 22202

Map indexes and maps for states west of the Mississippi, including Louisiana and Alaska, are available from:

Alaska Distribution Branch
U.S. Geological Survey
310 First Ave.
Fairbanks, AK 99701

Denver Distribution Branch
U.S. Geological Survey
Federal Center
Denver, CO 80225

Such maps are also available at the various Public Inquiries Offices maintained by the U.S. Geological Survey. Survey publications can be obtained at these offices, and you can also consult various open-file geological reports pertaining to the area. The addresses of the Public Inquiries Offices are shown here:

108 Skyline Building
508 Second Ave.
Anchorage, AK 99501

1C45 Federal Building
100 Commerce St.
Dallas, TX 75202

7638 Federal Building
300 North Los Angeles St.
Los Angeles, CA 90012

8102 Federal Building
125 South St.
Salt Lake City, UT 84138

504 Custom House
555 Battery St.
San Francisco, CA 94111

National Center, 302
Reston, VA 22092

1028 General Services Building
19th and F Streets N.W.
Washington, DC 20244

678 U.S. Court House
West 920 Riverside Ave.
Spokane, WA 99201

1012 Federal Building
1961 Stout St.
Denver, CO 80202

Offices of the National Cartographic Information Center also have U.S. Geological Survey maps available on an over-the-counter basis. Locations of the NCIC offices are as follows:

Western NCIC
345 Middlefield Rd.
Menlo Park, CA 94025

Eastern NCIC
Room 2Bs00
USGS National Center
Reston, VA 22092

Rocky Mountain NCIC
Room 2206
Building 25
Federal Center
Denver, CO 80225

Mid-Continent NCIC
1400 Independence Rd.
Rolla, MO 65401

Last, U.S. Geological Survey maps are available from some fifteen hundred private map dealers. For the addresses of map dealers in your area, simply consult the Yellow Pages of your telephone directory under "Maps." Map dealers are also listed on the

National Cartographic Information Center regional boundaries. (*National Cartographic Information Center*)

Geological Survey's index maps.

Geological and topographic maps of Canada can be obtained from the:

Map Distribution Office
Department of Energy, Mines, and Resources
615 Booth St.
Ottawa, Ontario K1A OE9

Topographic maps of Mexico do not exist in large quantities. However, for information regarding those that are available, write:

Instituto de Geologia
Universidad Nacional Autonoma de Mexico
Ciudad Universitaria, Mexico 20, D.F.

Geological Maps

Geological maps can also be of value. What a geological map does is give one a portrait of what lies beneath the surface, the kinds of rocks that are there, their distribution and structure.

Geological maps frequently accompany survey reports and, in such cases, are likely to relate to an individual mine or ore deposit. Geological maps are also available from the U.S. Geological Survey. They are published in various scales from 1/24,000 to 1/125,000. Geological maps published by the Geological Survey of Canada are available in scales ranging from 1 inch to 1 mile and 1 inch to 4 miles.

Geological maps are valuable to anyone planning placer mining because they reveal the composition of streambeds you may want to work. For a stream to give up gold flakes or particles, the bedrock must contain fissures and cracks that will act as gold traps. There are three types of rock that contain fissures and cracks in abundance: slate, granite, and phylitte (a slaty rock containing mica). If you can establish a streambed composed of one of these, you've found prime territory for a dredging operation.

Some geological reports and their related maps concern ancient rivers, known as Tertiary channels. The term *Tertiary* refers to a geological period during the Cenozoic era, a period which lasted about 70 million years and ended about one million years ago. Tertiary channels flowed in different streambeds from today's rivers. Before the volcanic eruptions that formed the Sierra Nevada, all rivers in the area ran north and south. During the course of the upheaval, the Tertiary channels were buried beneath enormous quantities of volcanic debris. New channels for the rivers and streams that flow today were formed. Generally speaking, these rivers flow east and west.

Wherever a Sierra Nevada stream cuts through an ancient river, the gold in the old streambed is likely to have been redistributed. Indeed, perhaps the richest source of gold-bearing material occurs wherever a modern-day river knifes through a Tertiary channel. You can obtain geological reports that describe these Tertiary channels. It's then up to you to pinpoint any unprospected site within or near a particular channel.

How to Preserve Your Maps

As part of your geological investigation and overall planning you may plot areas for follow-up investigation, campsites, or trails on your maps. That way you can enhance their value, turning them into documents that you'll want to preserve.

Mounting your map on cloth will keep it fresh looking for an extended period of time. It's easy to do. The cloth most commonly used as backing is bleached muslin sheeting. You need a piece of muslin that exceeds the size of the map by two inches on each side. Then you mount the map to the center of the cloth, using home-made flour-and-water paste or commercial wallpaper paste.

The first step is to stretch the cloth over a board or work table, fastening it in place with tacks. Next, carefully fold the map to a convenient size, and then cut the map along the fold lines. Or you can draw the fold lines on the reverse side of the map, and then cut along the lines.

The actual mounting follows. Begin with one of the map's corner sections. Float the section in a pan of water for a minute or two. Then, using a large, flat brush, apply the paste to the back of the section. Wait until the paste is partially dry, and then place the section on its proper place upon the cloth. Apply pressure gently with a damp cloth or sponge.

Follow the same procedure with an adjacent section. Leave a ⅛-inch space between sections (to allow for the folds). The final step is to cleanse the map of excess paste with the damp cloth or sponge.

Allow the map to dry thoroughly before removing it from the board. Drying takes at least 24 hours. Use a straightedge and razor blade to trim the edges of the cloth.

Art supply or photographic supply stores sell a commercial cloth backing to which map sections can be bonded by means of a household iron. Map mounting is easier when you use this cloth.

Other Maps and Surveys

You also may be able to derive helpful information from the Mineral Investigation Resources maps published by the U.S. Geological Survey. These maps show the areas in the United States from which most of the commercial metals and minerals have been obtained. While they do not show mines or quarries, these maps do indicate the major mining districts for various mineral commodities, including gold.

Printed in two colors, Mineral Investigation Resources maps are published at a scale of 1/3,160,000, in which 1 inch equals 50 miles.

Data compiled from aeromagnetic surveys of the United States have been published by the U.S. Geological Survey as Geophysical Investigations maps. These employ isometric contours, overprinted on sheets depicting cultural features, to show variations of magnetic intensities. Some of these maps also show geological characteristics for an area, and thereby facilitate the correlation of geological features and magnetic anomalies.

For a listing of Mineral Investigation Resources maps and Geophysical Investigations maps, write to:

> Geological Inquiries Group
> U.S. Geological Survey
> 907 National Center
> Reston, VA 22092

Results of aerial radioactivity surveys conducted by the U.S. Geological Survey on behalf of the Atomic Energy Commission (now the Nuclear Regulatory Commission) are also available in map form. For information about these maps, use this address:

> Library
> U.S. Geological Survey
> 950 National Center
> Reston, VA 22092

Information on satellite imagery and aerial and space photography can be obtained from these sources:

> National Cartographic Information Center
> U.S. Geological Survey   EROS Data Center
> 507 National Center   U.S. Geological Survey
> Reston, VA 22092   Sioux Falls, SD 57198

Aerial Photographs

Aerial photographs of most parts of the United States are available from several government agencies, and they're not expensive, costing only a few dollars apiece. General information concerning aerial photographs and a price list for reproductions can be obtained at this source:

> National Cartographic Information Center
> U.S. Geological Survey
> 507 National Center
> Reston, VA 22092

Because of the great array of photographs available, requests for information should define the area in which you're interested by means of a detailed description, sketch, or listing of the latitude and longitude coordinates. A topographical map with the area of interest outlined is always helpful.

Information regarding aerial photography can also be obtained from the regional offices of the National Cartographic Information Center. These offices also sell photographic reproductions on an over-the-counter basis. The offices are as follows:

Eastern Mapping Center
U.S. Geological Survey
536 National Center
Reston, VA 22092

Mid-Continent Mapping Center
U.S. Geological Survey
Box 133
900 Pine St.
Rolla, MO 65401

Rocky Mountain Mapping Center
U.S. Geological Survey
Box 25046
Federal Center
Denver, CO 80225

Western Mapping Center
345 Middlefield Rd.
Menlo Park, CA 94025

Other Research Sources

In your effort to gather information on mines and mining activity in your area, you're likely to find that there are as many as a dozen local, state, and federal agencies that you can call upon for assistance. Searching old records is never an easy task. One thing to do at the outset is put the information you're seeking into proper historical context. Take the case of California. California became a U.S. Territory in 1847 when Mexico surrendered to John C. Frémont. On January 24, 1848, James W. Marshall discovered gold at Sutter's Mill, triggering the gold rush which brought settlers by the tens of thousands to California. On September 9, 1850, California entered the Union.

You can't drop into the offices of the California Division of Mines and Geology in Sacramento and expect to riffle through the records of whatever mining has been carried on since gold rush days. Records simply weren't kept in the earliest times. Some of those that were kept have been lost. The California State Mining Bureau, the agency that was the forerunner of the Division of Mines and Geology, wasn't even founded until 1880. Even records compiled since that time aren't complete.

As a researcher, you also have to be aware that in 1850 California was made up of 27 counties. There are 58 today. A claim in Mariposa County in 1850 might now be located in Merced, Madera, Fresno, Tulare, or San Bernardino county.

Be sure to visit local libraries in your search for historical background. There are countless books available on the history of mining in California and the West. Don't overlook local historical societies, many of which maintain libraries and other sources of information. In California, the most noted organization of this type is the California Historical Society in San Francisco, but there's also the Society of California Pioneers, which specializes in the period before 1860.

To find out the names of historical societies in your area (there may be several), along with information as to the research programs in which each is engaged, and the time period covered by its collections, consult the *Directory of Historical Societies and Agencies of the United States and Canada.* A publication of the American Association of State and Local History, it is the best handbook available concerning historical associations, listing more than four thousand of them. Your local library is sure to have a copy.

Once you have some general information about a claim, you can visit the recorder's office in the county in which the claim is located. Although you may become involved in a time-consuming search, the results can be worthwhile. Different documents pertaining to mining claims filed with the county recorder often include the location notice, deed, quitclaim deed, affidavit of proof of annual labor, and various options, contracts, and leases.

In California, the county recorder's office maintains the official records of mining districts, which were organized by the miners in frontier times, before statehood, before federal mining laws had come into being. The miners established the territorial jurisdiction of the districts and defined the ownership and size of claims, and the amount of assessment work required in each case. The importance of mining districts diminished with the passage of the Federal Mining Acts of 1866 and 1872.

In some cases, it is possible to write to a county recorder and request the information you're seeking. But even if you're successful in getting help by mail, there's likely to be a long wait involved.

The county assessor's office also contains important records. These include plats showing the location of patented claims, the names and addresses of property owners, and all tax records. The county assessor's office is also likely to be the best source for information pertaining to the current owner (his name and address) of a patented claim.

There are many state agencies that can be helpful. About one half of the states have agencies devoted to mines, mineral resources, and geology (they too are listed in the Appendix). Records of the California Division of Mines and Geology, for instance, give the location and geology of some gold mines, the name of the operator, and general production statistics. Publications of the agency contain many reports on the mines and minerals of individual counties.

In the case of a mining company incorporated within a state, consult the office of the Secretary of State for information concerning the corporation's current legal status. The Secretary of State, or the Department of Corporations, should also have records as to whether the corporation listed stock.

The various Land Offices of the Bureau of Land Management (listed in the Appendix) may also have useful records concerning patented mining claims. These can include the applications for patents, with legal descriptions and plats. Survey plats, notes, and title certificates may also be available.

Many of the famous mining towns of the West offer small museums in which maps and documents relating to mining activity are available. Mining photographs and artifacts are on display. In the Amador County Museum in Amador County, California, scale models of several famous historic mines are displayed. (From the mines of Amador County, came more than half the gold mined in the Mother Lode.) The rare book division of the Amador County Library, located in Jackson, offers one of the most comprehensive collections of mining books, articles, and documents in the West. Librarians and museum curators are often rich sources of local mining lore.

Forty-Niners pan for gold on California's Mokelumne River. (*New York Public Library*)

CHAPTER 3

All About
Panning

PANNING IS A centuries-old mining technique for separating particles of gold from lighter-weight materials. While many things have changed about gold hunting in the past hundred years or so, panning has remained pretty much the same.

You shovel the gold-bearing material into the pan, add water, then agitate the pan, and let the lighter materials wash over the pan edge. The heavier black sand containing the gold particles settles to the bottom. That's what the Forty-Niners did; that's what you do today.

You can use a gold pan as your one and only piece of gold-hunting equipment, and spend the entire day at stream-side panning out sand and gravel. The problem is that a pan is not a very efficient way of doing this. Either a sluice box, a rocker, or a surface dredge (discussed in succeeding chapters) is better. According to a study conducted by the Montana Bureau of Mines and Geology, an experienced adult working with a standard-size gold pan should be able to wash out up to 60 pans of material in a 10-hour day. At 20 pounds per pan, he will handle about 1,200 pounds of material. To put it another way, it would take him about three days to pan one cubic yard of material. At that rate, not much gold is going to be found.

Veteran prospectors use a pan for one of two reasons. Usually they're prospecting; they're sampling. They're probing up and down a stream, seeking a site to establish operations for a day. When they find a spot where they get a significant amount of color in their pan, they'll set up their sluice, rocker, or dredge.

The other time the pan gets used is at day's end when they're cleaning up the black

Panning is inefficient. To wash out a cubic yard of sand and gravel, it takes three days.

Steel pans of various sizes; 16-inch pan is at left.

sand they've accumulated. Some miners carry the black sand home and pan it out at their leisure.

Types of Pans

Miners of the nineteenth century sometimes used frying pans or pie plates for panning. Theoretically, you could use a bowl, jar, a plastic dishpan, Frisbee, or just about any other type of flat-bottomed container. California motorists have been known to use automobile wheel covers as gold pans when nothing else was available. As this suggests, gold pans cover a fairly wide range in size and style. Usually, however, the only decision you have to make concerning pans is whether to use steel or plastic.

Steel Pans In the earliest days of prospecting in the West, the pan was called a "gold dish." It was common for a prospector to carve a pan out of a solid block of wood, then sand the inner surface with gravel from a streambed. But once steel pans were introduced and came into general use, they replaced all other types.

The standard steel gold pan is 16 to 18 inches in diameter at the top, 10 to 12 inches in diameter at the bottom, 2 to 2½ inches deep, with sides that slope about 30 degrees. It can hold a good-sized shovelful of sand and gravel, about 20 pounds of it.

Does this sound as if the standard pan can be a bit heavy and unwieldy? It's true. That's why some panning instructors advise novice prospectors to consider using a smaller pan, one with a top diameter of 12 inches and a 7½-inch diameter at the bottom. A pan of this type, sometimes called a half-size pan, has a depth of 2 inches. Filled with gravel, a 12-inch pan weighs only about 9 pounds. Don't try anything smaller than 12 inches. "Anything less than 12 inches is a toy," says veteran prospector Bob Erwin.

Because you can swirl around the 12-inch pan at a faster rate, it's possible to wash out two pans of gravel in the time it takes you to wash out only one pan of standard size. Also, you won't tire as quickly.

Some steel pans offer riffles. These are either narrow ridges or indentations that run along the inner surface of the pan parallel to the top edge. Riffles are meant to aid in trapping the black sand and gold particles and thus aid in speeding up the panning process. Some riffles are short; others extend all the way around the inside of the pan. (More will be said about riffles later.)

In gold country, mining equipment firms and most hardware stores sell gold pans. They range in price from $5 to $10, depending on the size.

Any brand-new steel pan is coated with oil, grease, or some other type of rust preventive. This must be removed before the pan can be used. Even the slightest trace of oil—even oil from your fingers or palms—will cause the gold dust and the finer particles to ride out and over the pan edge during panning.

The best way to clean a new pan is to pass it back and forth over the flame of a gas-stove burner. At the same time, the steel turns dark blue in color. You must apply the heat evenly. Excessive heat in any one spot can warp the pan, rendering it useless. Once blued, the pan will be free of grease.

Plastic Pans Many veteran miners prefer the steel pan. They like it for its durability, its sturdy feel. But some experienced hands recommend the plastic pan for the beginner, however grudgingly. "It's probably simpler to use," says Bob Erwin. "If you don't understand the basics of panning, the plastic pan, with its built-in riffles and gold traps, can compensate for whatever deficiencies you might have."

The standard plastic pan is 14 inches in diameter. Also available is a 10½-inch finishing pan in plastic, used in the final panning stages, when separating the gold flakes and particles from the black sand. (A circular ridge in the bottom of the standard pan makes finishing difficult.) With longer and sharper sloping sidewalls, and a slightly textured surface, the finishing pan permits the user to separate efficiently even the smallest amounts of the finest gold under either wet or dry conditions. Since it's of fairly good size and riffled, the finishing pan can also be used as an all-purpose pan. Plastic pans cost about the same as steel pans, between $5 and $10.

Another advantage of plastic pans is that they weigh less. Neither does the plastic pan need to be flame-treated to cleanse it. If it needs washing, you just use soap and water. And plastic never rusts.

Some processing operations can be performed with greater ease in a plastic pan. You can use a magnet to remove the black sand (which, as mentioned above, is magnetic) from the bottom of a plastic pan. When you try this with a steel pan, the magnet clings to the metal.

Certain types of plastic pans offer riffles that help trap the black sand and gold particles. The gravity-trap pan, for instance, has three molded riffles, or ridges, molded into the pan's inner surface, and the heavier materials collect in these ridges during the panning operation, the same way the riffles trap the black sand and gold particles in a sluice box. Some old-timers refer to these as "cheater riffles," because they allow a novice to pan with almost the same proficiency as an experienced hand. The pan's forest green color is said to enhance the ability to see particles of gold amidst the black sand.

The Grizzly Pan Using a grizzly, which is essentially a sifter, enables you to pan faster. The grizzly can take several forms. A plastic grizzly, circular, 10½ inches in diameter, is available for use with the gravity-trap pan. There are also steel grizzlies. Other types are wood framed, with ¼- or ½-inch mesh across the bottom.

The material to be panned is first shoveled into the grizzly, then sifted through the

The 14-inch plastic gravity-trap pan and 10½-inch grizzly.

As in a sluice box, pan riffles act to trap heavy materials.

grizzly screen into the gold pan. The big stones retained in the grizzly are thrown away, and only the finer material is panned.

To start, use your grizzly by shaking it, letting the sifted material fall into the pan. But keep one eye on the contents of the grizzly; there may be a nugget there. Although it's unlikely that you'll find a nugget, don't throw away any of the heavier material without visually inspecting it.

With grizzly, you presort material you plan to pan.

There's also a "wet" method of sifting material through the grizzly. After the grizzly is filled with sand and gravel, it's nested in a gold pan of the same size and shape and the two pans are submerged together. After the mixture is well soaked, the grizzly pan is lifted a few inches and agitated underwater. The finer material passes through the grizzly into the standard pan. The larger material in the grizzly is discarded. The finer material is then panned.

How to Pan

Successful panning is based upon the principle that gold has a specific gravity of 19, meaning that it is 19 times heavier than water. Ordinary sand and gravel are only two to three times heavier than water. Thus, the gold sinks to the bottom during the panning operation.

However, this principle only applies when the material to be panned is thoroughly wet and shaken vigorously. If you placed a gold nugget weighing several ounces atop a rock-filled pail, and left it there, and came back and looked at the bucket years later, the nugget would be still in the same place, right on top of the pile. For the gold to work its way to the bottom, the rocks would have to be agitated. That's the "secret" of successful gold panning. You have to agitate the material in the pan. Every particle has to be made to move.

You have to work when you pan. You must shake the pan, really shake it. Many novices think that all you have to do is get the material covered with water, and then swirl the water about. Somehow, they believe, the gold will then work its way to the bottom. Not at all. You have to work.

No two individuals pan in exactly the same way. Some stress side-to-side movement of the pan. Others may put the emphasis on to-and-fro motion. Still others prefer moving the pan in a circular fashion most of the time. It really doesn't make too much difference. The whole idea is to keep agitating the gravel and sand in the pan until the heavier material is concentrated at the bottom. You spill the unwanted lighter material over the pan sides. Use any method you want, so long as it gets results.

Nevertheless, certain fundamentals apply. Begin by picking out a shallow place along the bank of a stream where the water is just deep enough to cover the pan and its contents. The water should be moving fast enough so it will stay clear as you work, but it shouldn't be sluicing so rapidly that it washes any material out of the pan. (If you're not near a stream, you can use a tub of water.)

Fill the pan from one third to two thirds full of material.

Don't try squatting on the shore and reaching into the water with the pan. Your back isn't likely to be able to take the strain for very long. Instead, wear boots or waders, or, if it's summer, don a swimming suit and sneakers, and wade into the water. The sneakers are vital, incidentally, to prevent your feet from getting bruised by the stream's gravel bottom.

Submerge the pan until it's resting on the stream bottom. Working carefully so as not to spill any material, begin mixing the contents of the pan with your hands, permitting the water to thoroughly soak the mixture.

Pick out any large rocks and, after washing them clean, toss them away. Sand clinging to plant roots should be washed into the pan, and then the plant life can be discarded. Break up any clumps of soil with your fingers. This applies especially to clayey lumps. Keep squeezing them until they're completely dissolved. What you want to do is end up with a soupy mixture. When you've achieved this, you'll have the gold particles in a state of liquid suspension. Then, by agitating the pan, you can send them to the bottom.

First step is to submerge pan; then, using your hands, allow the water to thoroughly saturate pan's contents.

Break up any clayey lumps.

The next step is to grasp the pan on opposite sides while it is submerged. Be sure to keep it level, not letting any material spill over the side. Start shaking the pan vigorously from side to side. Then swirl the pan in a circular fashion, then from side to side again. What you're doing, of course, is separating the heavier material, causing it to sink to the bottom, while the lighter gravel and sand come toward the surface. At this point, again rest the pan on the stream bottom and pick out, wash, and discard any large stones and other debris.

Now you can begin washing the lighter material over the side. Raise the pan out of the water, and simply tilt the pan's far edge downward, allowing the lighter sand and gravel to ride over the edge to be swept away in the stream's flow. (If you're using a plastic pan with riffles, or any other type of riffled pan, adjust the pan so the lighter materials pass over the riffles before washing over the pan edge.) Once you've released some light material, bring the pan level again, getting the heavier material concentrated at the bottom. Shake the pan from side to side again.

Tilt the pan forward to permit lightweight materials to wash over pan edge.

Keep swirling the material around in the pan, tilting it to allow lighter materials to flow out.

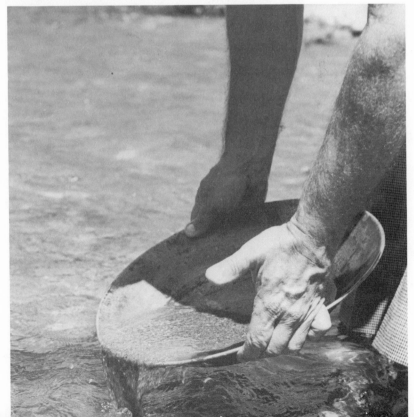

Repeat the process. Cover the material with water, level the pan, shake it from side to side, swirl it with a circular motion, then tilt the pan's far edge downward, letting the lighter sand and gravel flow over the edge.

After you've repeated the process several times, you should have only about two cups of material left in the pan. The accumulation of black sand concentrate in the pan bottom will have begun to show through the lighter material.

Horizontal strip of black sand shows through lightweight materials.

When panning is completed, only black sand remains. This accumulation contained several colors.

Now the operation is entering the final stage, which involves getting rid of the lightweight sand. You need only small amounts of water, you don't have to shake the pan quite so vigorously, and much less material gets washed over the edge. But the process is basically the same. You swirl the material around in the pan and dip the far edge downward to release the lightweight material into the stream. Anytime black sand starts to flow over the pan edge, bring the pan level, shake it, and resettle the material. By this time, you may be able to spot some color.

Eventually, only the mineral concentrates will remain in the pan. You can pour these concentrates into a storage container for final processing at home.

Practicing Panning You don't need a mountain stream at hand in order to be able to sharpen your panning skills. You can practice at home using your gold pan, a tub of water, and a couple of dozen air rifle BBs or pieces of buckshot. Or you can cut flakes from a lead sinker to simulate flakes of gold.

Fill the gold pan with sand and gravel. Mix in the BBs or chips of lead. Then soak the mixture in the water-filled tub and start panning.

If you get to the bottom of the pan and only about half the BBs or lead flakes remain, there's obviously plenty of room for improvement. If you had been working with gold, you would have washed half of it back into the water. Keep practicing until you can wash out a pan containing a dozen BBs or lead chips and have at least nine or ten of them remaining at the end.

Panning Schools While it takes only minutes to learn how to use a gold pan, it's no easy matter to become really skilled. The best way to learn is to watch an experienced miner at work. Once you're deft in the art, you should be able to handle up to six pans of sand and gravel in an hour.

You can also become skilled by attending a panning school. Such schools operate just about anywhere there are gold-bearing streams. Here are the names of some of the schools and their instructors:

Tenderfoot Prospectors, Inc.
6702 East Coronado Rd.
Scottsdale, AZ 85257

Bob Erwin
439 Sacramento St.
Auburn, CA 95603

George Massie, President
Gold Prospectors Association of America
P.O. Box 507
Bonsall, CA 92003

George Mroczkowski, President
San Diego Gem and Treasure Hunting Association
2493 San Diego Ave.
San Diego, CA 92110

E. Foley
R.F.D. 2
Woodsville, NH 03785

The Treasure Hunter
P.O. Box 5
Mule Creek, NM 88051

A. J. Haley
Lucky Leo's
   Prospecting School
709 West Columbia
Pasco, WA 99301

Prospector Ed
5263 Ranier Ave., South
Seattle, WA 98118

Dry Panning

There may be times when you'll want to pan out sand and gravel and no stream or other source of water happens to be nearby. Dry panning is the only recourse.

Dry panning is not as efficient as wet panning. And even if you become highly skilled in dry panning, you're still going to end up with a greater volume of concentrates than in wet panning. A riffled plastic pan is recommended.

In dry panning, the gravel and sand you work with must be *completely* dry. If the material contains even the slightest bit of moisture, the gold particles won't settle.

Don't put too much material into the pan. In your first attempts, fill the pan only from one-third to one-half full.

Set the pan down on a flat surface and, using both hands, mix the material thoroughly. Pick out any loose rocks or big pieces of gravel, cleaning them of sand before you toss them away.

Give the pan several vigorous shakes, either from side to side or forward and back. This helps to bring the lighter gravel to the surface so you can discard it.

Continue shaking the pan, settling the heavier concentrates and removing the lighter pieces of gravel with your fingers. If there's a great deal of gravel, you can simply rake it off over the pan edge. When you've reduced the amount of material in the pan to one or two cups, you're ready for the next step.

Adjust the pan so the riffles are facing you. Grasp one side of the pan with your left hand, and tilt the pan downward, away from you. The riffles will then be in the lower portion of the pan.

Using your right hand, tap the raised edge of the pan sharply. As you tap, the lighter material will begin edging its way up the side of the pan to tumble over the edge. Most of the heavier concentrates will collect in the pan bottom, but some will move in the opposite direction, that is, toward the hand that's doing the tapping. Still other heavier materials will move in the "wrong" direction, toward the lowered pan edge. But these will be trapped by the riffles and tumbled back toward the bottom of the pan.

Of course, all of this requires that you establish the proper amount of force with your taps while getting the pan tilted at exactly the right angle. It takes practice. But anytime the operation isn't going right, all you have to do is bring the pan level, resettling the concentrates, and then start again.

Go slowly at first, and develop speed gradually. An invidividual who is experienced in this dry panning can sometimes even recover flour gold.

Recovering the Gold

As you pan, you will eventually reach a stage in which the pan contains only a small quantity of black sand and gold particles. At this point, place a tiny amount of water in the pan and swirl it around, causing the contents to spread out over the entire pan bottom.

Use tweezers to pluck out the gold flakes or suck them out with a small eye dropper. Use a magnet to remove particles containing iron. Removing those ferrous particles from the magnet can be a chore. So first wrap the magnet prongs in plastic wrap, such as Saran Wrap. When you remove the wrap from the magnet, the particles will come with it.

If a considerable amount of black sand and minute gold particles remain, allow the sand to dry, and then spread it out on a big piece of cardboard. As you tap the

Amalgam is poured into glass vial for storage.

cardboard, blow gently across its surface. The sand will be carried away and the gold particles will remain.

Don't let yourself be taken in by what's called fool's gold—usually pyrite or mica. Under a magnifying glass, gold looks solid; fool's gold has a flaky appearance. If poked with a needle, gold will bend or dent; fool's gold shatters. If you have flakes of both in the bottom of a pan, cast a shadow in the pan. Gold will continue to glitter, while fool's gold loses its gleam.

Finally, there's a chemical test you can perform. Drop muriatic acid (sold in drugstores) on the sample. Gold is unaffected by the acid, while the pyrite or mica will foam and dissolve.

Mercury Separation You can also use mercury, the heavy, silver-white, metallic element, noted for its fluidity, in separating fine gold particles from black sand. It usually takes only a few drops of mercury to accomplish this. One rule of thumb states that you should first estimate the volume of gold in the pan, and then use twice as much mercury.

Be sure to work out of doors. Mercury vapor can be deadly, and good air circulation reduces the hazard of dealing with it.

Swirl the mercury through the black sand, getting it in contact with every gold particle. As it assimilates with the gold, the tiny globules of mercury and gold will join together in a single mass. If the two elements fail to puddle, it may be because you're not using enough mercury. Adding a drop or two should solve the problem.

Once you have a single mass of amalgamated mercury and gold, the next step is to recover as much of the free mercury as you can. Pour the metallic ball onto a square of chamois. Bring the four corners of the chamois together so as to form a pocket containing the amalgam. Squeeze all the mercury you can through the chamois into a dish.

What remains is still an amalgam, part gold, part mercury. Old-time miners used to separate the amalgam into its two components with some simple chemical wizardry. The amalgam was placed in a hole scooped out of a raw potato. A tin plate was then placed over the hole, the potato inverted, and the tin plate placed over a fire until the potato was partly cooked. The gold, in sponge form, would end up on the tin plate. The heat vaporized the mercury and it condensed on the inner surface of the potato. It could later be recovered by dunking the potato in a bucket of water.

Because of the lethal nature of mercury vapor, this is a hazardous experiment. Don't try it.

Retorting is the modern-day method of separating the amalgam into its two constituent elements. The process uses a cylinder-shaped vessel called a retort within which the amalgam is heated and the mercury removed by distillation. You can purchase a retort from a mining supplier.

The retort system you require has three principal parts. The recovery vessel for holding the amalgam usually takes the form of a small steel pot with a tight-fitting lid. The heat source is normally a portable propane gas torch. And the condenser, a length of tubing jacketed in a water-filled sleeve, is where the mercury vapors are converted back to liquid.

Apparatus for retorting. (*California Division of Mines and Geology*)

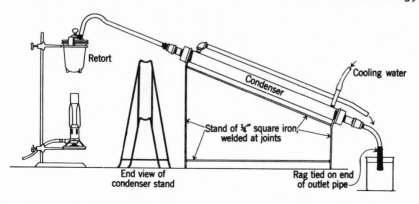

The amalgam is placed inside the steel pot and heated. At a temperature of 675° F, the mercury begins to vaporize and passes through a tube in the lid of the retorting vessel that leads to the condenser. As the mercury vapors pass through the condenser, cold water flowing through the jacket that surrounds the tubing reduces the vapors to a liquid state, and the liquid is collected in a vessel filled with cold water.

Because of the deadly mercury vapors, it's vital that the retorting operation be conducted outdoors. Leaks invariably occur in the network of retort pipes, tubes, and their connections. Individuals have died from attempting to retort amalgam in their homes.

Marketing the Gold   There are several different types of firms that buy gold, gold ores, nuggets, gold sponge, and other gold-bearing concentrates. Jewelers, refiners, and smelters are among them. Simply look in the Yellow Pages of your telephone directory under "Gold, Silver, and Platinum Buyers." As mentioned earlier, mining supply stores and gold-country dive shops also deal in gold in various forms.

In almost every case, the amount that you receive will be based on the current market price of gold or, at least, a formula that directly relates to the market price. For example, American Smelting and Refining Company once paid "98 percent of the gold content at the Daily London Final Gold Quotation, as published in *Metals Week,* averaged for the calendar month following the date of the delivery of the product."

Even though you may have cleaned your finds by washing, blowing, magnetic separation, and even mercury separation, some impurities still remain. These will be removed during the refining process. Because of the "shrinkage" involved, the final settlement may be a disappointment to you—unless you're aware in advance of the fineness of your gold.

You should also realize that smelter fees can include a minimum charge per shipment, a sampling and assaying charge, and a percentage discount for metallic content lost in handling.

The names of several refining companies are listed below. Be sure to contact any refiner before shipping gold or samples. Ask for a rundown on the company that handles purchases. Each has a different policy.

American Chemical & Refining Co.
Sheffield St.
Waterbury, CT 06714

Eastern Smelting & Refining Corporation
430 Montain Ave.
Murray Hill, NJ 07974

Associated Metals and Minerals, Inc.
11944 Mayfield Ave.
Los Angeles, CA 90049

Sabin Metal Corporation
316-34 Meserole St.
Brooklyn, NY 11206

Western Alloy Refining Co.
366 East 58th St.
Los Angeles, CA 90011

Wildberg Brothers Smelting & Refining Co.
349 Oyster Point Blvd.
South San Francisco, CA 95080

Cradle-rocking on the Stanislaus River in '49. (*New York Public Library*)

CHAPTER 4

Basic Placer Mining

MOST OF THE world's placer gold has been produced by two basic mining methods—sluicing or rocking. Either enables the miner to handle much more sand and gravel in a given day than when merely panning. Some sources say you can increase the amount of material handled by as much as a thousand percent. And sluicing and rocking are much easier than panning. The sluice box is powered by water, the rocker by a cradlelike rocking motion.

Some definitions, first. A sluice box is simply a shallow trough lined with raised obstructions called riffles. It is placed in a fast-moving stream, and gold-bearing gravel is fed into the upper end of the unit. The flowing water washes the material the length of the trough. The heavier material, the gold-bearing black sand, settles to the bottom of the sluice to be trapped in the riffles. Once the riffles have accumulated a sufficient amount of black sand, the flow of water is halted, the riffles cleaned out, and the black sand panned.

The rocker consists of a box about the size of a standard milk crate, mounted on two rockers. Once the unit is in place beside a stream, sand and gravel are shoveled into the rocker hopper—a grizzly, essentially—which retains the stones. At the same time, water is poured into the hopper and the unit is vigorously rocked. The finer material passes through the hopper screen onto a canvas apron, which traps some of the black sand. At the bottom of the rocker, there's a riffle system which catches any black sand the apron may have missed.

Sluice boxes can be purchased at mining supply outlets and most hardware stores in

Aluminum hand sluice has removable riffles. Slightly mo[re] than four feet in length, it costs less than $50. (*Kee[ne] Engineering*)

The rocker is an efficient, though underrated, means of gold recovery.

Tending a sluice box on a California stream.

the gold country. Rockers aren't quite so readily available. Bob Erwin (439 Sacramento St., Auburn, CA 95603) is one of the leading rocker manufacturers. The Erwin rocker, which costs about $75, is available at many mining supply stores. Sluice boxes are less expensive, beginning in price around $40.

Gathering Sand and Gravel

You can own the very best sluice or rocker and operate either with the greatest proficiency, but unless the sand and gravel you're working are of high quality, your time is going to be wasted. Always take your time selecting the material you plan to work. It's the key to success.

As mentioned earlier, gold flakes and particles lodge anywhere the stream's flow slackens, that is, on the inner side of a bend in the stream or where the stream widens. Gold lodges on the downstream side of boulders and in pockets and potholes in

Sluice is easy to carry, even over rough terrain.

the streambed. A fallen log or other natural breakwater can provide a spot where a fast-flowing stream drops gold. Don't overlook the roots of thick grass that may line the stream banks.

Gold flakes and particles are likely to be found in the cracks and crevices of the bedrock over which a stream flows or has flowed in the past. Crevice mining is the way to get this gold. It involves removing the accumulated material, then panning it or washing it in a sluice or rocker.

You don't need expensive equipment to be a crevice miner, although specialized tools are required. A long-handled teaspoon or ordinary tablespoon can be used in cleaning out bedrock cavities. An old screwdriver, with the blade bent over 90 degrees at a point two or three inches from the end, makes a good crevice digger. Or you can bend over the tang of a three-sided file in the same fashion, and dig with that. Mining-supply equipment firms have available special crevice tools made of case-

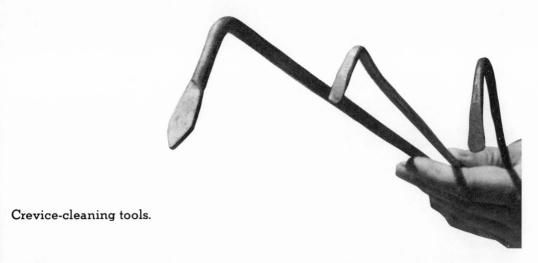

Crevice-cleaning tools.

hardened steel, as well as small chisels, gads, and gad pry bars for opening crevices and raking them clean.

You should also obtain a mineralogist's pick-hammer or rock-cracking hammer if you plan to do much crevice digging. With the former, the pick end can be used to open crevices, while the hammer end can be used to shatter rock. The all-steel pick-hammer gets high marks for durability, but most miners prefer the wooden-handled type because it is better balanced, light in weight, and transmits less shock and vibration when you're hammering rock.

You need a trowel and shovel for scooping up the debris generated by your crevice-cleaning tools, chisels, and hammers. A simple garden trowel may suffice, or, if you feel a shovel is necessary, consider a U.S. Army trenching shovel, the fold-up kind, available in camping supply stores. It's light in weight, can be used in cramped quarters, and has a pointed blade, necessary for scooping up rock pieces.

Use one of the digging tools to loosen the material in the crevice, then scoop it out with the spoon. Be sure to dig all the way down to the bottom. That's where most of the gold will be found. Don't discard any plants that may be growing in the crevice sand. Their roots will have penetrated to the very bottom of the opening, maybe even beyond, and the root hairs may have latched onto flakes or particles of gold. These will come loose when the plant is swirled about in your gold pan.

Even the narrowest of bedrock cracks should be probed. For this operation, a bulb snifter may be practical. This consists of a narrow tube fitted with a rubber bulb at one end. By squeezing the bulb, you draw in a quantity of silt; then pour the extract into your storage container for panning later.

Don't overlook any tufts or sods of moss growing on the rock formations. "You can't help but succeed as a miner when you dig moss," says one veteran prospector. "Moss on rocks acts exactly like matting in a sluice box in trapping gold particles."

Scrape the moss from the face of the rock with a knife or small chisel. Then, with a small brush, sweep it into a dustpan. The final step is to wash and pan the moss. Another, and perhaps better method, is to dry the moss, burn it, and then pan the ashes.

Wooden-handled pick-hammer is rated superior to all-steel type.

Small penknife is used to scrape moss from slatey rock.

Probing a rock crevice.

Many individuals gather their sand and gravel from bench placers, that is, areas often adjacent to and immediately above the present streambed. These formations consist of gravel deposits on top of bedrock along what was formerly the stream bed. The material to be worked is carried by bucket to the sluice or rocker.

If you're operating in an area where the gravel covers a wide range in size, you may want to use a grizzly to separate the larger stones from the bulk of the material you plan to sluice. Although it adds another step, screening the material provides a significant boost in efficiency.

The Sluice Box

In years past, sluice boxes were built of wood planks. Some modern-day sluices are made of wood, but most are constructed of aluminum and plastic. Light in weight, weighing from 5 to 10 pounds, they're much easier to store and carry. Sluice boxes vary in length from 3 to 5 feet, and from 10 to 16 inches in width.

Recreational miners use sluice boxes to process virtually every type of gold-bearing gravel. There are even instances of sluices being used to process gold-bearing minerals in desert areas. Of course, this means that water has to be trucked in, then used over and over again. Even prospectors who operate suction dredges keep a lightweight sluice available to sample gravel bars that might prove worthwhile to dredge. If it develops that the bar offers little potential, the labor of carrying in and setting up the dredge has been avoided.

Operating the Sluice Once you've located a promising deposit of gold-bearing gravel, search for a spot where the stream is flowing swiftly and the water is deep enough to almost cover the sides of the trough. To test the water's ability to wash the gravel efficiently, scoop up a handful of gravel from the stream bed and drop it onto

Take your time getting the sluice positioned. Use rocks to hold it in place.

the upper end of the trough. If the current is able to wash the lighter gravel components through the trough in just a few seconds, then you've found a good location.

If, on the other hand, the stream isn't flowing fast enough, you may be able to compensate by adjusting the slope of the sluice. Place some rocks beneath the upper end of the sluice, making it higher than the discharge end. You can also use rocks to brace the sides of the sluice should the current be so swift that it changes the box's position. It can take as long as half an hour to get the sluice operating properly.

When you begin feeding gravel into the upper end of the sluice, do so in carefully regulated amounts. Too much gravel will overload the riffles. When this happens, you no longer have a sluice box, merely a free-flowing trough, and the gold particles will ride over the material that's clogging the riffles and out the discharge end.

So keep your eye on the riffle system. If you can't see the top or uppermost portion of each riffle bar at all times, you're feeding in too much gravel too fast. Slow down!

When rocks get trapped in the riffle section, pick them out immediately. They can

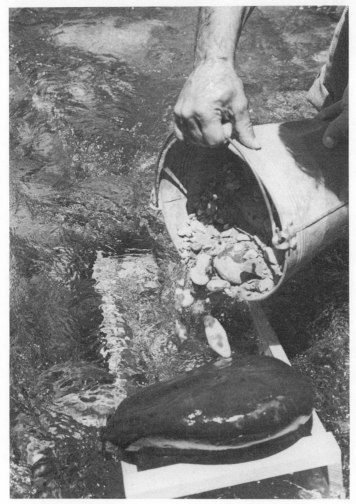

Feed gravel in carefully regulated amounts.

disrupt the water's flow, and cause the current to wash away much of the concentrated material that lies adjacent to the rocks. This is not a big problem should the rock be lodged in the upper portion of the sluice; the washed-out concentrate will simply be trapped by the riffles beyond. But if the wash-out occurs toward the lower end of the trough, the concentrate may be sluiced out of the box and lost.

Also be watchful of the tailings. They can pile up just beyond the discharge end of the box, and act as a dam to slow down the flow of the water. Eventually, the back-up water can flood the lower part of the trough. The solution is simple—just shovel the tailings away from time to time.

Although the sluice box is much more efficient than the gold pan in concentrating gold, it doesn't work automatically. It needs constant tending. That's why sluicing is frequently a two-person operation, with one individual obtaining the sand and gravel to be washed, while the other minds the sluice.

Removing the riffle section.

Cleaning the Riffles When the riffles have collected enough black sand, it's time to interrupt the operation and perform the cleanup. If you're using a sluice box with conventional bar riffles, you should clean out the riffle system whenever the sand accumulated at the bars extends more than halfway up the trough to the next higher bar.

Be very careful when you lift the sluice box from the stream. Keep it as level as you can so that none of the collected sediment spills out. Set it down on the stream bank. Remove the riffle section. You'll notice that there will be sand clinging to each of the riffle bars. Stand the section in a bucket, tub, or gold pan, and wash it off with water from another bucket. The material you collect should be panned along with whatever has been collected by the sluice.

Roll up the matting that lines the bottom of the trough and wash it in a water-filled bucket. Be sure to wash off every bit of concentrate.

Tiny particles of concentrate will have worked their way beneath the matting and have lodged in the bottom of the sluice, so wash the trough, collecting the water in your concentrate bucket. In other words, every part of the sluice should be carefully washed.

The final step is to pan the concentrate to recover the gold.

Two men can sluice as much as a ton of gravel in a day. One man digs the gravel and carries it to the sluice. The other feeds and tends the unit. Besides the sluice itself, you need several metal buckets and a sturdy shovel—and the willingness to work diligently.

The Rocker

Of all the gold-concentrating devices that depend on manpower exclusively for their operation, none is more efficient than the rocker. Widely used by early miners, it is not

Washing the riffles. Use gold pan to collect concentrates.

Removing the matting.

seen today as frequently as either the sluice box or the suction dredge. It undoubtedly deserves more popularity than it's achieved.

The rocker pictured on these pages is quite typical, consisting of two separate collecting systems that work in combination with the riffles of a sluice box. One person can operate the unit, but with two people it's much more efficient. As one man keeps filling the rocker hopper with sand and gravel, the other bails water into the hopper and rocks the unit in cradlelike fashion. Since the device manages to retrieve even the finest gold particles, some miners use the rocker to concentrate materials collected in sluice boxes and dredges.

The hopper takes the form of a screen classifier—a sifter—trapping the coarse pieces of gravel. The finer gravel, sand, and gold particles pass through the screen to fall upon a canvas apron stretched over a wooden frame. Here most of the black sand concentrates are trapped. Beyond the apron and lining the bottom of the unit, there's a riffle system that also catches heavy concentrates.

The rocker has many advantages over other gravity concentrators. It's light in weight and can be dismantled for easy transport. It can be set up just about anywhere, even in a desert context, where it can function as a dry washer. No pump or power equipment is involved and, thus, the rocker is quiet and nonpolluting; yet it's capable of handling truly large amounts of material.

Operating the Rocker In most cases, you'll want to set up the rocker as close as possible to a stream. Position the back edge of the unit about three inches higher than the front edge. If, however, the material you're handling contains an abundance of black sand, you'll want to increase the rocker's slope. If there is more than the usual amount of clay in the material, and the clay tends to bind the gravel, less slope is what's needed. Very fine gold particles also demand less slope.

Be sure the rocker's two support boards lie flat. Adjust the position of the unit so the metal pins within each of the rockers line up with the related holes in each of the support boards. As you rock the unit, the pins plunge into the board holes, and this prevents the unit from slipping from side to side.

Washing the matting. Material recovered then panned.

Rocker hopper traps coarse gravel.

Wood-framed canvas apron fits beneath rocker screen to snare fine gold particles.
It's angled so as to thrust coarse material onto unit's riffle system.

Metal pins in each of the two curved rockers fit into holes in support boards, preventing slippage.

Material is dumped into hopper.

Once the unit is in place, begin shoveling material into the hopper. Or you can do as many miners do: Collect the material in buckets, and then dump the buckets into the hopper. Once the hopper is loaded, buckets of water are bailed in and the unit is shaken vigorously. The shaking sifts material through the hopper and onto the canvas apron. All but the finest particles then drop into the riffle system.

With experience you'll learn how much water to use. Too much water can carry the finer materials right through the rocker system, and the black sand and fine gold particles will never be caught. Too little water can create a muddy mess.

Once the fine gravel and sand have passed through the hopper screen, and the gravel that remains in the hopper appears to have been washed clean, examine the hopper contents for nuggets. The hopper, being removable, can then be lifted out and dumped.

Whenever you remove the hopper, inspect the canvas apron. If it's filled with so much material that it can't handle any more, remove the apron and wash its contents into a bucket or pan for panning later. You may want to screen this material through a classifier to reduce the amount of material to be panned. In days past, miners used to save worn apron material, dry it thoroughly, and then burn it. The ashes were then panned to recover any fine gold particles they might contain.

The rocker's riffle system is not unlike the riffle system in the conventional sluice box, and should be cleaned in much the same way and with the same thoroughness. After removing the riffles, wash off any material clinging to them into a bucket or gold pan. Again, you may want to sift the material before panning.

Remove the carpet by rolling it up into a loose cylinder, and dunk it in a bucket of water to clean it. Last, flush out any material that remains in the bottom of the rocker down the trough into a bucket or tub.

According to one estimate, in one hour of operation, a gold rocker can handle an amount of material equal to that contained by one thousand gold pans. That estimate may be a bit enthusiastic, but it gives some idea of the rocker's high performance rating.

Working Dry Placers

Placer deposits in the desert regions of southeastern California and elsewhere are being worked with increasing frequency, even though the methods involved in recovering gold are slower and less efficient than in the case of streambed placer deposits.

Ideal ground for dry concentrations is not easy to find. Once you put a shovel in the ground, and penetrate the surface, you're almost certain to encounter soil with sufficient moisture to prevent the efficient separation of the gold from the unwanted sand and gravel. In other words, what you have to find is a really *dry* placer.

In days past, dry "washing" was performed by winnowing. The gold-bearing sand would be placed in a blanket, and then two men, holding the blanket at the ends, would keep tossing the concentrate into the air in a strong wind. The lighter particles would be blown away, while the gold particles would fall back onto the blanket. The weave of the blanket would trap the finer gold particles. Or, in the cases where a wool blanket was used, an electrostatic charge would build up on the blanket surface, which helped the blanket to hold the tiny gold particles.

One of the most successful of dry-placer operations dates to a time many years ago, when Mexicans and Indians, working in the El Paso Mountains, Chocolate Mountains, and Picacho and Potholes areas of Imperial County, California, used hand-operated dry washers to recover gold on a large-scale basis. The dry washer operates, at least to some extent, on the same principles as a sluice box, except that there's no water involved. Sand and gravel to be concentrated are shoveled onto the screen (*A*). The finer material passes through the screen into the hopper (*B*). From the lower end of the hopper, the material drops onto the riffles (*C*). Air from a bellows system (*D*), operated by turning a crank (*E*), lifts the lighter

Water goes in and box is rocked vigorously by means of upright handle.

Cleaning the rocker's riffles.

Removing the rocker carpet.

Forty-Niners winnowing gold near Chinese camp. (*New York Public Library*)

The dry washer. (*California Division of Mines and Geology*)

operated by turning a crank (*E*), lifts the lighter particles over the riffles and off the lower end of the washer, while the heavier material remains trapped behind the riffles.

Dry Concentrator The operating principles common to the dry washer have been incorporated in the dry concentrator, a modern system for handling dry placer deposits which plays about the same role as the sluice box does in working stream gravels. In the separation and recovery of gold, the concentrator employs the principles of gravity and also electrostatic attraction.

The Model DW-1, a product of Keene Engineering, weighs about 19 pounds and folds to carry in a back pack. It consists of a hopper with a classifying screen mounted at about a 90-degree angle to the recovery tray, within which a riffle system is contained. The bottom is covered with heavy-duty synthetic cloth.

That's not all. Mounted beneath the recovery tray is a vinyl-reinforced bellows.

The dry concentrator works like this: You, as the operator, shovel material into the steel hopper. The classifier screen permits only material ½ inch in diameter or larger to pass into the riffle system.

Once the hopper is filled, you keep jerking a pull cord that operates the bellows. Air is forced through the cloth that covers the bottom of the recovery tray, and sets up an electrostatic charge in the cloth. Even though gold is nonmagnetic, it has an affinity for any electrostatic charge; thus the tiniest gold particles adhere to the cloth in just the way tiny bits of paper are attracted to a comb after you pass it through your hair.

The recovery of black sand is also accomplished by the riffle system, of course. As the gold-bearing material passes over the riffles, the heavier sand and gold particles are trapped for panning.

The slope of the hopper and classifier screen is adjustable so you can change the rate at which the material flows. The slope of the riffle-system tray can also be adjusted.

To clean out the riffle system, a task that must be performed every hour or so, you simply lift the riffles out of the tray, and let the concentrate slide into a bucket.

For best results, the material handled by the dry washer should be thoroughly dry. The unit can process up to a ton of "bank run" gravel in an hour. The Model DW-1 costs about $125.

For speed and efficiency in handling large amounts of sand and gravel, nothing equals the suction dredge.

CHAPTER 5

All About
Dredges

SHOVELS, PANS, SLUICES, rockers, and the other equipment mentioned up to this point come right out of gold rush days. And the methods involved in using these are the same. From a standpoint of technology, the one big change in placer mining is represented by the suction dredge. It has made underwater prospecting and mining a reality.

The suction dredge is almost always operated by two individuals. One wields the hose that sucks up underwater sand and gravel, while the other tends the sluice box, motor, and pump at the surface, all of which are mounted on a flotation assembly. When the sluice box becomes filled with black sand, the unit is shut down, the riffles cleaned, and the concentrates recovered for panning later.

If you can run a power mower, you'll be able to operate a suction dredge. Although a knowledge of scuba diving is essential if you're going to be operating in deep water, there's no necessity to equip yourself with the breathing apparatus, wet suit, and all the rest. Many dredge operators work the shore line and seldom venture into deep water. Most dredging, in fact, is done in water that is less than 10 feet in depth.

The suction dredge has two important advantages over other types of mining. The dredge enables you to process much more gold-bearing gravel than it is possible to do otherwise. Suppose you've been relying on a sluice box to do your processing. Well, with even the smallest dredge, one with a 2-inch nozzle, you'll double or triple the amount of sand and gravel you can handle.

The dredge's second advantage is that it makes it much easier for you to thoroughly clean out bedrock crevices, some of which might be impossible to empty in any other way.

There's really nothing new about dredges in gold mining. During the 1880s, a number of crude, steam-powered suction dredges were put into use on the Feather, Stanislaus, and Yuba rivers. In these operations, divers would plunge to the stream bottoms to loosen packed gravel and clear boulders away from the dredge's intake pipe. In the early 1900s, diving operations were conducted on the American River, with divers attired in deep-sea diving suits and heavy metal helmets.

It wasn't until the introduction of scuba equipment in the late 1940s that suction dredges achieved widespread popularity. With a couple of air tanks strapped to his back, breathing through a regulator, and clutching a vacuum hose, a diver could work streambeds that had been untouched for many decades. Enormous amounts of placer gold were recovered by the first scuba miners. Reports of their finds triggered a great deal of private development work in dredge manufacture.

Dredges today are much more efficient and portable than those of the 1950s and 1960s. One of today's smaller units, equipped with a 2½-inch nozzle and weighing about 70 pounds, can perform the same amount of work that a 300-pound unit did 15 years ago.

Types of Dredges

There are two types of dredges—the surface dredge and the submersible. The surface variety is the sort you're most likely to use. As its name suggests, it rests on the surface of the water, usually riding on a big inner tube, and dredges gravel from below. Besides the flotation assembly, the unit consists of a suction hose, suction nozzle, a gasoline-driven jet pump, hopper, and sluice box.

Typical suction dredge functions much as a vacuum cleaner. (Drawing, *California Division of Mines and Geology***)**

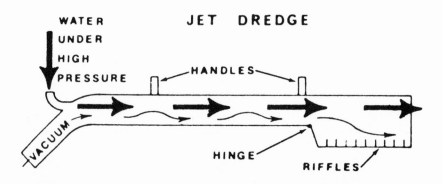

It works like this: The suction hose connects to a device called a suction jet or eductor. Water pumped down the pressure hose to the jet under high pressure creates a powerful vacuum in the suction hose, which causes it to suck up sand and gravel. As the material is drawn toward the sluice box, it first enters a hopper where the larger pieces of gravel are screened out, while the finer material passes into the sluice box.

The submersible dredge is simpler, consisting of a metal tube which is curved at one end. It can be from 4 to 8 feet long and weigh as much as 20 pounds. At the other end of the tube are a series of riffles. High-pressure water is pumped through a connection at the bend in the tube, creating suction which picks up the gold-bearing gravel and passes it over the riffles.

Although the submersible dredge weighs less, it is not nearly as popular. This is largely because the riffle system in the submersible dredge is not as efficient as the surface dredge's sluice box. In other words, with a surface dredge you're much more likely to get finer gold. According to Jerry Keene of Keene Engineering, the largest supplier of dredging equipment, a well-designed surface dredge should be able to recover 100- to 150-mesh flour gold.

Submersible dredges are generally used for removing overburden.

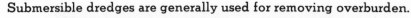

Another advantage of the surface dredge is that you're able to discard your tailings a good distance from your work area. With the submersible unit, tailings can easily accumulate in the area being worked.

Submersible dredges are used mostly for removing overburden. Once this has been accomplished, the area is then worked with a surface dredge.

As you've undoubtedly gathered by now, each dredge is classified by the inside diameter of its suction hose. Thus, a 2-inch dredge, the smallest size in general use, has a suction hose that is 2 inches in diameter. Some novice miners believe that when a dredge is described as being 2 inches in size, the figure refers to the diameter of the suction nozzle or the intake pipe. Not at all. The suction nozzle or intake pipe is always smaller in diameter than the suction hose, the difference ranging from ½ inch to a full inch. This is to prevent big rocks from being drawn through the nozzle and into the hose, clogging it. Plug-ups occur in the nozzle, of course, but they're easy to remedy.

There is a direct relationship between the diameter of the suction hose and the amount of gravel that can be moved in any given period of time. By implication, a law of hydrodynamics states that when you double the inside diameter of a hose or pipe, you increase its carrying capacity by a factor of four. Suppose you're able to move one cubic yard of material per hour with a 1½-inch dredge. If you traded that dredge in for a 3-inch model, you should be able to move about 4 cubic yards of gravel an hour. (Of course, there would have to be a related increase in the size of the pump and the engine.)

A word about yardage-per-hour figures: An example is a dredge described as "having the capability of processing up to one cubic yard of loosepack gravel per hour." This description and others like it refer to the maximum amount of gravel the dredge can handle under ideal conditions. "Ideal conditions" refers to gravel that has been screened and sorted until it resembles a pile of soy beans. The gravel that you encounter in any given streambed is not going to be sorted and screened. Much of it is going to have to be moved by hand or with a pry bar. Some pieces are going to momentarily clog the intake pipe.

Another figure you'll encounter is the one describing the unit's potential dredging depth. A typical 3-inch dredge is described as being "capable of processing up to 8 cubic yards of loosepack gravel per hour from depths as great as 30 feet." But to be able to pull gravel from 30 feet you have to be able to operate the engine at full speed, and do so continually. This puts undue strain on the engine.

The thing to do, of course, is to purchase a dredge with a capacity to operate well beyond your anticipated requirements. If you feel there's much of a chance that you're going to be operating in 20 feet of water, then you should buy a dredge that's capable of handling gravel from depths down to 30 feet.

The power generated by the engine and pump is not only affected by the operating

Portability is important feature to consider in buying a dredge.
(*Treasure Emporium*)

speed of the engine. The temperature and altitude are other factors that must be considered. For every 1,000 feet of altitude above sea level, engine power decreases by 3½ percent. For every 10 degrees of temperature above 60° F, engine power is reduced by 1 percent. Thus, if you're working in a mountain stream on a hot summer afternoon, the power loss your engine suffers can amount to as much as 15 percent.

Buying a Dredge

Picking out a surface dredge is something like choosing a family automobile. Everyone has a different set of requirements. How much money you want to spend is an important consideration. Surface dredges range in price from about $500 to several thousand dollars.

Your aims are important, too. Maybe you only want to recover enough gold particles and flakes to put into a vial and show your friends. On the other hand, maybe you want to seek out "big stuff" in remote mountain streams.

Where you're going to be working has to be considered. Is it a shallow stream or a deep river? Portability—how much the rig weighs—is another factor. Are you going to be able to pull your automobile or van right up to the water's edge, or are you going to have to tote the rig over hill and dale to get to the stream?

No matter what your ultimate decision, it's wise to buy dependable and durable equipment, quality equipment. A breakdown in the river is seldom less than a catastrophe, ruining your day and perhaps an entire vacation. You want a dredge system that's going to function faithfully in the rugged terrain you're going to be encountering, and this implies you should avoid mass-produced equipment.

Increased nozzle size and engine power mean increased weight, too.

What Size Dredge? When it comes to deciding what size dredge to buy, several factors have to be weighed. Most dealers recommend that the novice begin with a small, inexpensive unit. If you find you're successful in its operation, and your interest continues to build, you may then want to move up to one of the larger models.

Naturally, the bigger the unit, the more material you're going to be able to process in any given period of time. But as the capacity of the dredge increases, so do its size, weight, and cost. What one manufacturer classifies as "small" dredges are those that range from 2 to 6 inches in hose size and that have a capacity of from 4 to 20 cubic yards of material per hour. It's not likely that you, as a recreational prospector, would want to be involved with anything bigger.

Dredges that are 2, 2½, and 3 inches in size are meant primarily for shallow-water work and for areas where the bedrock is easy to reach. Unless you have a 4-, 5-, or 6-inch dredge, you can't handle more than 4 or 5 inches of overburden with any efficiency.

As the size and horsepower of the engine increase, it becomes practical to add a

compressor to your equipment package. The compressor provides air through a long length of hose to an underwater diver. Actually, the air hose is connected to a check valve at the diver's chest and is routed to the diver's regulator. This is the *hookah system* of underwater diving. While the compressor air also passes through standard air tanks, the diver does not have to wear them, increasing his mobility. The tanks are for emergency use in case of compressor breakdown. With the compressor pumping air to him, the diver can stay underwater for as long as the engine keeps running.

The hookah system significantly increases one's range. But, naturally, as the amount and size of your equipment increase, so does the weight of that equipment. You may need additional people to help you carry the equipment to your mining site.

For underwater mining, equipment package includes air compressor. Tanks are for emergency use.

This table gives an idea of the relationship between the hose diameter and the weight of the related equipment package:

Hose diameter	Equipment weight
2½″	70 pounds
3″	100 pounds
4″	200 pounds
6″	400 pounds

Flotation System   When you're choosing a dredge, you'll also have to make a decision as to how you're going to float the rig in the water. Sometimes the flotation gear is included in the basic equipment package, but at other times you'll have to purchase it separately, then mount the engine, pump, and sluice box on it.

Inner tubes, almost always *truck* inner tubes, constitute the most popular type of flotation system. Since they can be deflated very quickly, they're easy to pack and transport. Of course, you have to carry along an air pump, too.

The principal disadvantage in using inner tubes is that they can puncture and deflate. If that happens in fast water or deep water, your engine and pump could be irretrievably lost.

Inner tubes are also susceptible to pinhole leaks. While a small leak may not be as disastrous as a big one, it can prove troublesome by altering the plane of the sluice box, and thereby disrupting its operating efficiency.

Before you start your dredging operation for the day, inspect the tubes carefully for punctures. Fill the tubes with air and then submerge each one in still water and watch for air bubbles. Be sure to carry a tube-patching kit with you so that you can repair punctures on the spot.

Inner tubes are popular as flotation assembly. This is double-truck tube assembly.

Styrofoam pontoons are said to lack durabililty.

If your dredge floats on styrofoam pontoons, you don't have to worry about punctures, but there are precautions you have to follow. When filling the engine with gasoline, be careful not to spill any of the gas on the styrofoam, for it causes the substance to disintegrate. Also, styrofoam lacks durability: It doesn't stand up well to rugged terrain, to being dragged down hillsides or over rocks. The best way to protect syrofoam pontoons is to wrap them with ducting tape.

Plastic-module floats, the newest innovation in flotation gear, are the most durable of the three types, but there also can be problems associated with them. The plastic can dry out in the hot sun and thereby become susceptible to cracking. A collision with a rock can splinter the dried plastic, causing the unit to flood and sink.

If you have a flotation system of this type, the upper surface of the unit is likely to be fitted with a small ventilation hole. Never let the hole get clogged with sand. It's meant to allow for air expansion on very hot days, preventing the module from rupturing.

One of the more sophisticated flotation systems, developed by Rainbow Mining, involves the use of a 10- to 14-foot aluminum boat hull, which is fitted with a fiberglass deck. Polyurethane billets within the hull make the unit virtually unsinkable.

Types of Dredges

The pages that follow describe in detail suction dredges of several different sizes.

2-inch Companies often refer to models of this size as "mini-dredges" or "back-packer dredges." Portability is one of their leading features. They're very popular among novices. Gold Grabber manufactures a 2-inch model that is equipped with a

Aluminum boat hull, fitted with polyurethane billets and fiberglass deck, is latest word in flotation gear.

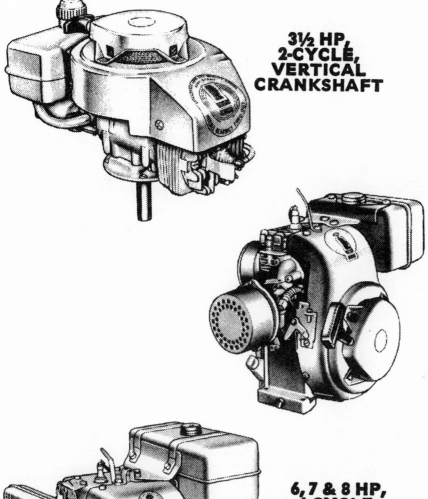

**3½ HP,
2-CYCLE,
VERTICAL
CRANKSHAFT**

**3½ HP,
4-CYCLE,
HORIZONTAL
CRANKSHAFT**

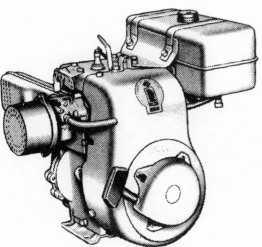

**6, 7 & 8 HP,
4-CYCLE,
HORIZONTAL
CRANKSHAFT**

Dredge engines of various sizes and styles. (*Gold Grabber Manufacturing Co.*)

Keene Engineering's 2-inch suction-nozzle backpack dredge. It weighs a mere 38 pounds. (*Keene Engineering*)

padded shoulder sling. The unit is powered by a 2-hp Fuji/Robin engine and pump. The sluice system is incorporated in the flotation assembly.

Other models of this size are powered by a 1½-hp, 2-cycle engine that delivers approximately 75 gallons of water per minute to the power-jet eductor. These rigs can operate for 5 hours on a gallon of gasoline. Most manufacturers furnish 10 feet of suction hose with such units.

2½-inch Frequently seen on the streams and rivers of the Western states, the 2½-inch dredge is practical wherever you expect to encounter a fair amount of overburden. The rig can move as much as 4 cubic yards of gravel an hour from depths as great as 10 feet. Most models weigh between 75 and 100 pounds. The typical unit of this size is powered by a 3- or 4-horsepower engine, and equipped with a pump that puts out approximately 100 gallons of water per minute. The newer 2½-inch surface dredges are usually furnished with power-jet eductors, while the older rigs employ the suction vacuum system. Some firms, however, are still producing

Treasure Emporium's 3-inch surface dredge. (*Treasure Emporium*)

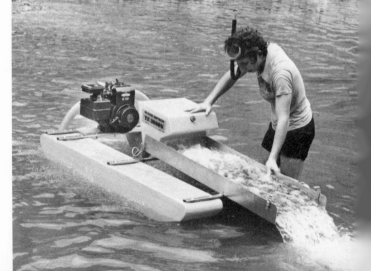

breed, recommending them for use in shallow water. The 2½-inch power jet dredge is furnished with a sluice box that is 52 inches long and 10 inches wide. The rig will run for up to 3 hours on a gallon of gasoline.

Gold Grabber's 2½-inch dredge is fitted with a grizzly screen that prevents rocks from entering the riffle system, and a built-in recovery drawer for quick sampling. The sluice system can be purchased as a separate unit for installation with other dredges.

Gold Vac's 2½-inch rig is powered by a 3½-hp, 4-cycle engine. The unit boasts a direct-coupled submersible pump that eliminates manual priming. Weighing only 65 pounds, the unit can operate for 2 hours on a gallon of gasoline.

3-inch Models of this size and larger are often classified as workhorse rigs, since they're capable of moving substantial amounts of overburden and gravel from deeper-than-normal depths ("deeper than normal" being more than 10 feet). Keene Engineering's Model 3501, for example, can process up to 8 cubic yards of gravel per hour from depths of up to 30 feet. Its power unit is a 5-hp engine, and it offers a pump that delivers approximately 175 gallons per minute to the power-jet eductor. The sluice box is 52 inches long and 14 inches wide. The unit weighs 87 pounds. You get 2½ hours of dredging time per gallon of gasoline.

A 5-hp engine has enough reserve power to run an air compressor to support underwater diving operations. The compressor is driven by a V-belt from a pulley on the engine shaft. Two divers can work off this system, penetrating to depths down to 50 feet. Including the air compressor, the unit weighs approximately 110 pounds.

The Model 3501 also has the versatility to be used in combination with a 5-inch submersible dredging tube for removing overburden. Fitted with a detachable riffle tray, the 68-inch tube is capable of moving up to 12 cubic yards of material per hour. Once you've used the 5-inch dredge to remove the overburden, and the gold content of the gravel starts improving, you switch over to the 3-inch surface dredge, with its bigger sluice box. Total weight of the unit, including both dredge systems, air compressor, hoses, nozzles, and flotation device, is 155 pounds.

4-inch With suction dredges that are 4 inches and bigger, you enter the realm of high-capacity work. A 4-inch dredge employing a power-jet eduction system can move as much as 12 cubic yards of gravel per hour from depths as great as 30 feet. The engine is usually rated at 7 or 8 horsepower; the pump delivers approximately 225 gallons of water per minute to the eductor. The sluice box is 63 inches long and 16 inches wide. The unit can operate for 2 hours on a gallon of gasoline. It weighs about 180 pounds.

The 4-inch surface dredge can easily be fitted with a compressor that will supply the air requirements of two divers operating at a maximum depth of 50 feet. It can also be used in combination with a 6-inch submersible dredge. This 6-inch tube, 80 inches in

Dredging with Nevada-Rainbow Mining's 4-inch unit.

length, can move up to 16 yards of overburden per hour. Once you begin nearing the bedrock, you switch over to the 4-inch surface dredge. The total weight of the "four-six combination," as it's usually termed, is 225 pounds.

Rainbow Mining calls its 4-inch the Streamliner. It features a 10-foot aluminum hull with a fiberglass deck that doubles as a motor mount and into which the sluice system is recessed. The hull is also fitted with polyurethane-foam billets which render the rig unsinkable under normal conditions. Grizzly bars over the riffle system keep out rocks. Another feature is a nugget trap at the sluice head to capture the coarse gold. Concentrates caught in the nugget trap can be panned to give you an idea of the richness of the stream or area you're working. In other words, you don't have to pan all of the concentrates trapped by the riffles. The Streamliner operates efficiently in shallow water as well as deep water.

5-inch While there are a few firms that offer 5-inch surface dredges, it is not considered a standard size. With most manufacturers, the jump is from 4 inches to 6 inches. The 5-inch size offered by Keene Engineering, the Model 5010, can process up to 16 cubic yards of gravel per hour from depths of up to 30 feet. The rig is powered by a 10-hp engine and its pump delivers up to 250 gallons per minute to the power-jet eductor. The sluice box is 64 inches in length and 18 inches wide. A compressor can be added to the rig, supplying air to two divers. The unit operates 1¾ hours per gallon of gasoline. It weighs about 260 pounds.

6-inch A high-powered, high-volume unit, the 6-inch surface dredge is meant for commercial mining operations and for those individuals involved in large-scale operations over extended periods of time. The rig is capable of moving up to 20 cubic yards of gravel per hour from depths of up to 30 feet. But forget portability; a 6-inch

Grabber calls this 4-inch model the Trapper. (*Gold Grabber, Inc.*)

is Keene Engineering's 4-inch sur-dredge.

5-inch unit provides powerful flow over riffle system.

Grizzly in Nevada-Rainbow Mining's dredge keeps rocks out of riffle system.

surface dredge weighs about 500 pounds. The unit is usually powered by a 4-cycle industrial engine that delivers 16 hp, but some manufacturers offer a 20-hp engine when greater operational depth is required. The unit's high-volume pump is capable of delivering well over 500 gallons of water per minute to the power-jet eductor. The sluice box is 8½ feet long and 22 inches wide.

The rig's compressor hook-up can furnish air to two divers to depths of 500 feet. While the overall weight is substantial, the unit breaks down in several pieces for transportation. The heaviest component is the pump and engine assembly, which weighs approximately 135 pounds. The 6-inch dredge consumes one gallon of gasoline per hour.

Rainbow Mining's 6-inch dredge, called the "Gold Factory," is mounted on a 14-foot boat assembly. It offers many of the same features as the company's 4-inch model (see above).

8-inch You may hear of 10-, 12-, and even 14-inch dredges, but the 8-inch surface dredge is the biggest size readily available. Like the 6-inch rig, it's designed for commercial use. The 8-inch unit can handle up to 30 cubic feet of gravel per hour in depths of up to 30 feet. Keene Engineering's 8-inch surface dredge, the Model 8032, is powered by a pair of 16-hp, 4-cycle, single-cylinder engines. The high-volume pumps deliver 1,100 gallons per minute to the power-jet eductor system, which is fitted with a special "twin eductor." The unit requires two gallons of gasoline per hour. Total weight of the rig, including the air compressor to supply air to a pair of underwater divers, is 750 pounds.

Electrostatic Concentrator

When it comes to handling gold-bearing material in dry form, the motor-driven electrostatic concentrator is the counterpart to the surface dredge. One of Keene Engineering's concentrators uses a high-volume blower assembly, powered by a 3½-hp gasoline engine, and 10 feet of 4-inch vinyl ducting hose, to deliver partially heated air to the unit's plastic recovery tray. There the air becomes electrostatically charged. The tray is covered by a synthetic cloth which helps to increase the intensity of the charge, and a hinged set of riffles. Of course, the material is also classified as it travels downward through the tray's riffle system.

All you, as the operator, have to do is shovel material onto the unit's aluminum trommel screen, which permits only gravel ½ inch in size or smaller to pass into the plastic tray. A simple turnbuckle arrangement enables you to adjust the tray angle to accommodate gravel of various sizes.

Under normal operating conditions, the riffle system should be cleaned about once every hour. This is easy to do. You simply raise the riffles, and the concentrate slides down the tray into a bucket. The cloth should be cleaned of all excess material before the riffles are replaced.

The total weight of the unit (the Keene 150) is 78 pounds. It runs about 2¾ hours on a gallon of regular gas, and is capable of handling as much as two tons of "bank material" per hour.

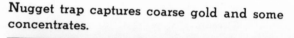

Nugget trap captures coarse gold and some concentrates.

This electrostatic concentrator is powered by a 3½-hp Briggs & Stratton engine that drives its blower assembly. (*Keene Engineering*)

Manufacturers of Dredges

Dredge dealers, located in abundance almost anywhere gold-bearing streams are to be found, are too numerous to list. Consult the Yellow Pages of your telephone directory under "Mining Equipment and Supplies." The principal dredge manufacturers, along with a handful of nationally known suppliers, are listed below. Most will be happy to supply you with a free catalog.

ATC Wet Washer
10092 Soquel Dr.
Aptos, CA 95003

Eureka Gold &
   Mining Co.
325 North First St.
San Jose, CA 95113

Exanimo Establishment
Box 448
Fremont, NE 68025

Gold Divers
P.O. Box 2848
Carson City, NV 89701

Gold Dredge
   Marketing Co.
1744 West 22nd Ave.
Eugene, OR 97405

Gold Grabber
   Manufacturing Co.
P.O. Box 3255
Boise, ID 83703

Keene Engineering
9330 Corbin Ave.
Northbridge, CA 91324

Lo-Sierra Mining
   Equipment
179 Palm Ave.
Auburn, CA 95603

Lucky Leo's
709 West Columbia
Pasco. WA 99301

Nevada-Rainbow Mining
8680 Lincoln
   New Castle Highway
New Castle, CA 95603

Oregon Gold Dredge,
   Ltd.
50 Grimes
Eugene, OR 97440

Precision Machine
21316 Gilber Rd.
Otis Orchards, WA 99027

Spartan Dredge Co.
210-5 Main St.
Ames, NE 68621

Thurman Engineering
545 Swan Hill Dr.
Big Gork, MT 59911

Treasure Emporium
6507 Lankershim Blvd.
North Hollywood,
   CA 91606

V.R.E. Inc.
P.O. Box 1086
Walla Walla, WA 99362

Western Sports
3725 South Stone
Spokane, WA 99203

Dredging Permits

Before you take your brand-new dredge out onto a river, be sure to check whether a dredging permit is required in the state in which you're going to be operating. Many states now have laws regulating the use of dredges, chiefly because they wreak havoc in streambeds, seriously disrupting fish life. These state regulations generally set a time period for dredge use, which usually falls between June 1 and September 15.

Dredge permits must usually be obtained from the state Department of Ecology or Department of Fisheries. In California, it's the Department of Fish and Game. The permit is good for a year and applies anywhere in the state. There is no charge for such permits—as of this writing.

CHAPTER 6

Dredge Operation

INDIVIDUALS WHO HAVE never seen a suction dredge are usually surprised at the ease with which these units operate. This is especially true of the 2-, 2½-, and 3-inch surface dredges, some of which can be manned by one person.

It shouldn't take any longer than half an hour to get the dredge set up. Once it's in place on the river, the sluice properly adjusted, and the engine churning away, it's simply a matter of guiding the suction nozzle into the sand and gravel.

If you've purchased a quality dredge and you continually have operational problems with it, the fault may be with you, the operator. The most common failing is expecting the dredge to do too much. You wouldn't try to cut a field of wheat with a power mower. It's the same with a dredge; it's a machine with certain limitations. You can't, for instance, plunge the intake nozzle haphazardly into an accumulation of gravel and expect the dredge to operate without difficulty. Do that and a plug-up is almost certain to occur. Instead, guide the nozzle, keeping it a few inches away from the gravel, so it sucks the gravel toward it.

Always keep an eye on what's happening below the surface. When you spot a big rock that could possibly cause a plug-up, move it out of the way.

Don't force the dredge to work too hard. After sucking in sand and gravel for five or six seconds, lift the nozzle from the gravel and allow plain water to run through the system for a couple of seconds. This is comparable to allowing a distance runner to take a breath of air. You're permitting the vacuum system to move several feet of sand and gravel up the suction hose and into the sluice system before it has to cope with

Key to successful dredging is skillful handling of intake nozzle.

Setting up the dredge for operation shouldn't take more than 30 minutes. Here, operator connects foot valve to pump.

additional material. It will thus be able to handle the sand and gravel efficiently, and the chances of a plug-up are greatly reduced.

Always plan your operation so that you begin dredging on the downstream side of any deposit, working your way upstream. When you follow this method, your tailings don't cause you any difficulty. They fill up any holes you create. But when you work in the reverse direction, from upstream toward downstream, you'll find yourself spending a good amount of time dredging your own tailings.

If you're operating in a fast-moving current, it may be worthwhile tying the dredge to a fixed object on the stream banks. This means, of course, you'll have to pack along rope, specifically, a pair of heavy-duty waterproof nylon ropes, each up to 50 feet in length. Tree trunks make the best anchor points. Tie one end of one rope to the left upstream corner of the dredge, and the other end to a tree on the left bank. Do the same on the right side. You can then work the suction hose without worrying where the dredge and pump are going to go.

Coping with Cold Water

If you have ever plunged a hand or a foot into a mountain stream, you're aware of a major problem that confronts all prospectors—cold water. If you happen to be merely panning or sluicing, the temperature of the water isn't likely to be too much of an inconvenience because you don't have to be immersed in it for very long. But should you be sniping or doing any serious dredging, the water can seriously interfere with the amount of work you're able to do.

Water temperatures can vary over a very wide range, depending on the time of the year. In California's Mother Lode country, the water temperature is sometimes as low as 30° F and as high as 70° F. It's not easy for an individual to endure very low water temperature for very long.

The solution is a diving suit. There are two types—the wet suit and the dry. The dry suit, made of paper-thin sheet rubber, fits loosely, except at the ankles, wrists, and neck, where the fit must be so tight that it seals the water out. Forget the dry suit. It is wholly impractical for mountain-stream diving because the fabric tears easily on rocks or other sharp objects. Once a leak develops, the suit is virtually worthless.

Get a wet suit. Made of ¼-inch neoprene, the wet suit fits tightly from chin to toes, like a second skin. Yet the material allows water to slowly seep through to become trapped between the suit and the diver's skin, forming a warm, insulating layer. A diver can work in water as cold as 30° F in a wet suit, and remain comfortable for hours.

Wet-suited diver primes the pump.

Adjusting the Sluice Box

Before you begin purposeful operation of the dredge, make sure that the sluice box slopes correctly for the type of gravel you're processing. If it's fine gravel, you need less slope than normal. Coarse or heavy gravel demands a steeper slope.

Adjust the level of the sluice box so that when the sand, gravel, and water are running through the sluice, you're still able to see the top half of the riffles. If you're unable to see the top half of the riffles, then the sluice is too level. The lighter waste will back up behind the riffles and clog them. Additional gravel that then pours into the sluice flows right over the waste, never even coming in contact with the riffles. You're no longer classifying material; you're simply operating a conveyor.

Too great an angle is just as bad. If you look into an operating sluice and see each riffle in its entirety, the angle of the sluice is too steep. Gravel is undoubtedly pouring out of the sluice without making any significant contact with the riffles.

You must keep adjusting the sluice, positioning the leveling rod in higher or lower holes, until you've established the right angle. "It has been my experience," says Matt Thornton in his informative handbook, *Dredging for Gold,* "that a slope of one-half inch per linear foot of sluice box will usually do the trick." But, as Thornton himself declares, there will always be special circumstances. Be guided by the general rule that the coarser the gravel, the steeper the slope; the finer the gravel, the shallower the slope.

Operating in Shallow Water

You don't need deep water for successful dredging. Most rigs are capable of operating in any area where the water is deep enough to support the flotation assembly and provide clearance for the pump intake nozzle (in the case of a suction-nozzle type of dredge) or jet tube (common to power-jet units). This can be as little as two feet of water.

Slope of sluice box should be about one half-inch per linear foot.

If you do happen to be operating in relatively shallow water, you may have to contend with flow sand that threatens to clog the intake. Clogging itself isn't the only problem. Flow sand serves to reduce the amount of water that is taken into the pump and passed on to the eductor, and the result is less suction. That's not all. While the pump is engineered to cope with sand in small amounts, sand in quantity can raise havoc with the pump's working parts.

Operating in shallow water can also mean you'll have to fend off various types of vegetable matter, chiefly leaves and roots, that lurk and travel along the stream bottom. Vegetable matter is drawn to the pump intake pipe like black sand to a magnet, and can quickly clog the intake screen, significantly reducing the flow of water into the pump. You have to shut down the engine, and then peel away the leaves and other debris from the intake screen.

There are, however, more efficient methods of coping with both of these problems. Perhaps the simplest is to place the intake-pipe foot valve in an ordinary water pail. First, dig a hole in the bottom in the streambed, and place the pail in it so the top half protrudes. When the dredge is operating with the foot-valve intake nozzle inside the bucket, it will pull in only clear water. Flow sand, leaves, roots, and other debris pass right around the pail.

Another method is to use rocks and boulders to construct a small dam on the upstream side of the foot valve. Flow sand and vegetable debris will be diverted to the right and left of the dam, and the intake pipe will pull in clear water.

Dredging to Bedrock

How you conduct the actual dredging depends on your knowledge of the stream you're working and the gold-bearing deposits it contains. Usually you'll have to make your way through six to ten feet of overburden before you reach bedrock. Of course, there may be times you'll excavate 20 feet of overburden and still not encounter the bedrock. Keep going; it's there. If you're not certain as to the depth of the overburden or the type of bedrock you're going to find, it's best to dredge a steep-sided vertical hole. This procedure gets you to the bottom fast, to make an advance inspection.

The excavation might be four feet on each side at the top and three feet on each side at the bottom—as long as the walls don't cave in. As this set of dimensions suggests, you won't have much working room once you reach it, but at least you'll be able to inspect the bedrock and determine what it holds.

Most dredgers, however, go toward the bottom in stages, excavating material in wide swathes on each side of the hole they create. Once they reach the bottom, the excavated area may be as much as 35 or 40

Most dredges are designed to operate efficiently in shallow water.

feet on each side. A hole of this size gives plenty of working room and makes it easy to move big boulders. Since the walls of the hole slope gradually, you don't have to worry about cave-ins.

An excavation of this size is dredged in stages. You might begin with a hole that is five feet on a side and five to ten feet in depth, depending on the condition of the bedrock and how much caving-in you encounter. The next step is to enlarge the hole by five feet on each side. When you've finished this effort, each side of the hole will measure 15 feet. Enlarge the hole by five feet on each side again, and you'll have an excavation that is 25 feet on a side. Do it yet again, and the excavation expands to a pit with 35-foot sides.

If the bedrock is still several feet beyond, move to the center of the excavated area and start dredging a hole within the hole. Again, make your way toward the bedrock by first excavating a hole with five-foot sides, and then enlarge it by extending the sides by five feet at a time.

Whacking and shaking the suction hose will usually serve to unplug it.

About Plug-Ups

The most frequently heard complaint voiced by novice miners has to do with plug-ups. Gravel and rocks get jammed in the intake nozzle or suction hose, the dredge stops sucking, and operations have to be curtailed until whatever causing the obstruction is removed.

Sometimes plug-ups are caused by faulty use of the intake nozzle. As mentioned earlier in this chapter, you can't heedlessly plunge the nozzle into an accumulation of gravel and expect the dredge to pull it all in without faltering. The proper method is to hold the intake nozzle within a few inches of the gravel; the suction will pull the gravel toward the nozzle.

And you can't force the dredge to pull in gravel without letup. After it has sucked in gravel for five or six seconds, pull the nozzle away from the gravel deposit, and let clear water flow through the system for two or three seconds.

It's always wise to keep your eye on the intake nozzle at all times. Watch what it's pulling in. Be careful of rocks that are too big for the nozzle, and kick them out of the way or shove them aside with your free hand. Watch for unusually formed rocks that can prove troublesome. For instance, a long, thin rock may pass through the intake nozzle without any trouble, but can easily jam in the bent section of tubing just beyond the nozzle.

With experience, you'll be able to identify at what point within the hose system the plug-up has occurred. The simplest obstruction to recognize is the one that occurs just inside the intake nozzle. With this type of plug-up, there's no suction at all because the vacuum created by the pressure hose is unable to draw any water. It's also the easiest

Clearing a plug in a power-jet dredge with a jam pole.

obstruction to unplug. Simply thrust a long screwdriver or pry bar into the opening, and work it around until you're able to extract the rock.

With suction-nozzle dredges, the usual plug-up point is in the bent section of tubing just beyond the intake nozzle and orifice. Gravel and rocks lodged at this point cause a powerful backwash to stream from the nozzle.

To free a plug-up of this type, simply insert a screwdriver or a pry bar into the nozzle, and jab at the obstruction. But jab lightly—you don't want to damage the inside walls of the nozzle or pipe. If the obstruction is beyond the reach of the screwdriver or pry bar, disconnect the nozzle from the suction hose and unclog the pipe from the other end.

Plug-ups also occur within the suction hose and at the collar where the hose attaches to the sluice box. Plug-ups of this type can be identified by water flowing from the intake nozzle, but that flow is much less than occurs when the nozzle is plugged. First check the hose. Use your fist and whack the hose every few inches until you've pinpointed the obstructed spot. Then rap the spot a few times with your fist or a rock. If that doesn't do the trick, shake the hose vigorously, and then try a few more whacks. This seldom fails.

If the plug-up occurs at the sluice box collar, you're likely to have to shut down the engine, detach the suction hose from the collar, and visually inspect for the plug-up. It should be simply a matter of unplugging the opening with a pry bar. In some rigs, the sluice's baffle box is fitted with a hinged opening which enables you to inspect the collar and remove the obstruction without having to kill the engine.

Power-jet surface dredges are less troublesome in this regard simply because there are no curved sections of tubing through which gravel and rocks have to travel. When obstructions do occur, there's a standard method of unplugging them. When you purchase a power-jet rig, it comes furnished with a long steel probe rod. You insert this rod into the round, stoppered hole in the face of the sluice box's baffle box, through the baffle box itself, and then through the power-jet tube and into the suction hose. Once the probe has made contact with the obstruction, a few sharp pushes should dislodge it.

You can use the same set of clues as in the case of a surface-nozzle dredge in pinpointing the trouble area. If water is flowing out of the intake nozzle, the plug-up is located between the orifice and the intake nozzle, possibly in the suction hose. Use the same methods of dislodging the obstruction as outlined for the suction-nozzle dredge, that is, rap the hose with a rock until you've located the plug-up, and then whack the spot a few times. If the plugged-up area is located between the orifice and the sluice box, there will be no suction, or almost none.

Engine Operation and Maintenance

If you own or operate a power lawn mower, you'll recognize the engine on your dredging rig as being very similar. Actually, it's quite likely to be exactly the same. The engines used to power surface dredges *are* power-mower engines, with only the barest modifications.

Most engines of this type run on regular, low-lead or lead-free gasoline. But some require gasoline mixed with oil on a 24-to-1 basis, or thereabouts. If you're going to be operating a dredge with an engine of this type, mix a batch of fuel before you leave home to save time in the field.

These engines were designed to be run for an hour or so once every week or ten days.

In a dredging operation, you're going to be running the engine for six or seven hours a day and for two or three consecutive days at a time. Is this going to cause any mechanical problems? It won't if you're careful about engine maintenance. This means, first of all, that you must be meticulous about following the manufacturer's recommendations as to changing the oil. Most engines of this type require an oil change after the first five hours of operation; thereafter, the oil must be changed every 25 hours. To the individual operating an engine in a power-mower context, this means changing the oil every six months or so. To you, operating the engine for sustained periods every weekend, it means an oil change every three or four days.

Don't try to stretch the manufacturer's recommendations. If you have any doubts about this statement, simply examine the oil in the crankcase after the engine has operated for 25 hours. It will be black and sludgy. The engine can't deliver optimum power when run on oil that needs changing. In addition, the sludge can harm the engine's working parts.

Between oil changes, be sure to stay aware of the oil level in the crankcase. Check it at the end of each day's operations, and be ready to add oil if it's needed.

Incidentally, when you purchase your dredging rig, there's not going to be any oil in the crankcase. You have to fill it. Fail to do so before you operate it, and you'll ruin the engine.

Another important maintenance chore is keeping the sparkplug tip free of carbon deposits. You'll have to do this about once a month. Even if you're careful about performing this task, it's a good idea to carry into the field the tools necessary for cleaning the plug and resetting the gap. It pays to be prepared. Also, take along a couple of extra plugs.

About the only other regular maintenance chores you'll have to perform are keeping the air filter and the cooling system free of accumulated dirt.

Starting the engine shouldn't be any problem, unless the air temperature is low. But if you expect cold weather, switch to winter-grade fuel, which is more volatile. Use the proper oil for the temperature you expect. Setting the throttle at a "part-throttle" position and turning the carburetor needle valve an eighth of a turn in a counterclockwise direction also aid in starting the engine in cold weather.

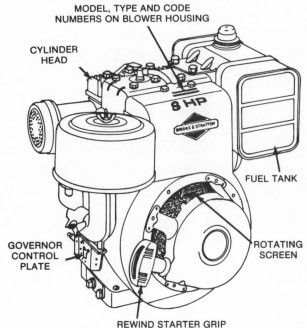

MODEL, TYPE AND CODE
NUMBERS ON BLOWER HOUSING

CYLINDER
HEAD

8 HP

BRIGGS & STRATTON

FUEL TANK

GOVERNOR
CONTROL
PLATE

ROTATING
SCREEN

REWIND STARTER GRIP

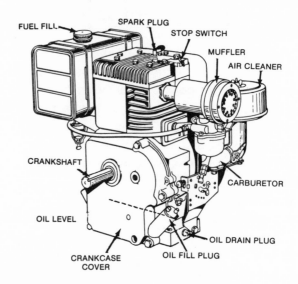

FUEL FILL · SPARK PLUG · STOP SWITCH · MUFFLER · AIR CLEANER · CARBURETOR · CRANKSHAFT · OIL LEVEL · CRANKCASE COVER · OIL FILL PLUG · OIL DRAIN PLUG

The gasoline engine that is included in most dredging rigs is an efficient and quite durable piece of equipment. It will give you years of service—provided you follow the manufacturer's recommendations in its operation and give it the required care.

Operating the Sluice

The riffle system in most surface-dredge sluice boxes is of "Hungarian" or "Lazy L" design. As the water pours over each riffle, a rotary or whirling motion is created, which causes the heavier material to settle into the corners formed by the junction of the riffle with the sluice box sides. Gravel that is light in weight also circulates in this eddying water, but eventually washes out of the sluice.

Throughout the day, keep an eye on the riffle system. If too much black sand concentrate is allowed to build up behind the riffle bars, the eddying effect is destroyed, and you'll start losing flakes and particles that should otherwise be trapped.

When should you clean out the riffle system? Anytime the accumulation of black sand concentrate approaches the midway point between riffle bars. This may mean you have to perform a cleanup every couple of hours. It depends on the type of black sand concentrate you're processing, that is, whether it's largely magnetic (magnetite) or nonmagnetic (hematite). Magnetite is considerably heavier than hematite. If it's mainly magnetite you're handling, there will be a relatively rapid accumulation of sand behind the riffle bars. But if the sand you're processing is richer in hematite, the water flowing through the sluice box will carry away much of the sand, and you'll be able to go for hours without performing any cleanup. Indeed, you may be able to operate for an entire day, and clean up only at the end.

Also keep an eye on the sluice box to see to it that no stones settle behind any of the riffle bars. A stone distorts the flow of water, causing accumulated black sand concentrate to swirl about, and frequently it washes over the riffle bar which had trapped it and sluices farther down the box. Pick out any stones that get trapped by the riffles and toss them away.

When you judge it's time to clean up the riffle system, the first thing to do before you kill the engine is to allow clean water to flow through the box to wash away any lightweight material behind the riffles. Simply place the intake nozzle where it won't pull in any additional gravel, only clear water.

After the lightweight material has been sluiced out, turn off the engine and make a careful inspection of the material in the box. Pay particular attention to the material trapped behind the first three or four riffles. This is where most of the gold particles settle. If you spot any particles, pick them out with tweezers and store them in your

Throughout the day, keep an eye on the sluice box.

gold-sample bottle. In general, follow the clean-up recommendations set down in Chapter 4 concerning sluice boxes.

In this dredge, long steel pin holds sluice box in place.

(Facing Page)
Concentrates from sluice box carpet go in gold pan.

CHAPTER 7

Electronic Prospecting

DURING THE LATE 1970s and early 1980s, scarcely a week passed without newspapers and television reporting on the finding of at least one gold nugget of spectacular size, a nugget that was invariably described in pounds, not ounces. Many of these stories carried an Australian dateline, but Canada, Mexico, California, and Oregon were other places where big nuggets were found.

What triggered the sudden and striking upswing in the discovery of native nugget gold was the introduction of a new electronic detection system. It endowed the novice prospector with professional expertise.

Of course, metal/mineral detectors have played a role in prospecting for more than two decades. Not only have they been used in nugget hunting, they have proved valuable in finding concentrations of black sand—mineral "hot spots"—in stream and river beds. They've been used successfully in prospecting old mines, in seeking ore-bearing veins or pockets that the original miners may have overlooked. And detectors have been used in working mine tailings, in finding valuable ore samples amid the residue produced by mine operations of the past.

The new detector system for prospecting features a detector with a ground-canceling feature. Known as the VLF (for very *low* frequency) detector, it permits the user, with the simple adjustment of a tuning knob, to neutralize, that is, cancel, the effect of iron mineral. Since ground minerals are invariably prevalent where gold is to be found, the VLF detector enables a prospector to search for nuggets and ore-bearing veins in areas that could not be touched with a detector only a few years ago.

Nugget hunting with a **VLF-TR** detector. (*Bill Sullivan*)

Electronic prospectors scan streambed of Mexico's Batopilas River for silver-bearing black sand deposits. (*Garrett Electronics*)

Types of Detectors

While a metal/mineral detector can be an important aid in helping you to spot a gold nugget as small as a grain of rice and find black sand concentrations, not every kind of detector is capable of performing these tasks with equal efficiency. From a technical standpoint, there are three different kinds of detectors: transmitter-receiver (TR); beat-frequency oscillator (BFO); and the newest type, VLF, which is actually a TR detector that operates on a lower frequency than the more conventional TR instrument. With two of these, the VLF and BFO detectors, it's possible to locate valuable gold specimens and deposits.

Almost all metal detectors have a family resemblance. The unit's electronic circuitry is contained within a metal case about the size and shape of a shoebox. On one face of the case are mounted the instrument's various switches and control knobs. These components are linked by a telescoping aluminum shaft to a circular search head, which can be as big as a dinner plate or as small as the bottom of a teacup.

Traditionally, detectors have been of two types—transmitter-receiver (TR) and beat-frequency oscillator (BFO). The TR type has two electronically balanced wire loops within the search head. One coil transmits signals; the other receives them. When something occurs to interfere with the signal, as when the search head is passed over a piece of metal, the interference is perceived by a user as a high-pitched sound; "*ooooheee,*" the detector signals.

The BFO detector is what is known as a frequency-change instrument. When adjusted to a specific frequency, it produces a constant audible response, a "*putt-putt*" sound. When the search head is passed over metal or iron mineral, the pitch and frequency of the putts increases or decreases. An experienced user knows how to "read" these variations.

One advantage of the BFO detector is that there is much more of a direct relationship between the character of the metal or iron-mineral deposit that the search head perceived and the quality of the sound produced. The user can thus make some judgment as to the size of the object being detected and the depth at which it is going to be found. When it comes to the hobby of treasure hunting, to the detection of metallic objects such as coins, rings, and similar artifacts that lie beneath the surface of the ground, both the TR and BFO detectors are effective. But when you're seeking gold nuggets, black sand concentrations, and ore samples, you should avoid the conventional TR detector.

Some manufacturers and their dealers continue to recommend TR detectors, and they instruct prospectors to tune the TR detector into the "mineral" mode of operation when seeking gold flakes or nuggets. This setting renders the instrument sensitive, they say, to ferrous metal objects, that is, those objects that contain a high percentage of magnetic iron. They're theorizing that the detector will respond to concentrations of black sand, and once these are located, some gold will probably be found, too.

But there are problems with this theory. Gold-country gravel deposits are so rich in minerals that almost anytime a TR search coil is swept anywhere near the ground,

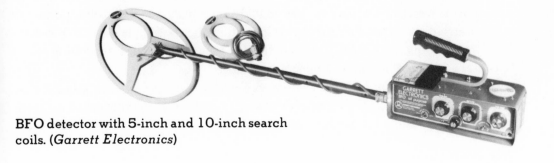

BFO detector with 5-inch and 10-inch search coils. (*Garrett Electronics*)

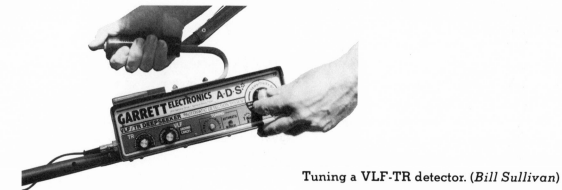

Tuning a VLF-TR detector. (*Bill Sullivan*)

there's a loud response. But there is no way of telling whether the instrument had detected an iron-rich chunk of gravel or gold-bearing black sand.

Tuning the TR detector to its "metal" mode of operation doesn't solve the problem. This makes the instrument respond "positively" to metal objects, including, of course, gold. But placer gold occurs in such small quantities—sometimes so small that it takes a magnifying glass to spot them—that they are almost impossible for the detector to perceive.

In the case of other metallic substances, the detector's sensitivity is enhanced by their oxidation. When they have been buried in the ground, especially damp ground, for any extended period of time, silver, copper, iron, and most other metals leach metallic ions into the ground. This zone of oxidized soil extends for several inches above, below, and to the sides of the object. A penny buried in 12 inches of soil is very likely to be perceived by a detector that is capable of operating to only a depth of 9 or 10 inches—because of the oxidized zone. But with gold there is no such advantage because gold does not oxidize, nor does it corrode or leach. Nothing but the gold itself is subject to detection.

Another problem with the TR detector is caused by the instrument's search coil. When a TR detector is tuned into the metal mode of operation, the coil will give different signals in response to the same specimen of metallic ore, depending on which part of the search disc it touches. A large gold nugget placed on the receiver portion of the search disc will produce a response indicating that the sample is metal. But when the nugget is placed on that portion of the disc that contains the transmitting coil, it causes a response indicating a mineral substance.

A small sample of high-grade iron ore—which is magnetic—also gets the dual reaction. On the receiver portion of the disc, it registers as something metallic. On the transmitter portion, it causes a mineral response.

Scanning mine-shaft entrance to locate conductive-ore vein.
(*Garrett Electronics*)

Quality headphones are a must. Important features are heavy-duty coiled cord, soft ear cushions, and adjustable volume controls. (*Garrett Electronics*)

This lack of reliability has caused knowledgeable gold seekers to reject the TR detector in favor of either the VLF or BFO instruments.

As mentioned above, the VLF detector is a special type of TR detector, one that operates on a much lower frequency. This lower frequency allows the detector to cancel out the disturbing effects of iron minerals, thus permitting detection to a greater depth.

But that's not the only reason that the VLF detector is becoming preeminent. Thanks to its sophisticated electronic engineering, the VLF detector is able to "discriminate," that is, the user can tune out troublesome interference, such as that caused by iron mineralization that might be present in a mine, streambed, or the ground.

While the VLF detector is the best all-around detector to use in prospecting, the BFO detector is superior when it comes to searching for black sand deposits. Although it is a more difficult instrument to learn to use, the sensitivity and reliability one gets with the BFO detector are worth the extra effort.

Choosing a Detector

Listed below are the principal companies that manufacture and distribute metal-/mineral detectors. They have dealers in all parts of the United States.

Bounty Hunter, Inc.
1309 West 21st St.
Tempe, AZ 85282

Fisher Research
  Laboratory
1005 I St.
Los Banos, CA 93635

Garrett Electronics
2814 National Dr.
Garland, TX 75041

Compass Electronics
  Corp.
3700 24th Ave.
Forest Grove, OR 97116

Gardiner Electronics Co.
4729 North 7th Ave.
Phoenix, AZ 85013

White's Electronics, Inc.
1011 Pleasant Valley Rd.
Sweet Home, OR 97386

Write to these firms, and request a catalog and sales promotion material from each. Point out that you're planning to specialize in gold prospecting. The reply you receive should explain in full how the equipment operates in various prospecting situations. The more detail, the better.

Be sure to check any advertising claims made for detectors in which you're interested. This is particularly true in the case of VLF detectors. Some detectors that are said to have ground-canceling capability actually perform that function poorly. If possible, test the detector in the field, or at least bench-test it in the store.

You should plan on paying more than $300 for a good-quality instrument. But price should not be your sole criterion. Examine the detector's construction features. The control knobs should rotate smoothly with only the slightest resistance. The machine should be sturdily built and have a solid "feel."

Check the audio signals for loudness. You should not have to operate the instrument at full volume in order to hear the signal clearly. If a slight adjustment of the tuning control causes an erratic or sizeable jump in the loudness of the signal, you are going to have trouble keeping the detector properly adjusted.

Once you have the instrument tuned and operating, test it for mechanical stability. Use the palm of one hand to sharply jar the control housing. This bumping should not cause any static or disruption of the signal. Raise the search coil over your head as if you were scanning the ceiling. Then lower it to the floor. Repeat the test. There should be no change in the audio signal.

Finally, seek out experienced detector users. Find out what they like and dislike about their detectors. Evaluate their comments as they apply to prospecting.

Bench-Testing

The use of metal/mineral detectors in prospecting for nuggets and placer deposits is a controversial topic. If you were to ask 12 gold seekers to describe their experiences with detectors, you would probably get a dozen different opinions as to the value these instruments have.

Some prospectors complain of signal drift. They say that the many mineralized rocks found in most streambeds cause audio signals to erratically fade or grow louder; the signal pitch is also subject to erratic behavior, they claim.

But experts in detector use say that this is not a real problem. "Sure, detectors are subject to drift, some more than others," says Charles Garrett, a treasure hunter and prospector for more than 35 years, and president of Garrett Electronics. "But most people who complain of drift don't know how to use their detectors. They simply don't know the difference between drift and mineral detection.

"Drift is a change in signal that doesn't disappear; it remains constant," says Garrett. "It's similar to what happens when you tune in your radio to a distant station and the signal fades, and then remains weak. That's drift.

"But abrupt changes in signal can't be called drift. They're not drift any more than the static you get on your car radio when you drive under power transmission lines.

"A quality detector won't lie to you," says Garrett. "If, for example, the coil passes over highly mineralized sand or rocks, you're going to get a signal reaction that's similar in nature to radio static. The detector is telling you something. You have to be able to interpret what it's saying. You can't dismiss it as drift.

"You need knowledge to be able to do this; you need experience."

One way to gain the necessary knowledge and experience is by bench-testing the detector at home. "Bench-testing can mean the difference between success and failure," says Charles Garrett.

Set the detector on a bench or table. In the case of a VLF detector, flip the mode-selector switch to the TR discriminate mode, then adjust the discriminator control knob to the zero setting. Adjust the tuning control to obtain a faint sound.

If you're using a BFO detector, tune it to the metal mode of operation and adjust the sound to a moderate rate of beats. You want to be able to hear even the faintest signal change.

Obtain samples of minerals and rocks of various types, including the type of streambed gravel and black sand that you expect to encounter. (Put a teaspoon or so of black sand in a small plastic bottle.) One by one, bring the samples to within an inch or so of the bottom of the search coil.

If there's more metal than mineral in the sample, the sound will get louder. A predominant amount of mineral or magnetic iron in a sample will cause the signal to

Charles Garrett conducts a bench test.

weaken. Testing the detector in this fashion enables you to become familiar with the types of responses you're going to be encountering in the field.

When bench testing, you must be aware of the type of coil you're using. Coils have traditionally been of the coplanar type. The transmitting and receiving units are mounted on the same plane and housed within a disc-shaped plastic case. The coaxial coil, which was introduced not long ago, is disc-shaped, too. But it's a thick disc because the transmitting units and receiving units are stacked atop one another, forming, in effect, a coil sandwich. This arrangement, incidentally, produces a more uniform response. In terms of pitch and intensity, you always get the same signal when using a coaxial coil, no matter from which direction the coil approaches the target.

If you're using a coaxial coil when bench-testing or ore sampling, you must pass the sample over the bottom of the coil; bringing the sample toward the top will give you a response that is the reverse of what you want.

In the case of a coplanar coil, it doesn't make any difference; you can pass the sample over either the coil top or bottom.

Field-Searching for Gold

Every quality VLF or BFO detector can be used in just about every type of prospecting situation. The VLF instrument, however, is superior in nugget hunting or when searching for the ore veins. But when seeking to locate pockets of black sand, the VLF detector may overlook some of the deeper concentrations. In addition, mineralized rocks can sometimes cause the VLF machine to give a "false" signal. That's why a BFO detector is the best to use when hunting for black sand deposits. This section explains how each instrument is to be used in each case.

Nugget Hunting Old placer diggings, old mines, mine tailings, gravel bars, the beds of dry washes and arroyos, exposed tree roots, and mountain streambeds—these are some of the areas in which to use a detector in prospecting for gold nuggets.

In almost every case, you'll want to operate the VLF detector in the ground-canceling mode of operation, tuning out interference from highly iron-mineralized soil. Using earphones will increase your ability to hear faint signals.

The size of coil to use depends on the size of nugget you're looking for. If only tiny nuggets were found in the area in the past, of only 1- or 2-pennyweight size, say, then a small coil, a 3½-inch coil, is probably the best. It has a "hot" response in depths as low as 3 or 4 inches.

But a larger coil, a 7½-inch coil, for example, could also work efficiently in searching for small nuggets. It not only would detect to greater depths, but it would enable you to cover more ground with each sweep because of its bigger size. In nugget

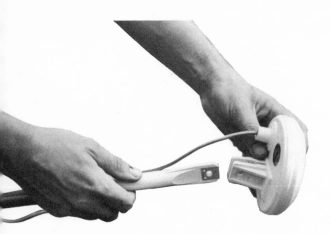

For nugget hunting, Garrett offers this 3½-inch coaxial coil. (*Bill Sullivan*)

hunting in Australia, prospectors use 10- to 14-inch search coils, which are effective in depths of up to 2 or 3 feet. But that's because some Australian nuggets are of such enormous size.

When working in a streambed, hold the coil from 1 to 4 inches above the bottom of the stream. If the gurgling of turbulent water is a problem, masking the detector's audio signal, wear earphones.

If you have a high-quality metal detector, you should be able to submerge the search coil and the shaft up to the point where the search-coil cable connects to the control box. This means that you can work in water that is 3 to 4 feet deep without any difficulty.

But many prospectors prefer the body-mount detector system. In this, the control box is strapped about the user's waist or chest. Two or three feet of extra search-coil cable connects the control box to the shaft. This not only enables you to attain greater depth when prospecting in a stream, but it also does much to increase your mobility.

The body-mount setup works best when you're using one of the smaller search coils, one that's 4 inches or less in diameter. As you increase the size of the coil, you're also increasing the weight that's concentrated at the end of the shaft. With a rod-mount detector, coil weight is not a problem because it is counterbalanced by the weight of the control box. But with a body-mount unit and big search coil, your arm and shoulder muscles can quickly tire because there's nothing counterbalancing the weight of the coil.

Compass Electronics' "Gold Probe." (*Compass Electronics*)

Bigger coils are deepest seeking. (*Bill Sullivan*)

Watertight search coils are a must. (*Bill Sullivan*)

•llect sand and gravel from target area in
ld pan, then scan it for "find."
ill Sullivan)

As you're working the streambed, have a gold pan nearby, perhaps on the bank. When you get the response you're waiting for, shovel out the sand and gravel from beneath the target area and place it in the gold pan. Use a hand trowel, which is similar to a garden trowel. When the pan is filled, scan the contents with the detector to see whether you've retrieved the metallic object. Of course, you must use a plastic gold pan; a metal pan will disrupt the detector's signal. If you haven't recovered the metallic object, keep shoveling out sand and gravel from the target area.

Sometimes, when you get a strong positive response with your VLF detector, you may dig down to find only a rock. A mineralized rock, it's what's sometimes called a "hot rock." The rock contains such a high concentration of iron mineral that it causes the detector to give off a false signal.

But there's a quick way to determine which of the signals you receive emanate from hot rocks and which from nuggets. When you get such a response, hover over the target area until you get the strongest possible signal. That maximum pinpoints the target.

Next, move the search coil two or three inches away from the target, keeping it a few inches from the ground or streambed. Then flip the mode-control switch to the TR discrimination mode (having preset the signal to bottle-cap rejection), and ease the coil back over the target.

If the audio signal decreases in intensity, you've found a hot rock. If the signal remains the same or grows louder, dig down. You may have found a nugget.

A BFO detector can also be used in nugget hunting. Tune the instrument to the metallic mode of operation. When the search coil is passed over a concentration of magnetic black sand or a "hot rock," the audio signal will slow down perceptively, or may even cut off entirely. Should the search coil detect a nugget, you'll hear a slight increase in the rate of beats.

The basic setting you use should be related to the amount of mineralization that's present in the area you're working. In the case of ground that is highly mineralized, increase the speed of the beats. But for ground that is relatively free of mineralization, adjust the tune to produce beats at a moderate rate.

Once in the field, sweep the coil moderately close to the ground. And don't be afraid to move the coil faster than normal. Rapid sweeps over highly mineralized ground produce the most distinctive signals. With a slow BFO movement, background noise has a greater chance of disrupting the signal.

Searching for Black Sand Deposits The BFO detector is superior to the VLF machine in detecting the deeper concentrations of black sand. Equipped with a good-sized coil, Garrett's 13-by-24-incher, for instance, the BFO detector is capable of finding black sand deposits at depths down to two feet or more.

In body-mount setup, control box is strapped to waist. (*Compass Electronics*)

In tuning a BFO detector to locate black sand concentrations, first find the detector's null point—the point where no sound is emitted by the speaker. (From the null point, you're able to rotate the tuning control in either direction and get sound. In one direction, you'll be in the metal mode; bring metal close to the coil and you'll get a positive response. Rotate the tuning control in the other direction, and you'll be in the mineral mode; iron minerals will give you a positive response.) Once you've found the null point, adjust the tuning control to the mineral side of null. This position causes the detector to give a loud and positive response every time a concentration of black sand is detected. If metal is detected, the sound will drop off.

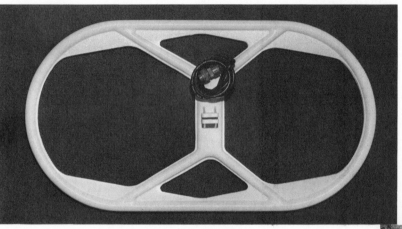

The 13- by 24-inch BFO search coil.
(*Garrett Electronics*)

Using a VLF-TR detector to search for black sand concentrations.
(*Garrett Electronics*)

When searching for black sand concentrations, hold the search coil out in front of you as you walk, keeping the bottom of the coil a few inches from the ground or stream bottom. Don't sweep the coil from side to side as you do in nugget hunting (or coin hunting). Walk in a back and forth pattern over the area to be covered; then walk at right angles. Hold the coil out in front of you, not sweeping it from side to side. This method is assurance that you'll get an abrupt, clearly defined response when black sand is detected. If you were to sweep the coil from side to side, the response would not likely be as distinct. You could even mistake it for an electronic aberration of one type or another.

Some BFO detectors cannot be used in cold water because of signal drift. The audio response either slows down or speeds up when the detector is plunged into cold water, and it never attains the stability required. Of course, even the best detectors require a few minutes to adjust to abrupt changes in temperature, but after a reasonable amount of time the audio signal should become stable.

If you're purchasing a new detector, test how it responds to temperature changes before you leave the store. Tune the detector and then walk outside the store into the hot sun. You should be able to tell within a minute or so whether the change in temperature causes the instrument's signal to drift excessively.

Gold Seeking in Old Mines In every part of the country where gold has been found, there are abandoned mines. Some of these were established more than a century ago and may not have been worked for fifty or sixty years. With the price of gold escalating, many of these abandoned mines are being reopened. If you've no accessible mine, you may have access to leftover ore or abandoned tailings. Modern milling methods have sometimes made it worthwhile to reprocess these materials.

Searching an old and abandoned mine with a metal/mineral detector can be exciting and challenging. But don't think you can enter a mine and confidently begin sweeping the search coil over the tunnel walls and floor and expect every response you get to be a rich ore vein. There's likely to be so much mineralization that signal distortion and false signals will prevent the instrument from properly reporting the presence of gold-bearing veins. Special techniques are required.

Tune the VLF detector to the ground-canceling mode. Once you have the instrument tuned, scan the mine walls and ceiling carefully. When the detector responds, indicating you've located an ore deposit, the next step is to identify it. Switch over to the TR discriminate mode. Scan the target area again. If the sound weakens or stops abruptly, the vein is predominately iron. Only iron will diminish the signal. If the signal remains unchanged and shows no sign of reducing in intensity, the vein is primarily conductive, that is, it may be gold or silver. By slowly adjusting the discriminator control knob, you can even determine the amount of conductivity the vein has.

Don't overlook the mine floor, likely to be one of the most productive areas. When the mine was being worked, all of the ore had to come through the main tunnel for dumping or milling. Whether it was moved by hand or by means of small ore cars, some pieces were dropped. These became covered by accumulated debris and now await discovery.

Lay the detector on the ground several feet outside the tunnel entrance. Using a hand pick, dig under the debris that covers the tunnel floor. After you've found some good specimens, carry them out of the mine and to the detector for testing. As Roy Lagal points out in his *Detector Owners Field Manual,* "Test small samples, keeping in mind that the large high grade samples would have been seen by the original miners, and only the smaller pieces would have been overlooked." Lagal adds, "Be mobile; test many different spots."

The mine dump, if there is one, can be as important as the mine itself. While some tailings may have been reworked, others may not have, and these can yield valuable ore samples mistakenly discarded by the original miners.

In investigating a mine dump, use the same technique you used in searching the mine floor. Piles of tailings also fall into this category. Lay the detector flat on the ground. You should be using a search coil that ranges in size from about 3 to 8 inches in diameter.

In the years in which the mine was in operation, it's likely that only a small portion of the dump received tailings from areas where veins were located. Most of the dump will be made up of debris that resulted when the mine shafts and tunnels were cut. That's why it's important to take samples from many different areas.

Be cautious about working old mines. They can be very hazardous. Many of the conventional hard-rock mines, typified by their vertical shafts, have been flooded in an effort to preserve the timbers used to shore up the mine walls and ceilings. Flooding is also meant to keep people out.

A mine with a horizontal shaft, perhaps following the course of an old streambed, can be just as perilous. Such mines are susceptible to cave-ins.

Two other dangers are snakes and various kinds of gases that build up in the

With a VLF-TR detector, Frank Mellish scans an embankment where gold concentrates have been found.

shafts. Snakes bed down in the shafts, especially in the summer. Bears and other animals have been known to wander into shafts. And in some shafts, there's an accumulation of noxious gases. "You won't know what hit you until it's too late," says Vic Renzoni, an Arizona gold-panning instructor.

And bear in mind that virtually all mines, particularly those in the West, are to be found on private property, or property held as a mining claim. You have just about as much right to enter a man's mine as you do his living room.

A claim may look inactive, but still be valid. All the claim holder has to do to maintain the claim's validity is to perform certain assessment work (explained in Chapter 8), and then file the appropriate affidavit with the county recorder.

A claim on which the annual labor has not been performed, or for which the affidavit has not been filed in the prescribed manner, can be challenged. Recent decisions in California courts have stated that failure to file the claim affidavit of annual labor "shall create a *prima facie* presumption of the act and intent of the owner to abandon such claim . . ." In other words, the burden of proof that the required work

Using a VLF-TR detector, Charles Garrett and Roy Lagal scan mine for gold veins earlier miners may have missed. (*Garrett Electronics*)

was, in fact, performed rests with the original claim holder. It used to be the reverse; the challenger had to provide the proof. What is likely to happen is that the original claim holder challenges the challenger and the matter winds up in court.

Treasure hunting and adventure magazines constantly feature articles about "lost" mines. These are invariably mines that are rich beyond one's dreams, but for some usually sinister reason cannot be located again by the prospector. While a handful of such legends have turned out to be true, never take lost-mine tales too seriously. They're meant to provide entertaining reading.

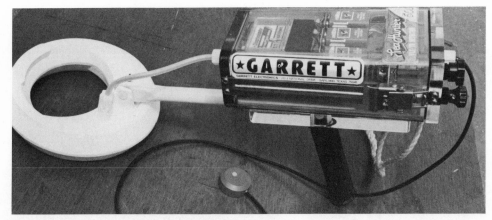

Sea Hunter is for underwater prospecting.

Underwater Gold Hunting A detector that can be used effectively for gold prospecting in a completely underwater environment was introduced by Garrett Electronics in 1980. Until then the better detectors made for underwater use were largely of foreign manufacture, and even these weren't suited for gold hunting because they couldn't be used successfully in a mineralized area.

The Garrett detector, called the Sea Hunter, is a VLF instrument. Its coaxial coil is linked by a telescoping metal rod to a control system housed in a watertight lucite case.

Instead of conventional earphones, the Sea Hunter is equipped with a receiver that operates on a bone-conduction principle. You simply place the small, circular device adjacent to one ear, slipping it under your diving hood. The Sea Hunter has two modes of operation, giving it the same operating characteristics as VLF-TR detectors.

For more information on the subject of electronic prospecting, you may want to consult one or more of the authoritative books published by Ram Publishing Company. The titles are listed under Additional Reading, in the Appendix.

Several films available on loan from Garrett Electronics (2814 National Dr., Garland, TX 75041) depict detectors being used in different types of prospecting situations. There is no charge for the films, but the borrower must pay return freight charges. Two of the films are *Treasures of Mexico* (16 mm, 18 min., color, sound) and *Gold and Treasure Adventures* (16 mm, 30 min., color, sound). The latter was filmed in California.

Have patience when learning to use a metal/mineral detector. It's a sophisticated piece of electronic equipment, and you can't expect to become skilled in its use overnight. Once you know the machine, once you're aware of all the subtleties of its operation, the detector can't help but produce results for you.

Tailings in Amador City, California, await investigation. Mine headframe looms in background.

CHAPTER 8

Staking a Mining Claim

ON FEBRUARY 26, 1980, not long after the announcement that indications of gold had been perceived in Alaska's Nelchina River Basin, about 145 miles northeast of Anchorage, nearly six hundred people crowded about the State Geological Office, waiting for authorities to disclose the exact location of the prime area.

When the announcement was made, pandemonium broke loose. Prospectors in cars, trucks, camping trailers, and mobile homes rushed to the Nelchina area to stake their claims. Some had been poised at the ready, near the area and in communication with associates at the geological office. When the reports came, they hurried into the Nelchina Basin on skis, snowshoes, and by snowmobile. Telephone lines became jammed, but some individuals had been clever enough to reserve lines in advance.

Seven big helicopters joined in. From at least one helicopter, stakes with concrete-weighted bases were dropped to mark claim boundaries (a procedure later deemed illegal by state authorities). One firm, the Geneva Pacific Corporation of Glenville, Illinois, had ten technical experts in the field awaiting the location announcement. They lived in mobile homes and had several different types of vehicles and a helicopter at their service. Walkie-talkies, side-band radios, and CB radios were used for communicating between the company's Anchorage office and the field. When confidential information was to be transmitted, coded messages were sent. Thanks to the preparation and planning, Geneva Pacific was able to stake forty-four claims in the prime area.

One reason for all the excitement was the fact that samples from the Nelchina area indicated a gold density of 0.4 ounce per ton mined. At the No. 1 producing mine in the United States, the Homestake Mine in South Dakota, gold density is the same 0.4 ounce per ton. State geologists also reported the Nelchina River Basin to have high concentrations of silver, lead, zinc, and copper.

While the real value of these claims may not be known for years, the great Alaska gold rush of 1980 is evidence that it's still quite possible to stake productive mining claims on public lands of the United States. The key term in that sentence is "public lands." Public land is administered for the most part by the Bureau of Land Management of the U.S. Department of Interior. If the land on which you're planning to file a claim is state-owned land, mineral rights must be obtained from the owner by lease or purchase.

Most public land on which claims may be staked is to be found in the states of Alaska, Arizona, Arkansas, California, Colorado, Florida, Idaho, Louisiana, Mississippi, Montana, Nebraska, Nevada, New Mexico, North Dakota, Oregon, South Dakota, Utah, Washington, and Wyoming. Records that can be consulted at the appropriate Bureau of Land Management or Forest Service state or regional office will inform you exactly which areas of these states are available for mineral exploration and the filing of claims. (The addresses of these offices are listed in the Appendix.)

Not all BLM and Forest Service land is available for mineral development. Certain areas are said to be "withdrawn from mineral entry or location." Attempting to prospect or remove minerals from "withdrawn" land is a violation of federal law.

Other federal lands closed to mining and related activities include all national parks, national monuments, Indian reservations, military reservations, reclamation projects, scientific testing areas, and some wildlife protection areas, such as the federal wildlife refuges.

Some of the regulations that now govern the staking of mining claims hark back to the gold rush days of the nineteenth century. Before the General Mining Law of 1872, the regulation of mining activities was carried out on a local or regional basis. General meetings were held by those miners who had staked claims on a particular creek or within a specified area of land. At these open meetings, a leader was selected and rules were set down to protect the property rights of all present.

Each area of land was clearly defined and assigned a name as a particular mining district. A prominent building in the largest settlement or town in the mining district was designated as the location where claims were to be filed and the dates of the filings recorded. Disputes that arose concerning claims were referred to the district leaders. These mining districts continued to function as governing authorities in the gold country until the organization of territories. The buildings housing the claim offices were often converted into county courthouses.

While mining districts have no political or governmental authority nowadays, the designation itself lingers on, the term being used to identify certain county or state subdivisions. More than that, when a present-day prospector records a mining claim and fills out a form called a Notice of Location, he or she must designate the name of the mining district in which the claim is located.

In addition, the Notice of Location requires that the claimant describe the location of the site in relation to some permanent and natural object within the township or county, which was the way claims were recorded in mining-district days. Of course, the claim also has to be delineated according to the current system of land surveying and described in current legal terms.

Questions and Answers on Mining Claims This section is devoted to questions on mining claims (most frequently asked of federal and state administrators) and to their answers.

What is a mining claim?

A mining claim is a portion of public mineral lands that an individual holds for the purpose of extracting minerals. The right is granted under the General Mining Law of 1872, which protects the claimant in "all lawful uses of his claim for mining purposes," and also establishes the limits of those rights.

How does one acquire a mining claim?

Mining claims may be staked, purchased, leased, or inherited.

When can one stake a mining claim?

Whenever valuable minerals are discovered on open public land, a claim may be staked.

May anyone stake a mining claim?

A mining claim may be staked by any citizen of the United States or by any person who has declared the intention of becoming a citizen. He may be a resident or nonresident of the state in which the claim is to be located. He may be a minor, if he is "of discretion and understanding."

What types of claims are there?

Most claims are either lode claims or placer claims. A lode claim refers to a vein of mineralized rock that fills a seam or fissure within a mass of unmineralized rock.

Deposits that are not classified as lode claims are registered as placer claims. These include "true" placer deposits scattered through masses of sand, gravel, or similar material that have been formed through the erosion of solid rock.

There are also mill-site claims and tunnel-site claims. A mill-site claim usually refers to land that is nonmineral in character, and which is to be used for the erection of a mill or reduction works. A tunnel-site claim refers to the tunnels necessary to gain access to a specific vein or lode.

How much land may a claim cover?

A single lode claim contains approximately 20 acres. A placer claim, for one person, must be limited in size to 20 acres. But an association of two people may file a placer claim of up to 40 acres in size; three persons, 60 acres; and so on, up to a total of eight people and 160 acres. Each of the individuals must have a bona fide interest in the claim.

What is the "prudent man" rule as applied to a mineral discovery?

You must do much more than speculate that gold is present on your claim site. For your claim to be valid, you must know that gold exists within the site, and be able to prove its existence.

Traces, isolated bits of mineral, or minor indications are not enough to satisfy the test. You have to be able to establish that the gold deposits are valuable deposits.

How valuable?

The courts have established the "prudent man" test in determining what constitutes the discovery of gold or any other valuable mineral. *Castle* v. *Womble,* 919 Land Decisions 455,457 (1897), defined the test as follows: "Where minerals have been found and the evidence is of such character that a person of ordinary prudence would be justified in further expenditure of his labor and means, with a reasonable prospect of success in developing a valuable mine . . ."

More recently, in 1968, the U.S. Supreme Court, in *United States* v. *Odeman* (39C, U.S. 599), sanctioned the marketability test, that is, that one be able to mine and market the mineral in question at a profit.

There is no limit to the number of claims an individual may hold. But in each case, there must have been actual physical discovery of the mineral on each claim, and each of these discoveries must satisfy the "prudent man" and "marketability tests."

If you're the least bit familiar with the public lands available for mineral development, you know that they offer some of the most scenically attractive areas in the United States. Be advised that using the Mining Law to appropriate public land for residential use or recreational purposes constitutes a fraud.

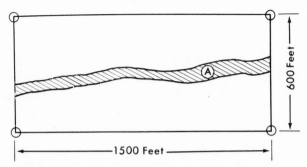

In this plan of a lode claim, the discovery post (*A*) that contains the notice of location is on the vein and within the claim boundaries. Common practice is to place one end line several hundred feet from the discovery post. (*Herb Field Art Studio*)

How does one actually stake a claim?

There are three steps:

1. Post the location notice.
2. Stake the claim corners and mark the boundaries.
3. Record the location notice.

Posting the Location Notice. After the discovery has been made, a notice of location must be posted at the discovery site. Location notices usually contain this information: the name of the claim, the date of the filing, the names of the individuals doing the filing (legally, they're referred to as "locators"), the amount of acreage being claimed, plus a legal description of the claim boundaries (see below). Printed location-notice forms can be obtained at many mining country stationery stores, printing shops, and other legal-form suppliers.

Staking the Claim. Federal law specifies only that claim boundaries should be distinctly and clearly marked, but most states have established detailed requirements for staking claims. In many cases, these mandate that corner posts be erected to mark the claim site, and a center stake between the corner posts may be erected on each side. Some states require that the name of the claim be painted on each stake, with a designation regarding direction, i.e., "southwest corner," "northwest corner," and so on.

Almost always, a discovery-post monument must be erected. This takes the form of a wooden or metal stake that projects several feet above the ground, or it can be a pile of rocks several feet in height. The notice of location must be affixed to the post or buried within the rock pile.

In the case of lode claims, the notice of location must be erected on or near the vein that is to be worked. Common practice is to place one end line several hundred feet from the discovery post.

Placer claims, wherever practical, must be located by legal subdivision. A legal subdivision is a section or a part of a section that lies within a designated township, range, and principal subdivision. (In most of the United States west of Ohio-Pennsylvania border, the term "section" refers to one of the 36 numbered subdivisions, each one mile square, of a township.) Should you be unable to find any section corners on surveyed land, or in cases where the land is unsurveyed, you're required to conform as nearly as possible to the government survey system.

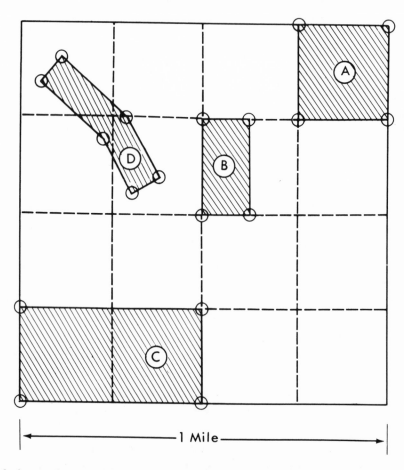

Several placer claims within a section of land, described as Section 2, T. 40 N., R. 8 E. The claims are described as follows: *A* (40 acres)—NW 1/4, NW 1/4, Sec. 2, T. 40 N., R. 8 E. *B* (20 acres)—S 1/2 SE 1/4 NW 1/4 Sec. 2, T. 40 N., R. 8 E. *C* (80 acres)—E 1/2 SE 1/4 Sec. 2, T. 40 N., R. 8 E. *D* Gulch claim; does not conform to legal survey description. (*Herb Field Art Studio*)

Recording the Location Notice. Within ninety days after posting a lode claim, a copy of the location notice must be filed with the office of the county recorder and the state office of the Bureau of Land Management or regional Forest Service office. In the case of a claim involving a Forest Service wilderness area, the applicant has only thirty days in which to file. Failure to file with the appropriate government agency within the specified time period can void the claim.

With placer claims, the locators must file a copy of the location notice with the county recorder within thirty days of staking. In addition, a copy of the location notice must be filed with the state office of the Bureau of Land Management or regional office of the Forest Service.

How can I tell whether a claim has been staked?

The presence of a location notice and marked boundaries, as well as evidence of recent mining activity, can indicate a valid mining claim. If the names of the claim and claim owners are found, check with the county recorder to find out whether the claim has been legally recorded and whether the annual assessment work has been performed (see below). If the claim has not been recorded, or if the assessment work has not been performed, the claim may be invalid and open to re-location.

What must be done to hold a mining claim?

Once a mining claim has been established, the owner must perform $100 worth of labor, or make improvements equal to that sum, each year. This is known as "annual assessment work," and must be performed within the assessment year, which begins on September 1. A statement that the work has been completed must be filed with the county recorder and with the appropriate BLM or Forest Service office. The work and related filing are necessary in order to verify your continuing interest in the claim.

Assessment can include the following:

Excavations to uncover mineralization
Drill holes
Tunnel work
Road building
Construction of mining-related buildings
Sharpening tools
Moving machinery and material to the claim
Removing tumber and overburden
Rental of mining equipment
Cost of powder, fuses, and drill steel.

These are among the types of work and expenditures that are not counted as assessment work:

Sampling and assaying
Surveying
Travel expenses to and from the claim
Fencing

Owners of mill-site or tunnel-site claims are not required to file evidence of assessment work. They must, however, file a notice of their intent to retain their claim to the site.

How does one patent a claim?

If your claim is properly filed and you continue to perform the annual assessment work, and meet the requirements of state and federal mining law, you'll maintain what's known as a "possessory right to the land." This limits you to developing and extracting the valuable minerals. No other person can mine the land without your consent. You are not likely to have to pay any county or state taxes. You can sell the claim or pass it on to your heirs.

This is known as an unpatented claim. You do not own title to the land.

One drawback of an unpatented claim is that the government can challenge its validity at any time, claiming that no bona fide mineral discovery was ever made, or that the claim's minerals have never been mined. Indeed, you can be challenged on any or all provisions of the mining law. If the government's challenge is upheld, your claim can be cancelled and you will be made to forfeit all your rights to the land. You can even be charged with trespass.

With a patented claim, you can avoid such a contingency. A patented claim is one for which the federal government has given a deed or whose title it has transferred to an individual. However, once you have obtained a patent for a claim, you must satisfy a number of legal requirements. This first step is to prove the discovery is a valuable one by satisfying the "prudent man and marketability" test.

You then must have the claim surveyed by a mineral surveyor. In the case of land administered by the Bureau of Land Management, the surveyor must be selected from a roster maintained by the BLM.

A copy of the survey and the notice of intention to apply for a patent must be posted at the claim site. Copies of these documents must then be filed with the BLM office.

The actual application must be made within sixty days of the filing of the notice of intention. The application must include evidence of your right of possession and basis of your right to patent. You have to explain in full why you believe the site contains valuable mineral deposits. Evidence in support of this contention might include descriptions of the geology of the area, the results of drilling and sampling, the mining methods likely to be used, anticipated mining and milling costs, factors involved in transportation, and even a market analysis.

In the case of a placer-gold claim, you cite the gold per cubic yard recovered as determined by your development work, indicate the thickness of the overburden and of the gold-bearing gravel, and state the extent of the deposit as accurately as you can.

In the case of a lode claim, you are required to submit a full description of the vein or lode, and to state the amount of ore that has been extracted and its value. You must also describe mine shafts, tunnels, cuts, and other improvements.

With both placer and gold lode claims, you must submit proof as well that you have performed not less than $500 worth of development work on the site.

Once the requirements have been fulfilled, and the application is approved, you then must pay the purchase price. For placer claims, the price is $2.50 an acre; for lode claims, $5 an acre.

The paragraphs above describe the patent application process in a general way. Processing a claim can vary from case to case. For more detailed information, write to the appropriate BLM state or Forest Service regional office.

When your claim is patented, it takes on status as real property and, hence, can become subject to real estate taxes. If you receive a tax bill and fail to pay it, the tax collector may do what is quite usual in such circumstances—seek to recover the value of the money due by condemning the property and selling it at auction.

When planning to buy a placer claim, how can one tell whether it is going to be productive?

The first thing to do is assure that the claim involves a gold-bearing stream, that is, a stream with a reputation for productive placer deposits. Some streams, even in the gold country, do not have any gold in them, or at least not a sufficient amount of gold to justify dredging. Look for piles of old tailings. Test the grass, roots, and moss by panning. If there are color and fine particles along the stream, there's likely to be gold at the bottom.

This presupposes, of course, that the bottom is capable of catching gold. Inspect the bottom carefully. If the rock is slaty, broken, cracked, and has alternating layers of gravel and clay, it's capable of trapping the finest particles. Also look for trees, roots, and blocky boulders. If these conditions prevail, there will be plenty of traps in which the gold particles can be caught. Avoid smooth, hard bottoms like those formed of igneous rock such as some granites.

Try to inspect the claim in the winter when the water is high. Then you'll be able to spot the eddies and back washes that contribute to the collection of gold. In the summer, these normally aren't apparent.

In buying a claim, how can one be sure it's a valid claim?

Obtain from the seller copies of the location notice, amendments, and documentation concerning each year's assessment work, and then check the county records to ascertain that the information has been properly recorded and indexed, and that the filing has been made within the time periods specified by federal and state law.

Also obtain from the seller copies of documents recorded with the local office of the Bureau of Land Management, the official claim number, the recording receipt, and the assessment work receipts for each year the claim has been worked.

Be sure to ascertain that the claim was recorded with the BLM within ninety days of the posting of the location notice, and that each year's assessment work was properly recorded. Remember, if the seller has not complied with BLM requirements, he does not have a valid claim.

After obtaining a map of the claim, walk over it with the seller, having him point out the boundaries, corner posts, location notice, evidence of work performed, adjacent claims, roads, and the like.

Make all transfers of money, deeds, receipts, and records through a title-company escrow account. Although the title company cannot guarantee that the claim is valid, they can prevent money from changing hands until you have completed your investigation.

Helpful Publications There are various booklets and brochures you can obtain from the federal government that will provide you with additional information:

> *Staking a Mining Claim on Federal Lands* (Bureau of Land Management, Department of the Interior, Washington DC 20240)
> *Patenting a Mining Claim on Federal Lands* (Bureau of Land Management)
> *Mining Claims Under the General Mining Laws* (Bureau of Land Management)
> *Mining in National Forests, CI-14* (Forest Service, U.S. Department of Agriculture, Washington, DC 20250).

There are also state mining laws of which you should be aware. Generally, these statutes do no more than support federal law, but there are exceptions. Be sure to become informed on this subject. Copies of the mining laws of the state in which you plan to stake a claim can be obtained from the state bureau of mines or natural resources, or are available at most larger public libraries within that state.

Glossary

alloy—a substance composed of two or more metals

alluvial—pertaining to any loose deposit of sand or gravel

alluvium—a deposit of sand formed by flowing water

assay—to analyze an ore in order to determine the quality of gold, silver, or other metal in it

Au—the chemical symbol for gold

bar placer—a placer deposit formed within a river or stream, and left exposed when the water level drops

base metal—any metal other than a precious or noble metal. Base metals include copper, lead, zinc, tin, iron, and others

bedrock—unbroken solid rock, overlaid in most places by soil or rock fragments

bench placer—a placer deposit in a former streambed, located anywhere from a few to several hundred feet above the present level of the stream

BFO metal/mineral detector—a beat-frequency oscillator detector, characterized by the "*putt-putt*" sound the instrument emits when the search coil is passed over a metal or mineral substance

black sand—grains of heavy, dark minerals, such as magnetite, limenite, and chromite, found in streams and commonly associated with gold and platinum

bullion—gold that is at least 995-fine

claim—a piece of public land for which a formal request is made for mining

coarse gold—particles of gold that will not pass through a 10-mesh screen color—any particle of gold that is large enough to be seen readily by the naked eye

concentrates—the mixture of heavy sand and gold particles that results from sluicing or dredging

contour line—a line on a topographic map that joins points of equal elevation

dry washer—a device for separating gold concentrates from dry sand and gravel, which consists of a hopper, screen, bellows, and riffle system

eductor—in a surface dredge, the jet nozzle that creates the high-pressure suction

electrostatic concentrator—a dry washer that incorporates electrostatic principles in separating gold concentrates from dry sand and gravel

fault—a break in the continuity of a vein or a body of rock, with discoloration along the plane of the fracture

fine gold—particles of gold that will pass through a 20-mesh screen, but not through a 40-mesh screen

fineness—the proportion of pure gold in an alloy

fissure—a narrow opening or crack in a rock

float—the loose and scattered pieces of ore that have broken away from an outcrop

flour gold—particles of gold that are powdery fine, smaller than very fine gold particles

fool's gold—iron and copper pyrites, and iron-stained mica, sometimes mistaken for gold

gad—a pointed, chisel-like tool used for opening crevices or breaking up mineral samples

gold-filled—an article of jewelry coated with a layer of at least 10-karat gold that constitutes at least 1/20th of the total weight of the article

gold flashed, gold washed—gold electroplate thinner than 7 millionths of an inch of gold

grain—originally the weight of a grain of wheat or a barleycorn; the smallest unit of weight in the troy and avoirdupois systems. 1 grain = 0.0648 gram; 24 grains = 1 pennyweight

gram—a metric unit of mass and weight equal to 1/1000 kilogram. 1 grain = 0.048 gram; 24 grains = 1 pennyweight

gravel—small stones and pebbles, or a mixture of these with sand

igneous—produced under intense heat, as rocks of volcanic origin

karat—a unit used for measuring the fineness of gold. Pure gold is 24 karats fine.

locating—marking and staking a mining claim

locator—an individual who files a mining claim

lode—a veinlike deposit or other body of ore set off from adjacent rock formations

Long Tom—a series of connecting troughs, each lined with riffles, used in separating gold-bearing material from streambed sand and gravel medium gold—particles of gold that will pass through a 10-mesh screen but not a 20-mesh screen

nugget—a lump of native gold that is more than one grain in weight

ore—gold-bearing mineral or rock

ore body—that part of the vein that carries the ore

outcrop—that portion of a mineral deposit or sedimentary bed at the surface of the earth

overburden—the valueless material that covers a placer deposit or an outcrop

patent—the legal instrument by which the federal government conveys title to public land

patented claim—a claim for which the federal government has given a deed or transferred title to an individual or corporation

pennyweight—originally the weight of an English silver penny; in troy weight, 20 pennyweight = 1 ounce

placer—an alluvial deposit of mineral-bearing sand and gravel

plat—a plan or map of a plot of ground

retort—a cylinder-shaped vessel within which an amalgam of mercury and gold can be heated so as to remove the mercury by distillation

riffle—the lining at the bottom of a sluice or rocker, made of slats of wood or metal and arranged in such a manner that grooves or openings are created for catching and collecting particles of gold

rocker—a box mounted on curved pieces of wood. The box is swayed from side to side in washing sand or gravel to separate gold or other minerals

section—in most of the United States west of Ohio-Pennsylvania border, one of the 36 numbered subdivisions, each one mile square, of a township

shaft—a vertical or sloping passageway in a mine

sluice box—a long trough used in placer-mining operations in washing and separating gold particles from sand and gravel

sniping—crevice mining in stream or river bedrock

specific gravity—the ratio of the density of any substance to the density of some other substance taken as a standard, water being the standard for liquids and solids

sponge gold—a type of porous gold that results when an amalgam of mercury and gold is retorted

strip—to remove the overburden from a placer deposit or the barren outcrop from an ore deposit

submersible dredge—a type of surface dredge that consists of a long metal tube through which gold-bearing sand and gravel are drawn from the streambed to be passed over a built-in riffle system at the tube end

suction dredge—a type of surface dredge in which gold-bearing sand and gravel are drawn from the streambed by a suction hose and passed over a sluice box at the water's surface

tailings—the rock residue that results from lode-gold mining

topographic map—a map that shows the actual form of an area by means of contour lines

TR metal/mineral detector—a transmitter-receiver detector, the chief characteristic of which is the constant high-pitched sound the instrument emits when the search coil is passed over a metal or mineral substance.

troy ounce—a unit in the troy system, used for precious metals. The system is based on a pound of 12 ounces and an ounce of 20 pennyweights or 480 grains

unpatented claim—a claim for which an individual or corporation has obtained a legal right to extract and remove minerals from the land but for which full title has not been obtained

vein—a distinct body of mineralized material

very fine gold—particles of gold that will pass through a 40-mesh screen

VLF metal/mineral detector—a very-low-frequency TR detector that permits the user to neutralize signals created by iron mineral

Appendix

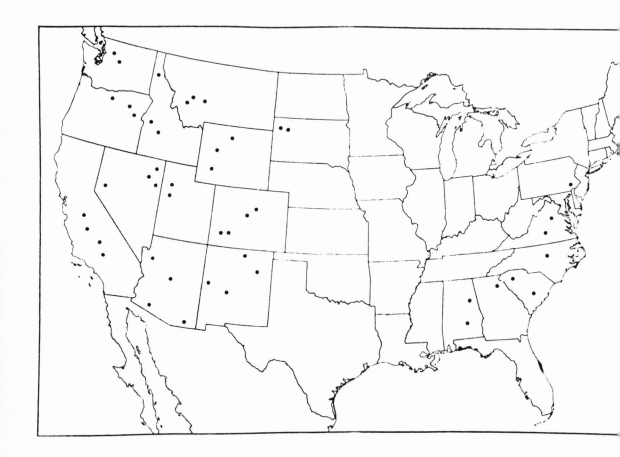

Significant gold-producing areas in the contiguous United States. (*U.S. Department of the Interior, Geological Survey*)

For More Information —
State-by-State, Province-by-Province

THE PAGES THAT follow are meant to help you get information and advice on placer mining sites in the United States and Canada.

While libraries, historical societies, and some museums are excellent sources concerning local mining activity of the past, a great deal of information can be obtained simply by writing for it.

The very best starting point is likely to be the State Geologist in the state in which you're interested—or the Office of Mines and Geology, or Geology and Mineral Research, or whatever name that office happens to bear. Branch offices of the Bureau of Land Management (listed below) and the U.S. Geological Survey (listed in Chapter 2) should not be overlooked.

The U.S. Geological Survey (119 National Center, Reston, VA 22097) has published reports of the principal gold-bearing districts of the country. These are listed in a booklet titled *Publications of the U.S. Geological Survey*. A similar publication, titled *Gold* (MCP-25), is available from the Bureau of Mines (4800 Forbes Ave., Pittsburgh, PA 15213).

STATE OFFICES

Bureau of Land Management

Note: Be sure to include "Bureau of Land Management" as the first line in each of the following addresses.

Alaska:
701 C Street
Box 13
Anchorage, AK 99573

Arizona:
2400 Valley Bank Center
Phoenix, AZ 85073

California:
Federal Building, Room E-2841
2800 Cottage Way
Sacramento, CA 95825

Colorado:
Colorado State Bank Building
1600 Broadway
Denver, CO 80202

States east of the Mississippi River, plus Iowa, Minnesota, Missouri, Arkansas, and Louisiana:
Eastern States Office
350 S. Pickett St.
Alexandria, VA 22304

Idaho:
Federal Building, Room 398
550 West Fort Street
P.O. Box 042
Boise, ID 83724

Montana, North Dakota, and South Dakota:
222 N. 32nd Street
P.O. Box 30157
Billings, MT 59107

Nevada:
Federal Building, Room 3008
300 Booth St.
Reno, NV 89509

New Mexico, Oklahoma, and Texas:
U.S. Post Office and Federal Building
P.O. Box 1449
Santa Fe, NM 87501

Oregon and Washington:
729 N.E. Oregon Street
P.O. Box 2965
Portland, OR 97208

Utah:
University Club Building
136 East South Temple
Salt Lake City, UT 84111

Wyoming, Kansas, and Nebraska:
2515 Warren Avenue
P.O. Box 1828
Cheyenne, WY 82001

Forest Service Regional Offices

Note: Be sure to include "U.S. Forest Service Regional Office" as the first line in each of the addresses below.

Alaska Region
Federal Office Building
P.O. Box 1628
Juneau, AK 99802

California Region
630 Sansome Street
San Francisco,
   CA 94111

Eastern Region
633 West Wisconsin Avenue
Milwaukee, WI 53203

Intermountain Region
Federal Building
324 25th Street
Ogden, UT 84401

Northern Region
Federal Building
Missoula, MT 59807

Pacific Northwest Region
319 S.W. Pine Street
P.O. Box 3623
Portland, OR 97208

Rocky Mountain Region
11177 West Eighth Avenue
Box 25127
Lakewood, CO 80225

Southern Region
Suite 800
1720 Peacetree Road,
                  N.W.
Atlanta, GA 30309

Southwestern Region
Federal Building
517 Gold Avenue, S.W.
Albuquerque, NM 87102

Alabama

Gold was discovered in Alabama in 1830, and a genuine "rush" took place in the years that followed, with thousands of people working the deposits at Arbacoochee in south-central Cleburne County and Goldville in northern Tallapoosa County. But the boom died with the discoveries in California, when Alabama miners departed for the Mother Lode country.

But some mineral production continued in the state until the 1860s, when the War Between the States shut down virtually all the mines in the Southeast. Gold mining in Alabama recovered in the 1870s, with some of the more prosperous mines continuing in operation until World War I. Most of the mining enterprises that were attempted after World War I were marginal in character, the Hog Mountain mine north of Alexander City being an exception. Numerous mine pits, trenches, and tunnels still remain in the major gold-bearing districts of the state.

Weekend miners and vacationing gold hunters now seek out the auriferous streams of the state's Piedmont area, a vaguely triangular piece of land in east-central Alabama, with one side of the triangle bordering Georgia. The following streams and their tributaries have yielded placer gold: Talladega, Dynne, Chulafinnee, Weogufka, Hatchet, Hillabee, Yellow Leaf, Chestnut creeks, and many small tributaries of the Coosa and Tallapoosa rivers.

The Geological Survey of Alabama advises us that practically all the land in the Piedmont area is privately owned, except for a National Forest preserve. Permission should be obtained from the Forest Service regional office before beginning gold-seeking operations in the preserve.

Contact:

State Geologist
Geological Survey of Alabama
P.O. Drawer 0
University, AL 35486

Source material:

Gold Deposits of Alabama, Bulletin 40, Geological Survey of Alabama, $2.50.
Gold in Alabama, Geological Survey of Alabama (free bulletin).

Gold-mining districts:

Cleburne, Randolph, Clay, Coosa, Tallapoosa, and parts of Elmore, Chilton, and Talladega. Small amounts of gold have been found in other counties of the Piedmont.

Alaska

Gold is what drew many of the first settlers to Alaska. Its presence was known as early as 1848, almost two decades before Alaska was purchased by the United States from Russia.

In 1869, miners who were disappointed in British Columbia's prospects discovered placer deposits at Windham Bay and Sumdum Bay southeast of Juneau. The state's major lode deposits were uncovered at Juneau in 1880, and the town quickly became the center of mining activity for the Territory. Encouraged by the successes at Juneau, prospectors began to range throughout southern Alaska, making important discoveries at Berners Bay and Eagle River on the mainland near Juneau, at Klag Bay on Chicagof Island, at Willow Creek near Anchorage, and on Unga Island, over a thousand miles to the west. As transportation facilities improved, new gold discoveries were made in some of the more remote areas of the Territory.

What about today? "The placer deposits are widespread," says the U.S. Geological Survey, "occurring along nearly all of the major rivers and their tributaries and in beach sands near the Nome area."

But Alaskan authorities are forewarning prospectors planning to visit the state. "Old-time prospectors were very thorough and active in their quest for placer gold," says a spokesman for the state's Department of Economic Development. "They were unlikely to have passed over many near-surface bonanzas. Colors or a few nuggets from beach or stream gravels, too lean to attract the original hand miners, may be your best reward.

"Although new discoveries are possible and even probable, most placer gold lies at or near bedrock, and may require much digging for discovery."

An overriding problem faced by present-day prospectors is Alaska's constantly changing land status. For that reason, the State Geological Survey recommends that prospectors follow this set of procedures:

1. Obtain a U.S. Geological Survey topographic map of the area in which you're interested.

2. Take the map to the office of the Bureau of Land Management in either Fairbanks or Anchorage to determine:

    A. Whether the land is open or closed for mineral location.

    B. Whether the land is federal or state property.

3. If the land is owned by the state of Alaska, check with an office of the Division of Lands, either in Anchorage or Fairbanks, to determine whether the land is open for mineral location or has been classified for other uses.

4. Last, check with one of the mining-information offices of the Alaska Division of Geological and Geophysical Surveys to determine whether any preexisting claims have been filed covering the area in which you're interested. These offices are located in Anchorage, Fairbanks, Juneau, and Ketchikan.

Contact:

State of Alaska
Department of Natural Resources
Division of Geological and Geophysical Surveys
Box 80007
College, AK 99708

Public Inquiries Office
U.S. Geological Survey
Room 108
Skyline Building
508 Second Ave.
Anchorage, AK 99501

State of Alaska
Department of Commerce and Economic Development
Division of Economic Enterprise
Pouch EE
Juneau, AK 99811

Source material:

The state Department of Natural Resources (address above) has available the following information circulars at no charge:

Circular No.	Title
1	Proper Claim Staking in Alaska
3	Hand Placer Mining Methods
6	Alaskan Prospecting Information
10	Skin Diving for Gold in Alaska
18	Amateur Gold Prospecting in Alaska

Other recommended sources include:

Anthony, Leo Mark, *Introductory Prospecting and Mining,* Mineral Industry Research Laboratory, University of Alaska, Fairbanks, AK 99701, $7.75.

Cobb, E. H., *Placer Deposits of Alaska,* U.S. Geological Survey Bulletin No. 1374, Superintendent of Documents, Government Printing Office, Washington, DC 20402, $3.10.

Madonna, J. A., *Guide for the Alaska Prospector,* 504 College Rd., Fairbanks, AK 99701, $4.95.

Wolff, Ernest, *Handbook for the Alaskan Prospector,* Mineral Industry Research Laboratory, University of Alaska, Fairbanks, AK 99701, $6.00.

Gold-mining districts:

See points 1-43 on accompanying map.

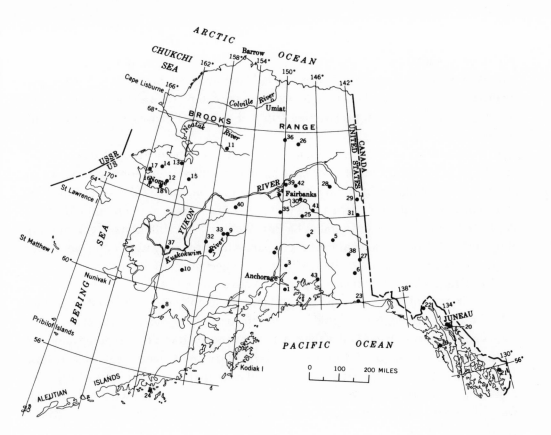

GOLD-MINING DISTRICTS OF ALASKA:

Gold-mining districts of Alaska:

Cook Inlet–Susitna region: 1) Kenai Peninsula 2) Valdez Creek 3) Willow Creek 4) Yentna-Cache Creek.

Copper River region: 5) Chistochina 6) Nizina.

Kuskokwim region: 7) Georgetown 8) Goodnews Bay 9) McKinley 10) Tuluk-sak-Aniak.

Northwestern Alaska region: 11) Shungnak.

Seward Peninsula region: 12) Council 13) Fairhaven 14) Kougarok 15) Koyu-kuk 16) Nome 17) Port Clarence 18) Solomon-Bluff.

Southeastern Alaska region: 19) Chichagof 20) Juneau 21) Ketchikan-Hyder 22) Porcupine 23) Yakataga.

Southwestern Alaska region: 24) Unga.

Yukon region: 25) Bonnifield 26) Chandalar 27) Chisana 28) Circle 29) Eagle 30) Fairbanks 31) Fortymile 32) Iditarod 33) Innoko 34) Hot Springs 35) Kan-tishna 36) Koyukuk 37) Marshall 38) Nabesna 39) Rampart 40) Ruby 41) Richardson 42) Tolovana.

Prince William Sound region: 43) Port Valdez.

Arizona

"A person not in robust health or who has not sufficient funds to finance his entire trip runs a splendid chance of starving to death if he tackles placer mining in Arizona," a booklet by the state Bureau of Mines in 1933 declared. "If, however, a man in good health is out of work, and has enough money to pay camp expenses for some time, and is willing to work hard, a prospecting trip will doubtless prove preferable to lying around and doing nothing."

From this rather cheerless statement, one would never deduce that gold deposits of at least some economic importance have been found in every Arizona county, except Apache, Coconino, Graham, and Navajo, and that the state ranks either fourth or eighth in gold production, depending on your source. More to the point, old placer deposits, such as those to be found at La Paz, Weaver-Rich Hill, Lynx Creek, and Big Bug Creek, still yield gold in almost every pan. And if you stay with it and work over some of the better gravels, a grain-of-wheat size nugget may be your reward. It's been a very long time since any miner is reported to have starved to death.

As early as 1774, when Arizona was still part of Spain's empire, a Jesuit priest and others worked rich deposits of gold in southern Arizona. The legendary period of the state's gold history dawned late in the nineteenth century. It was then that the fabulous Vulture Lode was discovered in northwestern Maricopa County near what is now Wickenburg. The first important underground mine in the state, the operation got its name when a prospector threw a rock to shoo off a vulture eating his lunch. The rock was solid gold. A gold strike in the Bradshaw Mountains in 1864 gave birth to the town of Prescott. The United Eastern Mine at Oatman proved to be Arizona's biggest gold producer. Between 1917 and 1924, United Eastern yielded about $15 million in gold. During the 1960s, Arizona's gold production outstripped that of California and Colorado.

Most of Arizona's lode and placer gold comes from deposits that occur in the mountain region of the state, a belt about 65 miles wide that borders the southwestern edge of the Colorado Plateau. Within this area, the deformed Precambrian rocks have been severely tilted, faulted, and intruded by masses of igneous rock. In the desert region to the southwest, gold deposits are widely scattered.

The Arizona Bureau of Geology and Mineral Technology (address below) provides informative booklets of value to both professional and amateur prospectors, updating the publications regularly. In addition, the Bureau classifies mineral and rock specimens found in Arizona at no charge, and provides advice for mining operators concerning techniques and economics of mine development, sampling, assaying, equipment selection, and similar topics.

Contact:

State of Arizona
Bureau of Geology and Mineral Technology
845 N. Park Ave.
Tucson, AZ 85719

Source material:

The Bureau of Geology and Mineral Technology (address above) has available a free booklet titled *List of Available Publications.* Among its more important listings are:

Arizona Lode Gold Mines and Gold Mining, Bulletin 137, $2.75.
Exploration and Development of Small Mines, Bulletin 164, 25 cents.
Gold Placers and Placering in Arizona, Bulletin 168, $2.

Gold-mining districts:
See points 1-42 on accompanying map.

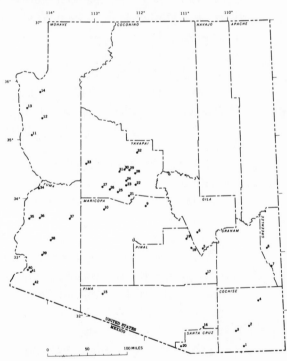

GOLD-MINING DISTRICST OF ARIZONA:

Gold-mining districts of Arizona:

Cochise County: 1) Bisbee 2) Turquoise 3) Tombstone 4) Dos Cabezas.
Gila County: 5) Banner 6) Globe-Miami.
Greenlee County: 7) Ash Peak 8) Clifton-Morenci.
Maricopa County: 9) Cave Creek 10) Vulture.
Mohave County: 11) San Francisco 12) Walapai 13) Weaver 14) Gold Basin.
Pima County: 15) Ajo 16) Greaterville.
Pinal County: 17) Mammoth 18) Ray 19) Superior.
Santa Cruz County: 20) Oro Blanco.
Yavapai County: 21) Tiptop 22) Black Canyon 23) Pine Grove-Tiger 24) Peck 25) Black Rock 26) Weaver-Rich Hill 27) Martinez 28) Agua Fria 29) Big Bug 30) Lynx Creek-Walker 31) Hassayampa-Groom Creek 32) Jerome 33) Eureka.
Yuma County: 34) Cienega 35) La Paz 36) Plomosa 37) Ellsworth 38) Kofa 39) Castle Dome 40) Laguna 41) Dome 42) Fortuna.

California

California is rightly called the "Golden State." It's there that tens of thousands of recreational miners pan, sluice, and dredge for gold on warm summer weekends. It's in California, according to the Bureau of Land Management, that more than 66,000 gold claims have been filed. Of the 58 counties in the state, significant quantities of gold have been found in 41.

The most productive areas are located along the western slope of the Sierra Nevada, says the California Division of Mines and Geology. Areas most favorable to placer mining extend along the Sierra Nevada streams from points where they enter the Great Valley of California to elevations of about 5,000 feet. As for lode deposits, there's the Mother Lode system, a strip of mineralized rock that is from one to four miles wide and that extends for a distance of about 120 miles along the lower western flank of the Sierra Nevada. From near Georgetown in El Dorado County, it extends southward to Mormon Bar, 21 miles southeast of Mariposa. The five counties it traverses—El Dorado, Amador, Calaveras, Tuolumne, and Mariposa—are called the Mother Lode counties. The rivers that drain the Mother Lode—the Feather, Mokelumne, Sacramento, American, Cosumnes, Calaveras, and Yuba—attract the bulk of the placer miners.

Another important placer-mining area is located in the Klamath Mountains in the northern part of the state. The relative inaccessibility and sparseness of population render this area attractive to only the more venturesome prospectors. The Smith, Klamath, Scott, Salmon, Trinity, New Hay Fork, and Sacramento rivers and their tributaries have traditionally been productive.

In southern California, gold has been recovered from placers along the Kern River; in the vicinity of Randsburg and Atolia; Placerita Canyon near Newhall, and the San Gabriel Canyon in Los Angeles County.

Although James W. Marshall, who plucked a gold nugget out of the American River near Coloma in 1848, is usually credited with being the discoverer of California's gold, there was mining activity there as early as 1775, when Mexicans were recovering placer gold from the Colorado River and lode gold from the Cargo Muchache Mountains. Gold was discovered in Los Angeles County in the 1830s. But these early strikes were in sparsely settled areas, far from lines of communication, and news of the discoveries did not spread

Suction dredges and divers attired in wet suits are commonplace on California streams and rivers today. But diving for gold is not something new. It's been going on for at least a century. As early as 1849, when miners realized that the deeper gravels in the river beds were rich in gold, divers were sent down to attempt to recover gold from the bed of the American River at Coloma. Some of the early divers may have been Hawaiian sailors who left their ships in San Francisco to try their luck in the gold fields. Later, in the 1880s and 1890s, when thousands of Chinese were brought to the gold regions, to remove the tailings and re-mine the rivers, it's certain that many of them were used as divers.

Today's divers and their suction dredges must operate within regulations established by California's Department of Fish and Game. A dredge permit must be obtained from that agency. It costs $5. While some rivers are open to dredging the year 'round, others may only be open during particular months of the summer, and, on still others, no dredging at all is permitted. A bulletin from the Department of Fish and Game (see below) lists open and closed areas.

If you're planning to prospect and mine in California, the State Division of Mines and Geology (address below) is your best starting point. Obtain a list of the Division's publications relating to gold and mining. The Division will also send you a rundown of the other state agencies that can be helpful. There are about a dozen of them. There's no charge for these lists.

Contact:

State of California
Division of Mines and Geology
Sacramento District Office
2815 O St.
Sacramento, CA 95816

State of California
Department of Fish and Game
1416 Ninth St.
Sacramento, CA 95814

Source material:

These are among the publications that may be obtained from the Division of Mines and Geology:

Geology of Placer Deposits, Publication No. 34, $1.

Basic Placer Mining, Publication No. 41, $1.

Gold Districts of California, Bulletin No. 193, $6.50.

"Diving for Gold," reprinted from *California Geology,* 25 cents.

Other publications of interest include:

Vacation/Gold Country: Touring Amador County, Amador Chamber of Commerce, P.O. Box 596, Jackson, CA 95642, 50 cents.

Gold-mining districts:

See points 1-97 on accompanying map.

GOLD-MINING DISTRICTS OF CALIFORNIA:

Gold-mining districts of California:

Amador County: 1) Mother Lode 2) Fiddletown 3) Volcano 4) Cosumnes River placers.

Butte County: 5) Magalia 6) Oroville 7) Yankee Hill.

Calaveras County: 8) Mother Lode, East Belt, and West Belt 9) Placers in Tertiary gravels 10) Jenny Line 11) Camanche 12) Campo Seco.

Del Norte County: 13) Smith River placers.

El Dorado County: 14) Mother Lode, East Belt, and West Belt 15) Georgia Slide 16) Placers in Tertiary gravels.

Fresno County: 17) Friant.

Humboldt County: 18) Klamath River placers.

Imperial County: 19) Cargo Muchacho.

Inyo County: 20) Ballarat 21) Chloride Cliff 22) Resting Springs 23) Sherman 24) Union 25) Wild Rose 26) Willshire-Bishop Creek.

Kern County: 27) Amalie 28) Cove 29) Green Mountain 30) Keyes 31) Rand 32) Rosamond-Mojave 33) Joe Walker mine 34) St. John mine 35) Pine Tree mine.

Lassen County: 36) Diamond Mountain 37) Hayden Hill.

Los Angeles County: 38) Antelope Valley 39) Acton 40) San Gabriel .

Mariposa County: 41) Mother Lode, East Belt 42) Mormon Bar 43) Hornitos 44) Merced River placers 45) Placers in Tertiary gravels.

Merced County: 46) Snelling.

Modoc County: 47) High Grade.

Mono County: 48) Bodie 49) Masonic.

Napa County: 50) Calistoga.

Nevada County: 51) Grass Valley-Nevada City 52) Meadow Lake 53) Tertiary placer districts.

Placer County: 54) Dutch Flat-Gold Run 55) Foresthill 56) Iowa Hill 57) Michigan Bluff 58) Ophir 59) Rising Sun Mine.

Plumas County: 60) Crescent Mills 61) Johnsville 62) La Porte.

Riverside County: 63) Pinacate 64) Pinon-Dale.

Sacramento County: 65) Folsom 66) Sloughhouse.

San Bernardino County: 67) Dale 68) Holcomb 69) Stedman.

San Diego County: 70) Julian.

San Joachin County: 71) Clements 72) Bellota.

Shasta County: 73) Deadwood-French Gulch 74) Igo 75) Harrison Gulch 76) West Shasta 77) Whiskeytown.

Sierra County: 78) Alleghany and Downieville 79) Sierra Buttes.

Siskiyou County: 80) Humbug 81) Klamath River 82) Salmon River 83) Scott River 84) Cottonwood-Fort Jones-Yreka.

Stanislaus County: 85) Oakdale-Knights Ferry 86) Waterford.

Trinity County: 87) Trinity River 88) Carrville.

Tulare County: 89) White River.

Tuolumne County: 90) Mother Lode 91) East Belt 92) Pocket Belt 93) Columbia Basin-Jamestown-Sonora 94) Groveland-Moccasin-Jacksonville.

Yuba County: 95) Browns Valley-Smartville 96) Brownsville-Challenge-Dobbins 97) Hammonton.

Colorado

Because Colorado has such rich lode deposits (it ranks second among the states in gold production), placer mining is often overshadowed. But the truth is that every gold district of the state has some placer production.

Colorado placers are normally confined to the narrow canyons below gold-mining areas in a belt that extends northeast across the western part of the state. Many of the streams emerging from the Front Range, the headwaters of the South Platte River, and the Arkansas River and its tributaries as far upstream as California Gulch, contain placer gold.

The Colorado Division of Mines offers a free map that shows the areas of the state that have been placer mined. However, no map exists to show which lands are open for prospecting, mining, and mineral claim. This information you must gather on your own, perhaps beginning by carefully examining the site in which you're interested, looking for location markers and monuments. If none can be found, check the county records maintained at the office of the county assessor, clerk, or recorder at the local county seat. Records pertaining to unpatented mining claims are maintained at the office of the county clerk and recorder. Records of patented land claims will be found at the office of the Bureau of Land Management in Denver (700 Colorado State Bank Building).

If no claim has been entered for the land, and it is not privately owned, not a state forest or park, not an Indian or military reservation, the land may be public land and open to mineral exploration. But state officials advise that a high proportion of the claims in the older mining districts, although entered years ago, are still valid.

There probably is no other state as environmentally conscious as Colorado, and ordinances abound that provide for the protection of domestic water supplies and limit the discharge of debris into streams. The regulations are too numerous to list here. But, clearly, any placer-mining operations you seek to conduct are likely to be subject to ordinances set down by a considerable number of agencies: the Reclamation Section of the Department of Natural Resources, The Colorado Division of Mines, Colorado Division of Wildlife, Colorado Division of Parks, Colorado Water Conservation Board, and the Colorado Water Quality Control Commission. They will also fall under the jurisdiction of various city and county statutes.

Contact:

Colorado Division of Mines
Department of Natural Resources
1313 Sherman St.
Denver, CO 80203

Source material:

Publications of the Colorado Geological Survey, Department of Natural Resources, State of Colorado, January 1980.

Location Regulations for Lode and Placer Mining Claims, Circular No. 3, Colorado Division of Mines, January 1976.

Gold-mining districts:

See points 1-44 on accompanying map.

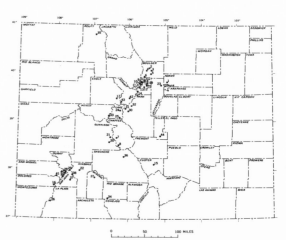

GOLD-MINING DISTRICTS OF COLORADO:

Gold-mining districts of Colorado:

  Adams County: 1) Clear Creek placers.
  Boulder County: 2) Jamestown 3) Gold Hill-Sugarloaf 4) Ward 5) Magnolia 6) Grand Island-Caribou.
  Chaffee County: 7) Chalk Creek 8) Monarch.
  Clear Creek County: 9) Alice 10) Empire 11) Idaho Springs 12) Freeland-Lamartine 13) Georgetown-Silver Plume 14) Argentine.
  Custer County: 15) Rosita Hills.
  Dolores County: 16) Rico.
  Eagle County: 17) Gilman.
  Gilpin County: 18) Northern Gilpin 19) Central City.
  Gunnison County: 20) Gold Brick-Quartz Creek 21) Tincup.
  Hinsdale County: 22) Lake City.
  Jefferson County: 23) Clear Creek placers.
  Lake County: 24) Leadville 25) Arkansas River Valley placers.
  La Plata County: 26) La Plata.
  Mineral County: 27) Creede.
  Ouray County: 28) Sneffels-Red Mountain 29) Uncompahgre.
  Park County: 30) Alma 31) Fairplay 32) Tarryall.
  Pitkin County: 33) Independence Pass.
  Rio Grande County: 34) Summitville.
  Routt County: 35) Hahns Peak.
  Saguache County: 36) Bonanza.
  San Juan County: 37) Animas 38) Eureka.
  San Miguel County: 39) Ophir 40) Telluride 41) Mount Wilson.
  Summit County: 42) Breckenridge 43) Tenmile.
  Teller County: 44) Cripple Creek.

Connecticut

E. C. Kennedy, "The Old Prospector," as he calls himself, once found a few small grains of gold in Enfield. He did it on a bet and it took an enormous amount of tenacity. Only minute quantities of gold have been found in the state, and what has been recovered is invariably of flour size. No commercial gold mines or placer deposits have been worked in Connecticut, according to the Geological and Natural History Survey.

Georgia

"Thar's gold in them thar hills" is, to most people, a familiar quotation from the writings of Mark Twain. But not to the residents of the town of Dahlonega in north Georgia. They know it as a phrase coined by the assayer of the Dahlonega mint. A phrase that was meant to dissuade local miners from straying off to California in 1849.

Georgia's gold rush had begun earlier, in 1828, when one Benjamin Parks kicked over an ore-laden rock near Dahlonega which contained a chunk of gold "as yellow as the yolk of an egg." Almost overnight, some four thousand prospectors were active in northern Georgia. Auraria, five miles south of what is now Dahlonega, in the heart of what had been Cherokee Indian Territory just a few years before, was the first gold mining town to be established.

In 1838, Dahlonega was selected as a location for one of the three branch mints in the United States. It continued in operation until the outbreak of the Civil War.

There's still "gold in them thar hills." Indeed, prospecting and mining have continued to this day. Such activity is concentrated in several mineralized zones amidst the crystalline schists and gneisses that underlie the northern third of the state.

There are no public lands in Georgia and, thus, prospectors must obtain permission from property owners before exploring streams and hillsides. Anyone planning to use a suction dredge or perform surface mining must adhere to guidelines established by the state's Environmental Protection Division (address below).

Contact:

State of Georgia
Department of Natural Resources
270 Washington St., S.W.
Atlanta, GA 30334

State of Georgia
Environmental Protection Division
North Georgia Regional Office
19 Martin Luther King Drive
Atlanta, GA 30334

Source material:

The Department of Natural Resources provides a free information packet for individuals planning to prospect in Georgia. One of the publications it contains, *General Information on Gold in Georgia,* describes and locates the distribution of gold deposits in the state.

Principal gold-producing counties:

Cherokee, Lumpkin, White.

Idaho

Either lode or placer gold is to be found in each of more than twenty counties of Idaho, one of the leading gold-producing states.

The first gold placers were discovered along the Pend Oreille River in 1852. Then, in 1860, Captain E. D. Pierce found rich placer deposits near what was to become the town of Pierce in Clearwater County. Other placers were discovered in Elk City, Orofino, and along the Salmon River in 1860 and 1861, and the following year discoveries were made at Florence, Warren, and Boise Basin.

The lode gold deposits of Idaho are related to the huge batholith, or enormous body of igneous rock, that makes up much of the central portion of the state. After 1900, when commercial interests had exhausted the state's placer deposits, they turned their attention to the lode gold.

There are vast areas of federal and state-owned land in Idaho that are open for prospecting, mining, mineral claim application, and mining claim location. The Idaho Department of Lands (address below) administers state lands and the beds of navigable rivers and streams.

To be sure that you're not trespassing on private lands, state officials advise that you use a good county map to help you in preparing a description of the land in which you're interested. The next step is to take the description to the county assessor's office. "The people there will help you in determining the ownership of the land," says a state spokesman. Mineral status plat books can be consulted at the office of the Department of Lands for information as to which lands have been leased or claimed.

State water quality standards permit no increase in turbidity or discoloration of any flowing stream. To you, as a miner, this means you must obtain a Stream Channel Alteration Permit if you're planning to use any equipment more sophisticated than a shovel and pan. Such permits can be obtained from the Idaho Department of Water

Resources. If your operation is going to be moving more than two cubic yards of material per hour, you will need a Dredge Mining Permit.

Contact:

State of Idaho
Department of Lands
Statehouse
Boise, ID 83720

State of Idaho
Bureau of Mines and Geology
Moscow, ID 83843

Source material:

A free information bulletin titled, "The Twentieth Century Gold Rush, Prospecting for Gold in Idaho," is available from the Department of Lands.

These publications are available from the Bureau of Mines and Geology:

A Metallurgical Study of Idaho Placer Sand, Idaho Bureau of Mines and Geology, Pamphlet 51, 1932.

Prospecting for Gold Ores, Idaho Bureau of Mines and Geology, Pamphlet 36, 1932.

Gold in Idaho, Idaho Bureau of Mines and Geology, Pamphlet 68, 1960.

Gold-mining districts:

See points 1-42 on accompanying map.

GOLD-MINING DISTRICTS OF IDAHO:

Gold-mining districts of Idaho:

Ada County: 1) Black Hornet.
Bingham County: 2) Snake River placers.
Blaine County: 3) Camas 4) Warm Springs.
Boise County: 5) Boise Basin 6) Pioneerville 7) Quartzburg.
Bonneville County: 8) Mount Pisgah.
Camas County: 9) Big and Little Smoky-Rosetta.
Cassia, Jerome, and Minidoka Counties: 10) Snake River placers.
Clearwater County: 11) Pierce.
Custer County: 12) Alder Creek 13) Loon Creek 14) Yankee Fork.
Elmore County: 15) Atlanta 16) Featherville 17) Neal 18) Pine Grove 19)

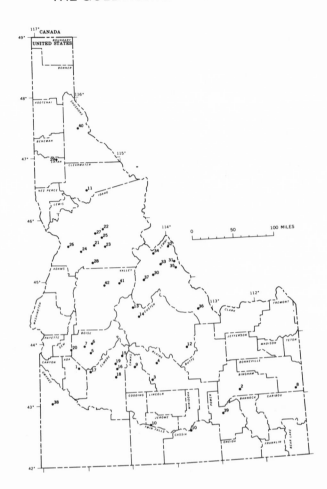

Rocky Bar.

    Gem County: 20) Westview.

    Idaho County: 21) Buffalo Hump 22) Elk City 23) Dixie 24) French Creek-Florence 25) Orogrande 26) Simpson-Camp Howard-Riggins 27) Ten-mile 28) Warren-Marshall.

    Latah County: 29) Hoodoo.

    Lemhi County: 30) Blackbird 31) Carmen Creek-Eldorado-Pratt Creek-Sandy Creek 32) Gibbonsville 33) Mackinaw 34) Mineral Hill and Indian Creek 35) Kirtley Creek 36) Texas 37) Yellow Jacket.

    Owyhee County: 38) Silver City.

    Power County: 39) Snake River placers.

    Shoshone County: 40) Coeur d'Alene region.

    Valley County: 41) Thunder Mountain 42) Yellow Pine.

Illinois

Minor amounts of gold have been found in Illinois amidst the sand and gravel associated with glacial deposits. But these occurrences, to quote a state geologist, "have been little more than curiosities." None has ever developed into a serious mining operation.

Claim staking is not valid in Illinois. Leasing is the accepted method for gaining access to property for mineral prospecting.

Contact:
Illinois Institute of Natural Resources
State Geological Survey Division
Natural Resources Building
Urbana, IL 61801

Indiana

While no veins bearing gold are known to occur in Indiana, small quantities of gold are to be found in the sand and gravel deposits in any of those counties whose surface makeup includes rock debris left by glaciers. W. S. Blatchley, who, in 1903, prepared a detailed list of the areas where gold had been found in the state, pointed out that gold accumulated in the greatest quantity "only at the edges of these . . . moraines, where the material composing the drift has been most weathered and washed, and where streams flowing from the moraines have deposited beds of gravel over the bed rock in their valleys."

To be specific, streams in Morgan and Brown counties have given up the most gold. That's where Hoosier prospectors of the past have concentrated their efforts.

Don't expect a "big" find. Flecks of gold found in Indiana rarely exceed 3 millimeters in diameter. The largest nugget ever found weighed 8.55 grains, less than one third of an ounce.

Contact:
Indiana Department of Natural Resources
Geological Survey
611 North Walnut Grove
Bloomington, IN 47405

Source material:
Minerals of Indiana, Indiana Geological Survey, Bulletin B18, 1961, 75 cents.
Gold and Diamonds in Indiana, reprinted from the 27th Annual Report of the Indiana Department of Geology and Natural Resources of 1902, 50 cents.

Principal gold-producing counties:
Brown, Morgan.

Iowa

There are no outcroppings of gold-rich rock in Iowa, but gold flakes and particles are scattered in small amounts throughout the glacially deposited sand and gravel found along many of Iowa's rivers. These materials have been derived from Minnesota and Canada, where igneous rock formations are abundant.

Historically, gold has been reported along the Iowa River near Steamboat Rock and Eldora in Hardin County; along Otter Creek near West Union, Brush Creek, and the Volga River in Fayette and Clayton counties; along the Des Moines River near Pella, Douds, and Farmington; along Vasser Creek in northeast Davis County; along the Little Sioux River near Correctionville in Woodbury County, and near the town of Cherokee. Along the Big Sioux River at Klondike in Lyon County, a small commercial placer mine operated during the early 1900s.

In 1904, a report concerning the geology of Fayette County stated that $1 to $1.50 in gold could be panned from Otter Creek by a "patient washer" in a day. An ounce of gold brought $20.67 at the time. If you now figure gold at $750 per ounce, a similar day's work could produce gold worth $37 to $56.

Contact:
Iowa Geological Survey
123 North Capitol St.
Iowa City, IA 52242

Kansas

"Kansas isn't really a prime area for gold mining," says a spokesman for the state

Geological Survey. Although there are anecdotal accounts of gold being found in Kansas from time to time, there are no verified cases of full-fledged gold mining having been conducted in the state. Geologists theorize that the best place to look for gold might be in the Arkansas River and the other rivers that drain the Rocky Mountains.

Maine

Although gold occurs in several different geological environments in Maine, most gold now being recovered in the state is placer gold. Gold panning is permitted in streams that traverse state-owned land, but prospecting or mining on state property requires prior authorization. The Maine Department of Conservation (address below) can provide more information on this subject.

The state Geological Survey names these streams as "gold-panning streams":

Stream	Township	County
East Branch, Swift River	Byron	Oxford
Sandy River	Madrid to New Sharon	Franklin
South Branch, Penobscot River	Sandy Bay; Bald Mountain; Prentiss	Somerset
Gold Brook	Bowman	Oxford
Gold Brook	Chain of Pounds, Kibby	Franklin
Gold Brook	Chase Stream Tract	Somerset
Gold Brook	T5-R6, Rangeley	Somerset
Nile Brook	Dallas, Rangeley	Franklin
Kibby Stream	Kibby	Franklin
St. Croix River	Baileyville	Washington
Black Mountain Brook	Rumford	Oxford

Interest in gold and gold panning is such that three "mineral shops" are now operating in Maine—in West Paris, East Winthrop, and Dresden—where mining equipment and supplies may be purchased.

Contact:

State of Maine
Department of Conservation
State House Station 22
Augusta, ME 04333

Source material:

Available Publications of the Maine Geological Survey, Department of Conservation, 1979.

Maine Geological Survey and Mining on State Lands, Statutes, Department of Conservation, 1977.

Maryland

Gold-bearing veins have been found through much of Maryland's Piedmont Plateau in the central part of the state, and gold panners are active on many Piedmont streams and creeks.

The most notable occurrences of gold have been pinpointed near Great Falls in Potomac, Maryland. Gold has also been mined in lesser amounts at the following sites: near Winfield, Carroll County (Costly Gold Mine); in Frederick County (Frederick-Clifton Mine); and near Catonsville, Baltimore County (Hays Gold Mine).

Maryland has strict property laws. To pan or prospect, you must first obtain permission from the landowner. To investigate many of the state's old gold mines, which are located on federal-park property, permission must be obtained from the Superintendent of Parks.

Contact:

Maryland Geological Survey
The Johns Hopkins University
Merryman Hall
Baltimore, MD 21218

Source material:
List of Publications, Maryland Geological Survey.
Reed, J. C. and Reed, J. C., Jr. *Gold Veins Near Great Falls. Maryland.* 1969, U.S. Geological Survey Bulletin 1286.

Michigan

It's believed that gold in Michigan was first discovered by Dr. Douglas Houghton, the state's first geologist, in 1842, in a streambed of the Upper Peninsula. The only significant output of gold in Michigan is also Upper Peninsula in origin, being derived from a small area about four miles northwest of Ishpeming on the northern side of the Marquette Range in Marquette County. It was in this area that the Ropes mine was located; it yielded gold bullion valued at more than $600,000. The mine closed in 1897.

Today, the state is better known for its placer deposits than any lode gold it might boast. These occur in stream gravel and moraine debris throughout a good part of the state.

Contact:
Geological Survey Division
Michigan Department of Natural Resources
Box 30028
Lansing, MI 48909

Source material:
The Geological Survey Division makes available a "Gold Open File Report," which contains reprints of selected out-of-print publications. The report costs $2.50.

Principal gold-producing counties:
Antrim, Charlevoix, Ionia, Kent, Leelanau, Manistee, Marquette, Newaygo, Oceana, Ontonagon, Wexford.

Minnesota

It's possible that occurrences of gold might be found almost anywhere amidst Minnesota's glacial drift. But the only substantiated reports of gold placers has come from panners along the Zumbro River between Rochester and Mazeppa. These discoveries have been invariably described as "small," and the locations of the lode gold from which the placers were derived is not known.

Reports of lode discoveries in the state refer to occurrences of insufficient size and richness to be mined profitably. A few attempts have been made in the late 1800s, however, most notably at the Little American Mine, and on the south shore of Lake Vermilion, but both operations were abandoned within a few years because the ore was too lean. Specific information on these and other occurrences of gold in Minnesota can be obtained from an article by Frank F. Grout which is cited below. It should be available at most university libraries in the state.

Contact:

Minnesota Geological Survey
University of Minnesota, Twin Cities
1633 Eustis St.
St. Paul, MN 55108

Source material:

Grout, Frank F., *Petrographic Study of Gold Prospects in Minnesota,* "Economic Geology," Vol. 32, No. 1, 1937.

Mississippi

While there are reports of gold being recovered in Jackson County, Mississippi does not offer great promise for prospectors. Gold particles that might be present in the sedimentary coastal plain would have been transported great distances, possibly from the southern Appalachians.

Contact:

Bureau of Geology
Mississippi Department of Natural Resources
P.O. Box 5348
Jackson, MI 39216

Missouri

Gold occurs in small amount within some of the glacial drift of northern Missouri, and occurrences have been reported in Macon, Adair, Livingston, Linn, Putnam, and Randolph counties. But no deposits of economic value are known in the state and there are no gold mines, active or abandoned. "Gold in Missouri," says the state geologist, "is considered little more than an interesting curiosity."

Contact:
Missouri Department of Natural Resources
Division of Geology and Land Survey
P.O. Box 250
Rolla, MO 65401

Montana

Gold seekers have been prowling the rivers and streams of Montana for more than a century, and although many worked with thoroughness, there is still placer gold to be found there. Montana's principal placer districts are clustered in the southwestern part of the state, with the Helena mining district and the many placers along the Missouri River in the vicinity of Helena being among the more productive areas. The headwaters and tributaries of the Missouri River in Madison County, particularly in the vicinity of Virginia City and Bannack, and the headwaters of the Clark Fork of the Columbia River, are also noted for their placer deposits.

Gold was first discovered in Montana in the sand and gravel of what is now Gold Creek in Powell County in 1852. But not until 1862, with the discovery of placers along Grasshopper Creek in Beaverhead County, did the influx of prospectors begin. Other discoveries, including the rich deposits along Alder Gulch, near Virginia City in Madison County, were made the following year, triggering an even greater inpouring of gold hunters.

Today, Montana ranks third in placer production of gold, behind California and Alaska. But more than half of the state's gold production comes as a by-product of base-metal refining, principally by Anaconda Copper's operations in and near Butte.

Contact:

Director
Information Service
Montana Bureau of Mines and Geology
Butte, MT 59701

Gold-mining districts:

See points 1-54 on accompanying map.

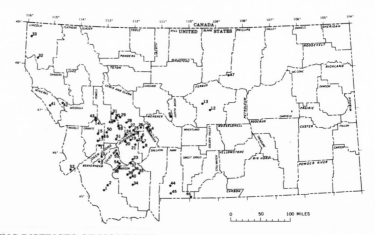

GOLD-MINING DISTRICTS OF MONTANA:

Gold-mining districts of Montana:

   Beaverhead County: 1) Bannack 2) Argenta 3) Bryant.
   Broadwater County: 4) Confederate Gulch 5) White Creek 6) Winston 7) Park 8) Radersburg.
   Cascade County: 9) Montana.
   Deer Lodge County: 10) French Creek 11) Georgetown.
   Fergus County: 12) Warm Springs 13) North Moccasin.
   Granite County: 14) First Chance 15) Henderson Placers 16) Boulder Creek 17) Flint Creek.
   Jefferson County: 18) Clancy 19) Wickes 20) Basin and Boulder 21) Elkhorn 22) Tizer 23) Whitehall.
   Lewis and Clark County: 24) Rimini-Tenmile 25) Helena-Last Chance 26) Missouri River-York 27) Sevenmile-Scratchgravel 28) Marysville-Silver Creek 29) Stemple-Virginia Creek 30) McClellan 31) Lincoln.
   Lincoln County: 32) Libby 33) Sylvanite.
   Madison County: 34) Virginia City-Alder Gulch 35) Norris 36) Pony 37) Renova 38) Silver Star-Rochester 39) Tidal Wave 40) Sheridan.
   Mineral County: 41) Cedar Creek-Trout Creek.
   Missoula County: 42) Ninemile Creek 43) Elk Creek-Coloma.
   Park County: 47) Little Rocky Mountains.
   Powell County: 48) Finn 49) Ophir 50) Pioneer 51) Zosell.
   Ravalli County: 52) Hughes Creek.
   Silver Bow County: 53) Butte 54) Highland.

Nebraska

There *were* goldfields in the western Nebraska Territory in 1858, with reports of panned gold coming from the Arkansas and Platte rivers and from such streams as Cherry and Sand Creek. The lure was so strong that several new communities were founded, and they exist to this day. What must be added, however, is that the Nebraska Territory of 1858 encompassed land far to the west and north of what is now the state of Nebraska. The Nebraska Territory's goldfields, nestled in the Rocky Mountains, now lie wholly within the borders of another state, Colorado.

The consensus among geologists is that Nebraska contains no native gold. The small quantities that have been found were carted in by the glacial drift in the eastern part of the state and alluvial deposits from the Rocky Mountains in the West.

"Placer operations in the Platte River have enjoyed an intermittent vogue during the past century," says the state's Institute of Agriculture and Natural Resources, "but always with uncertain or disappointing results. Some gold nuggets have been found in the gravels of the Platte, but only on a haphazard basis. No one has yet struck paydirt within Nebraska's present boundaries."

Contact:

Institute of Agriculture and Natural Resources
The University of Nebraska—Lincoln
Lincoln, NB 68588

Nevada

Although lode gold deposits are widely distributed throughout Nevada, the state has never produced placer gold in any important amounts. The problem is water—or lack of it. If you're placer mining in Nevada, your most important piece of equipment is likely to be a dry washer or an electrostatic concentrator.

The placer gold that has been found has been recovered chiefly in the western half of the state, an area that includes American Canyon and Spring Valley in the Humboldt Range, Pershing County, and the Manhattan and Round Mountain areas of Nye County. Placer deposits have also been worked below Virginia City and in northern Elko County near Charleston.

Commercial mining began in Nevada in 1850. The years 1859-1879 were a boom period, highlighted by the opening of the Comstock Lode and Reese River mining districts. In the latter part of the nineteenth century, the mining output of Nevada declined steadily, except in the Comstock Lode. Then, in 1900, came the discovery of silver in Tonopah and, in 1902, the gold bonanza at Goldfield. These finds helped to rejuvenate mining activity throughout the state.

Today, most of Nevada's mines and mining districts produce gold as a by-product in the processing of silver, copper, lead, and zinc ores. Partly because of the state's long mining history, its Bureau of Mines and Geology (address below) offers a wide range of bulletins, reports, maps, and other publications, including picture postcards. Write for a free publications list.

Contact:
Nevada Bureau of Mines and Geology
Mackay School of Mines
University of Nevada
Rene, NV 89557

Source material:
The publications listed below are available from the Nevada Bureau of Mines and Geology:

Placer Mining in Nevada, University of Nevada Bulletin, Vol. 30, Number 4, 1936, $3.

Mineral and Water Resources of Nevada, U.S. Geological Survey and Nevada Bureau of Mines and Geology, 1964, $3.

Gold-mining districts:
See points 1-71 on accompanying map.

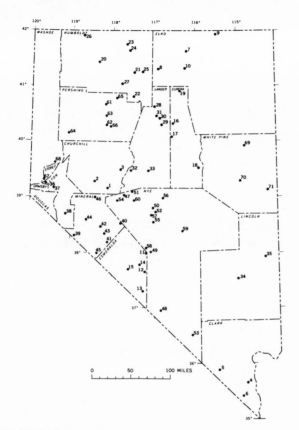

GOLD-MINING DISTRICTS OF NEVADA:

Gold-mining districts of Nevada:

   Churchill: 1) Fairview 2) Sand Springs 3) Wonder.
   Clark County: 4) Eldorado 5) Godsprings 6) Searchlight.
   Elko County: 7) Edgemont 8) Gold Circle 9) Jarbidge 10) Tuscarora.
   Esmeralda County: 11) Divide 12) Goldfield 13) Hornsilver 14) Lone Mountain 15) Silver Peak.
   Eureka County: 16) Buckhorn 17) Cortez 18) Eureka 19) Lynn.
   Humboldt County: 20) Awakening 21) Dutch Flat 22) Gold Run 23) National 24) Paradise Valley 25) Potosi 26) Warm Springs 27) Winnemucca.
   Lander County: 28) Battle Mountain 29) Bullion 30) Hilltop 31) Lewis 32) New Pass 33) Reese River.
   Lincoln County: 34) Delamar 35) Pioche.
   Lyon County: 36) Silver City 37) Como 38) Wilson.
   Mineral County: 39) Aurora 40) Bell 41) Candelaria 42) Garfield 43) Gold Range 44) Hawthorne 45) Mount Montgomery and Oneota 46) Rawhide.
   Nye County: 47) Bruner 48) Bullfrog 49) Ellendale 50) Gold Hill 51) Jackson 52) Jefferson Canyon 53) Johnnie 54) Lodi 55) Manhattan 56) Northumberland 57) Round Mountain 58) Tonopah 59) Tybo 60) Union.
   Pershing County: 61) Humboldt 62) Rochester 63) Rye Patch 64) Seven troughs 65) Sierra 66) Spring Valley.
   Storey County: 67) Comstock Lode.
   Washoe County: 68) Olinghouse.
   White Pine County: 69) Cherry Creek 70) Ely 71) Osceola.

New Hampshire

Robert I. Davis, New Hampshire's state geologist, calls gold a "rare commodity" in his state. And while placer gold has been recovered in some of the streams in the northern part of the state, it is always in very small amounts.

The best sources are the Wild Ammonoosuc, near Bath; the Baker River, near Warren; Indian Stream, north of Stewartstown, Tunnel Stream, in Easton; and the Gale River, in Franconia.

"I have tried to impress people with the fact that you can't earn a living panning for gold up here," says Ernest Foley, a prospector in New Hampshire for more than two decades. "It's strictly a fun thing or hobby."

Contact:
State of New Hampshire
Office of the State Geologist
Department of Resources and Economic Development
James Hall
University of New Hampshire
Durham, NH 03824

Department of Earth Sciences
Dartmouth College
Hanover, NH 03755

Source material:
Hitchcock, C. H., *Geology of New Hampshire,* Vol. 3, 1878.

New Jersey

Gold has been reported near Beemerville, Sussex County; south of Harmony, Warren County; and near Budd Lake, Morris County. Associated with copper ore, it has been found at the old Arlington Mine near North Arlington, Bergen County, and at the Griggstown Mine near Griggstown, Somerset County. In all cases, the gold has consisted of flakes about the size of the head of a pin.

Occurrences associated with pyrite have been reported at Fort Lee, Bergen County, and at the base of the Kittatinny Mountain in Warren and Sussex counties. At the base of the Shawangunk Conglomerate, which forms Kittatinny Mountain, there are some pyritiferous beds that have been assayed for gold. But commercial extraction of these deposits has not been deemed worthwhile.

It is not possible to stake a mining claim in New Jersey. Gold-mining operations require ownership of the surface land or a contractual arrangement with the landowner.

Contact:
State Geologist
State of New Jersey
Department of Environmental Protection
John Fitch Plaza
P.O. Box 1390
Trenton, NJ 08625

New Mexico

There's a rich tradition of placer mining in New Mexico, with the first operations beginning as early as 1828 in the Ortiz Mountains south of Santa Fe. Placer deposits were discovered farther south, in the San Pedro Mountains, in 1839.

Of course, a big drawback to placer mining on a widespread basis is New Mexico's lack of free-flowing streams and rivers. Nevertheless, there is more gold panning than one would expect in the state. The most productive streams have been Moreno Creek in Colfax County and the Old and New Placer districts in Santa Fe County. These locations are on private land, so panning requires permission from the owners.

Other streams that have produced gold include Percha Creek, east of Hillsboro, in Sierra County; Bear Creek and Rio de Arenas near Pinos Altos in Grant County; and the Rio Grande between Embudo and the juncture of the Red River and the Rio Grande. New Mexico's Bureau of Mines and Minerals (address below) has available a detailed description of the state's placer deposits.

Panning on state-owned land requires a general mining lease, which can be obtained from the State Land Office (address below).

Contact:

New Mexico Bureau of Mines and Mineral Resources
Socorro, NM 87801

State of New Mexico Land Office
310 Old Santa Fe Trail
Santa Fe, NM 87503

Source material:

The publications listed below are available from the New Mexico Bureau of Mines and Natural Resources:

Placer Gold Deposits of New Mexico, U.S. Geological Survey, Bulletin 1348, 1972, $1.25.

Gold Mining and Gold Deposits in New Mexico, New Mexico Bureau of Mines, Circular 5, $1.50.

Laws and Regulations Governing Mineral Rights in New Mexico, New Mexico Bureau of Mines and Mineral Resources, Bulletin 104, $1.50.

Gold-mining districts:

See points 1-17 on accompanying map.

GOLD-MINING DISTRICTS OF NEW MEXICO:

Gold-mining districts of New Mexico:

Bernalillo County: 1) Tijeras Canyon.
Catron County: 2) Mogollon.
Colfax County: 3) Elizabethtown-Baldy.
Dona Ana County: 4) Organ.
Grant County: 5) Central 6) Pinos Altos 7) Steeple Rock.
Hidalgo County: 8) Lordsburg.
Lincoln County: 9) White Oaks 10) Nogal.
Otero County: 11) Jarilla.
Sandoval County: 12) Cochiti.
San Miguel County: 13) Willow Creek.
Santa Fe County: 14) Old Placer 15) New Placer.
Sierra County: 16) Hillsboro.
Socorro County: 17) Rosedale.

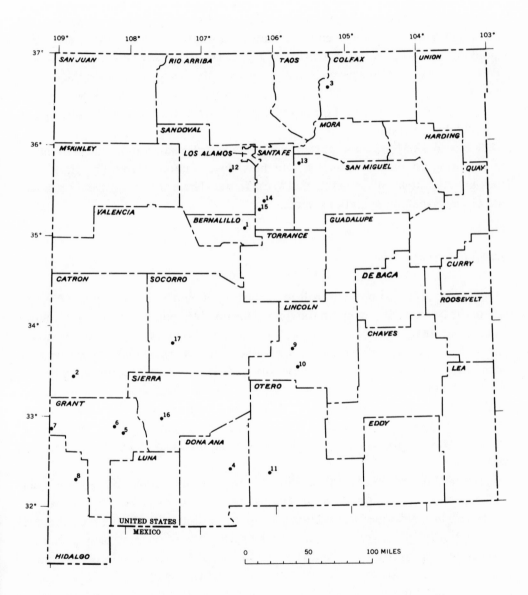

New York

Through the years, there have been frequent and invariably unjustified claims of gold in New York, often triggered by the state mining law, which held that discoveries of gold were the property of the discoverer, regardless of the ownership of the land, for a period of 21 years, after which a one-percent royalty had to be paid to the owner. That law was amended in 1948, requiring claimants to spend at least $250 every 30 months for exploration or exploitation. The number of claims being filed dropped substantially as a result.

The Bureau of Mines has reported that gold in small quantities has been found in the following counties: Allegheny, Dutchess, Erie, Fulton, Hamilton, Herkimer, Rockland, Saratoga, Washington, and Westchester. There are no active commercial-lode or placer-mining operations in the state.

North Carolina

Gold has played an important role in the history of North Carolina since 1799, the date of the first gold discovery in the state. This involved a lump of glittering yellow metal bigger than a softball; it weighed 17 pounds. Twelve-year-old Conrad Reed found the nugget in Little Meadow Creek, a stream that crossed his father's farm in Cabarrus County. It was used as a doorstop in the Reed home for three years, before Mr. Reed sold it to a jeweler in Fayetteville for $3.50. When he later learned of the terrible mistake he had made, John Reed demanded and received an additional $3,000. Reed and three helpers returned to the creek to find more gold nuggets, one of which weighed 28 pounds. In total, the Reed Mine provided 153 pounds of nugget gold.

From 1803 until 1828, North Carolina was the nation's number one state in gold production, and it continued as a leading producer until 1849, when gold was discovered in California. Production continued sporadically throughout the late 1800s and early 1900s, but output during the period was of only minor importance.

While deposits were mined in both the Piedmont and Mountain regions of the state, which today are pocked by more than 600 inactive mines and other workings, most of the early production was concentrated in the central Piedmont, particularly in the Carolina slate belt and Mecklenburg County. The slate belt includes the Gold Hill and

Cid Mining Districts and such mines as the aforementioned Reed Mine, the Gold Hill Mine in Rowan County, and the Silver Hill Mine in Davidson County.

Streams flowing through the former gold-producing districts are the best places to pan. These include Little Buffalo Creek in Rowan and Cabarrus Counties, Dutch River in Montgomery County, Cabin Creek in Moore County, and streams in the South Mountain area.

Most property in North Carolina is privately owned. What federal land there is takes the form of parks and forests. Although prospecting is not permitted in national parks, prospecting permits can sometimes be obtained for National Forests. Consult the District Forest Ranger. Prospecting is not permitted within the state parks of North Carolina.

All mining operations, including the use of surface dredges that discharge large volumes of water into streams, are affected by the State Mining Act of 1971. For information regarding dredge permits, contact the Division of Environmental Management, P.O. Box 27687, Raleigh, NC 27611. For a mining permit, contact the Land Quality Section at the same address.

Contact:

North Carolina Department of Natural Resources and Community Development
Division of Land Resources
Box 27687
Raleigh, NC 27611

Source material:

Gold Resources of North Carolina, Information Circular 21, Geological Survey Section, North Carolina Department of Natural Resources and Community Development, $1.50.

Metallic Mineral Deposits of the Carolina Slate Belt, Bulletin 84, Geological Survey Section, North Carolina Department of Natural Resources and Community Development, $4.

Gold-mining districts:

See points 1-15 on accompanying map.

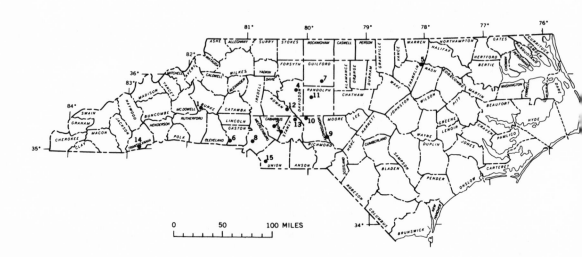

GOLD-MINING LOCALITIES IN NORTH CAROLINA:

Gold-mining localities in North Carolina:

1) Mills property 2) Phoenix Mine 3) Reed Mine 4) Cid District 5) Portis placers 6) Kings Mountain Mine 7) Gardner Hill, Lindsey, North States, Jacks Hill mines 8) Rudisil and St. Catherine mines 9) Iola and Uwarra mines 10) Russell and Steel mines 11) Hoover Hill Mine 12) Gold Hill District 13) Parker Mine 14) Fairfield Valley placers 15) Howie Mine.

North Dakota

There is no naturally occurring gold in North Dakota. The samples of that precious metal that have been found were brought into the state by the great continental glacier. These are invariably extremely fine, the largest recovered being about the size of a grain of wheat. They have been found over a very wide area, covering much of the northern half of the state, in fact. In North Dakota, according to one report, gold deposits should be regarded only as "curiosities."

Contact:
North Dakota Geological Survey
University Station
Grand Forks, ND 58202

Oklahoma

Of the many tales that have been told about gold in Oklahoma, only one appears to be authoritative. Dr. Edwin DeBarr, in an article written in 1904, titled "Report of Mineral Deposits in the Wichita Mountains" and published by the Oklahoma Department of Geology and Natural History, reported finding "a very small quantity of exceedingly fine gold in a limited area" after washing placer material from the creek southeast of Snyder and Kiowa counties. Prospectors worked the area until 1935. Other accounts of gold discoveries in Oklahoma are classified as folklore.

Contact:
Oklahoma Geological Survey
The University of Oklahoma
830 Van Vleet Oval
Room 163
Norman, OK 73019

Oregon

In 1850, when prospectors were still invading California in great numbers, a few of them stopped to sample the sand and gravel of the stream channels of southwestern Oregon, discovering the rich gold placers at Rich Gulch in Jackson County and Josephine Creek in Josephine County. There were also important placers discovered in northeastern Oregon in 1862.

Lode deposits were probably also found during those early days, but the difficulties associated with lode mining made the miners reluctant to abandon the easy pickings provided by the placers. However, lode mining developed rapidly in the last decade of the century. At one time, Oregon could boast 122 active lode mines and 192 active placer mines. Since World War II, however, only a handful of mines have operated in the state, and only one, the Buffalo Mine in the Granite District of Grant County, has produced gold consistently.

For recreational miners, the situation is more positive. Placers are located in the southwestern part of the state, in tributaries of the Rogue River, and in streams in the Klamath Mountains. The chief gold-producing areas are the Greenback District in Josephine County and the Applegate District in Jackson County. Placer gold also occurs in many of the streams that drain the Blue and Wallowa Mountains in the northeast of Oregon. The Sumpter area and the upper Powder River have produced important amounts of gold. Other notable areas include the Burnt River and its tributaries and the John Day River valley.

Most of the land under the jurisdiction of the Bureau of Land Management and the Forest Service is open for prospecting. However, most of the better placer and lode areas have been staked. No permits are required for the use of surface dredges, but the Department of Fish and Wildlife urges miners to read its bulletin, "Recreational Mining Can Be Compatible with Other Resources." It can be obtained from the environmental division of the Management Section, Department of Fish and Wildlife, 506 S.W. Mill St., Portland, OR 97208.

Oregon's Department of Geology and Mineral Industries maintains an assay service at its Portland headquarters where samples of sand and gravel ore can be analyzed. Specimens should weigh from one to five pounds. Fees range from $2 to $12.

Contact:

State of Oregon
Department of Geology and Mineral Industries
1069 State Office Building
Portland, OR 97201

Source material:

The Department of Geology and Mineral Industries has available a "recreational gold packet" containing information on the location of the state's placer deposits, a mineral localities map, and other bulletins of interest. The packet costs $1.50.

Gold-mining districts:

See points 1-31 on accompanying map.

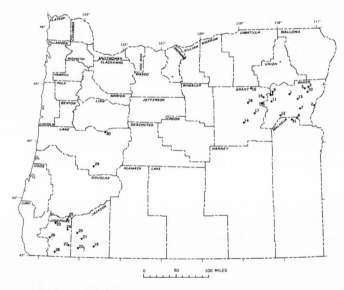

GOLD-MINING DISTRICTS OF OREGON:

Gold-mining districts of Oregon:

Baker County: 1) Baker 2) Connor Creek 3) Cornucopia 4) Cracker Creek 5) Eagle Creek 6) Greenhorn 7) Lower Burnt River Valley 8) Mormon Basin 9) Rock Creek 10) Sparta 11) Sumpter 12) Upper Burnt River 13) Virtue.

Grant County: 14) Canyon Creek 15) Granite 16) North Fork 17) Quartzburg 18) Susanville.

Jackson County: 19) Ashland 20) Gold Hill 21) Jacksonville 22) Upper Applegate.

Josephine County: 23) Galice 24) Grants Pass 25) Greenback 26) Illinois River 27) Lower Applegate 28) Waldo.

Lane County: 29) Bohemia 30) Blue River.

Malheur County: 31) Malheur.

Pennsylvania

The Cornwall Iron Mine, a few miles from Lebanon in the northern part of Lebanon County, is the only location in Pennsylvania to have given up any gold; and this occurrence was a by-product of the mine's iron and copper production. Mining ceased at that site many years ago. Other reports of gold in Pennsylvania are unverified, according to the state geologist.

Contact:
State Geologist
Department of Environmental Resources
Commonwealth of Pennsylvania
P.O. Box 2357
Harrisburg, PA 17120

South Carolina

Like North Carolina and Georgia, South Carolina offers placers in some of its streams that drain the eastern slopes of the southern Appalachian Range. Placer gold was first discovered in 1827 at Haile Mine which, within a few years, became the most productive of all Southeastern mines.

A long gold belt, known as the Carolina Slate Belt, a continuation of a North Carolina geological formation, and running in a northeast-southwest direction, cuts across several counties of the state. Many streams within this belt contain placer gold.

Contact:
South Carolina Geological Survey
Harrison Forest Rd.
Columbia, SC 29210

Source material:
Gold Resources of South Carolina, Bulletin No. 32, South Carolina Geological Survey, 1966, $1.25.

Principal gold-producing counties:
Chesterfield, Lancaster, McCormick.

South Dakota

In total gold production, South Dakota ranks third among the states. As a placer producer, the state is eleventh.

The first documented discovery of gold in South Dakota was made in 1874 by a pair of miners attached to General George A. Custer's expedition. Assigned to reconnoiter the Black Hills, they found gold in the gravel bars along French Creek in what is now Custer County. Miners worked their way northward gradually, with the big discovery of placer gold in Deadwood Gulch occurring in 1875. Prospectors working upstream from Deadwood discovered the Homestake Lode in 1877. That mine, still in operation, produces about a quarter-million ounces of gold per year. There is only one commercial placer-mining operation currently active, located in Whitewood Creek just below Deadwood.

Panners can still get colors in French, Spring, Rapid, and Castle creeks in the Central Black Hills, and in Deadwood, Whitewood, Gold Run, Potato, and Bear creeks in the northern hills. But much of the old placer ground is now privately owned, and the more promising stretches of Forest Service land have been staked.

Detailed information concerning the mining districts of South Dakota and their geological makeup can be obtained from the "Reports of Investigations" cited below.

Contact:
Department of Geology and Geological Engineering
South Dakota School of Mines and Technology
Rapid City, SD 57701

South Dakota Geological Survey
Department of Water and Natural Resources
Science Center, University
Vermillion, SD 57069

Source material:

Report of Investigations, No. 15, South Dakota Geological Survey, $2.63.
Report of Investigations, No. 16, South Dakota Geological Survey, $2.53.

Gold-mining districts:

See points 1-7 on accompanying map.

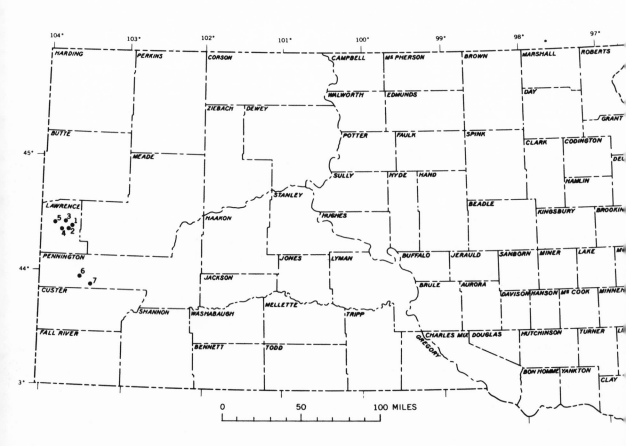

GOLD-MINING DISTRICTS OF SOUTH DAKOTA:

Gold-mining districts of South Dakota:

  1) Deadwood-Two Bit 2) Lead 3) Garden 4) Bald Mountain 5) Squaw Creek
6) Hill City 7) Keystone.

Tennessee

The only deposit of precious metal in Tennessee lies in a narrow belt in Monroe County in the Cherokee National Forest in the southeastern part of the state. It has been studied in detail, and a report based on that study is available from the state Division of Geology (see below).

Contact:
Division of Geology
Tennessee Department of Conservation
G-5 State Office Building
Nashville, TN 37219

Source material:
The Gold Deposits of the Coker Creek District, Monroe County, Tennessee, Division of Geology, Bulletin No. 72, $6.50.

Texas

The amount of gold produced in Texas has been described as "insignificant" in *Texas Mineral Resources,* a publication of the Bureau of Economic Geology. Nearly all gold that has been obtained has come as a by-product of the silver ores of the Presidio Mine in Presidio County and the copper-silver ores of the Hazel Mine in Culberson County. Smaller amounts have been derived from the Heath Mine and other sources in the Llano region.

In Trans-Pecos Texas, occurrences of gold have been reported in the Shafter, Van-Horn Allamoore, and Quitman mountain districts. Gold has also been found in the sand and gravel of Howard and Taylor counties, and from limestones in Irion, Uvalde, and Williamson counties. In the Llano region, gold is known in small quantities in quartz veins, in metamorphic rock, and in streambed gravels.

Most of the land in Texas is privately owned, necessitating the arrangement of a lease with the owner. There are no large tracts of federal land in Texas that are open to prospectors, nor is prospecting permitted in state parks. The state, however, owns surface and mineral rights to several million acres. These lands are administered by the Texas General Land Office (Stephen F. Austin State Office Building, Austin, TX

78701). Information about prospecting, mining, and leasing can be obtained from that agency.

Contact:
Bureau of Economic Geology
The University of Texas
Austin, TX 78712

Source material:
Gold and Silver in Texas, Mineral Resource Circular No. 56, Bureau of Economic Geology, $1.50.

Utah

Second only to South Dakota in total gold production, Utah is said to have vast areas of stream and terrace gravels that have yet to be prospected. But lack of water hampers placer miners.

The Kennecott Copper Corporation's Bingham mine, the world's largest open-pit copper mine, produces the lion's share of Utah's gold as a by-product of its copper-mining operations. Placer deposits have long been associated with the Bingham operation, as have the streams of other major mining districts. These occur along the Colorado River and its principal tributaries, along the Green River, in the La Sal Mountains, and in the Henry Mountains near Marysville. There's also the aforementioned Bingham district and the Oquirrh Range, which includes Bingham Canyon.

Contact:
Utah Geological and Mineral Survey
606 Black Hawk Way
Salt Lake City, UT 84108

Source material:
Gold Placers of Utah, Geological and Mineral Survey, Circular 47, $1.75.

Gold-mining districts:

See points 1-13 on accompanying map.

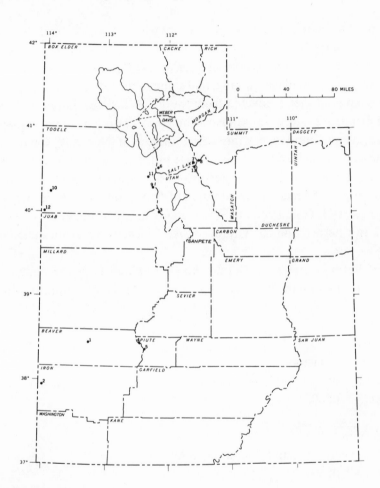

GOLD-MINING DISTRICTS OF UTAH:

Gold-mining districts of Utah:

    Beaver County: 1) San Francisco.
    Iron County: 2) Stateline.
    Juab County: 3) Tintic.
    Piute County: 4) Gold Mountain 5) Mount Baldy.
    Salt Lake County: 6) Cottonwood 7) Bingham.
    Summit and Wasatch Counties: 8) Park City.
    Tooele County: 9) Camp Floyd 10) Ophir-Rush Valley 11) Clifton 12) Willow
Springs.
    Utah County: 13) American Fork.

Vermont

While the state would never be ranked anywhere near the leaders in total gold production, Vermont has a long history of panning and sluicing. President Calvin Coolidge's forebears—Coolidge himself was born in Plymouth, Vermont—are said to have been very active in placer mining, and some of the family's heirloom jewelry was fashioned from gold they recovered from the state's fast-rushing mountain streams.

Buffalo Brook in Ludlow, in Windsor County, is a popular source of placer gold. "I have found gold in abundance in moss on both sides of the brook, wherever it widens out and the water velocity drops," says Ernest Foley, a long-time Vermont prospector. "The brook is not easy to find," he adds. "Get directions from local residents." Gold has also been panned recently in Shady Rill Brook north of Montpelier, Minister Brook in Worcester, Gold Brook in Stowe, Baldwin Creek and Lewis Creek in Bristol; Jail Brook in Chelsea, and the Ottauquechee River not far from the Killington ski area.

Many of Vermont's streams flood in the spring and large amounts of sand and gravel are deposited on the bedrock. A surface dredge is the standard solution to the problem.

Contact:

Office of the State Geologist
Agency of Environmental Conservation
Montpelier, VT 05602

State of Vermont
Agency of Development and Community Affairs
Montpelier, VT 05602

Source material:

Publications of the Vermont Geological Survey, 1979, free publication available from State of Vermont, Department of Libraries, Montpelier, VT 05602.

Virginia

Virginia was the first of the gold-producing states. One of the earliest printed references to gold in America occurred in 1782, when Thomas Jefferson reported on the finding of a four-pound gold-bearing rock on the north side of the Rappahannock River a few miles below the falls. Several placer and lode mines were opened in Virginia in 1825.

The gold-bearing regions of Virginia are found east of the Blue Ridge Mountains in a belt that is 15 to 25 miles wide and 200 miles long. Placer gold can be found in the gravels of some streams that cross gold-bearing rock within this area. Much of the present-day panning is done on such streams as Wilderness Run and Mine Run near the abandoned Melville, Vaucluse, Wilderness, and Grasty mines in Orange County, and Byrd Creek and Little Byrd Creek in Goochland County.

On state-owned land in Virginia, prospecting and leasing are a matter of negotiation between the prospector and the appropriate state agency. A permit must be obtained to prospect on county- or city-owned property. Information concerning prospecting in the George Washington or Jefferson National Forests can be obtained from the Forest supervisors.

Contact:
Virginia Division of Mineral Resources
Box 3667
Charlottesville, VA 22903

Source material:
Gold in Virginia, by Palmer C. Sweet, Division of Mineral Resources, 1980, $4.68.
"Gold Mines and Prospects in Virginia," *Virginia Minerals,* August, 1971, Vol. 17, No. 3, Division of Mineral Resources.

Principal gold-producing counties:
Albemarle, Buckingham, Culpeper, Cumberland, Fauquier, Fluvanna, Goochland, Orange, Louis, Spotsylvania, Stafford.

Washington

Placer gold was discovered in Washington in the 1850s and, during the next decade, with other discoveries at Peshastin Creek in Chelan County and Swauk Creek in Kittitas County, commercial placer mines began operating. While many individuals and a handful of mining companies earned a good return on their labor, the deposits, compared to those of the other gold-producing states of the West, were never considered notable.

Present-day prospectors and panners have worked profitably on the Swauk, Williams, and Boulder creeks in Kittitas County, where virgin deposits are sometimes found; and on the bars of the Similkameen and Snake rivers, where new supplies of placer gold are deposited each year. Placer deposits also occur along the banks of the Columbia River between Grand Coulee Dam and the Canadian border. Many of these deposits, however, are now covered by the waters of Roosevelt Lake, and cannot be worked because they fall within the boundaries of the Roosevelt Lake National Recreation Area. Says a bulletin issued by the Washington Department of Natural Resources: "Continued prospecting along the streams of the northern counties of the state will doubtless uncover areas where good returns can be made."

The fact that miners of the past did not recover all of the gold in the streambeds is substantiated by a recent placer operation at Liberty, where the reworking of a small section of a streambed resulted in the recovery of more than 100 *pounds* of gold.

A permit is required for large-scale sluicing or suction-dredge operation. In most cases, dredging is permitted only from June 1 to September 15. Work must be limited to areas around large rocks, along vertical rock faces, and cracks in bedrock. No spawning gravels or gravel bars can be disrupted in any way. To obtain a permit, contact the Washington Department of Fisheries, Stream Improvement and Hydraulics (3939 Cleveland Ave., Olympia, WA 98501)

Contact:
State of Washington
Department of Natural Resources
Division of Geology and Earth Resources
Olympic Olympia, WA 98504

Source material:

Placer Gold Mining in Washington, Division of Geology and Earth Resources, August 1979, free bulletin.

Inventory of Washington Minerals, Part II, Metallic Minerals, Washington Division of Mines and Geology, Bulletin 37, $4.50.

Gold in Washington, Washington Division of Mines and Geology, Bulletin 42 (out of print, but available in some public libraries).

Gold-mining districts:

See points 1-15 on accompanying map.

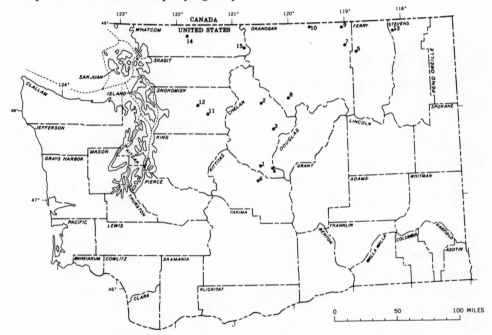

GOLD-MINING DISTRICTS OF WASHINGTON:

Gold-mining districts of Washington:
   Chelan County: 1) Blewett 2) Entiat 3) Chelan Lake 4) Wenatchee.
   Ferry County: 5) Republic.
   Kittitas County: 6) Swauk.
   Okanogan County: 7) Cascade 8) Methow 9) Myers Creek 10) Oroville-Night-
hawk.
   Snohomish County: 11) Monte Cristo 12) Silverton.
   Stevens County: 13) Orient.
   Whatcom County: 14) Mount Baker 15) Slate Creek.

West Virginia

"There are no substantiated records of commercial gold deposits in West Virginia," says a spokesman for the state Geological and Economic Survey, "and such deposits are unlikely." Nevertheless, geological reports for the counties of Braxton, Clay, Grant, Mercer, Mineral, Monroe, Pocahontas, Randolph, Summers, and Tucker, published during the years 1906-1939, mention several reputed gold mines and prospects.

Contact:

Geology Division
West Virginia Geological and Economic Survey
P.O. Box 879
Morgantown, WV 26505

Wisconsin

Through the years, there have been sporadic accounts of gold discoveries in Wisconsin, but the amounts involved have never been significant. "It may be possible to find gold in the glacial debris which covers much of the state," says William Scott of the Mineral Resources Section, "but I have no records showing this has been done."

Contact:

Mineral Resources Section
Geological and Natural History Survey
University of Wisconsin
1815 University Ave.
Madison, WI 53706

Wyoming

Placer deposits of gold were discovered along the Sweetwater River in Wyoming in 1842, and commercial mining operations began in that area during the next decade, continuing until the 1860s. Small amounts of gold have been mined in the state almost continually until fairly recent times, but only the South Pass-Atlantic City (the Sweetwater River) District and the Douglas Creek District in Albany County have yielded gold in any significant amounts.

Most of Wyoming's streams carry small amounts of placer gold. Panners are active on the Wind River near South Pass City; on Douglas and Beaver creeks about ten miles west of Foxpark; on the headwaters of the Little Bighorn River; and along the Snake River in the southern part of Jackson Hole.

Contact:

Geological Survey of Wyoming
University of Wyoming
Box 3008
University Station
Laramie, WY 82071

Source material:

Gold Districts of Wyoming, Geological Survey of Wyoming, 1980.

Mineral Resources of Wyoming, Bulletin 50, Geological Survey of Wyoming, 1966, $3.50.

Canada

SINCE 1858, WHEN commercial mining began, Canada has been one of the world's leading producers of gold. Although virtually all her provinces and territories have contributed to Canada's total output, the largest producers, in decreasing order of importance, have been Ontario, Quebec, British Columbia, Yukon Territory, and the Northwest Territories.

Placer gold is said to have been found in the Chaudiere River, Quebec, as early as 1823. But the first significant find did not occur until 1858, the year that placer gold was discovered in the Fraser Valley in British Columbia. As many as 25,000 people participated in the Fraser gold rush. Many of these individuals moved on to the Cariboo fields following discoveries made there in 1860. In 1862, lode gold was discovered in Nova Scotia. In 1896, a huge find was reported on the Klondike River in the Yukon. During this century, the emphasis has been on lode mining. Rich deposits of ore-bearing veins were found at Porcupine, Ontario, in 1911. These discoveries were followed by finds at Rouyn, Quebec, in 1924.

Other important fields include San Antonio and others in the Rice Lake district of eastern Manitoba, Nor-Acme at Snow Lake in northern Manitoba, the Premier near Stewart and the Bralorne and Pioneer at Bridge River, British Columbia, and the Giant Yellowknife and others in the Yellowknife district of the Northwest Territories.

The greatest opportunities for amateur gold seekers are offered by Canada's Rocky Mountains, for it's there that placer deposits are to be found. But these are often in remote country, requiring backpacking, camping, and perhaps canoeing. The transporting of surface dredges to mountain streams by helicopter is not uncommon.

Each of Canada's provinces and federal territories has its own legal requirements for staking claims and other aspects of mining. The pages that follow mention some of these regulations in general terms, but be certain to write to the agencies listed for specific information.

Any serious Canadian prospector should consult two publications available from the Geological Survey (at address below). They are:

Lang, A. H., *Prospecting in Canada,* Fourth Edition, 1970, Geological Survey of Canada, Economic Geology Report No. 7, 308 pages, $24.95. It's regarded as essential for the beginner.

Boyle, R. W., *The Geochemistry of Gold and Its Deposits,* Geological Survey of Canada, Bulletin 280, $54. Amply illustrated, this publication describes Canada's

best-known gold deposits and also contains information on prospecting techniques.
For more information, contact:

Information Officer
Department of Energy, Mines, and Resources
588 Booth St.
Ontario, Canada K1A OE4

Alberta

There is placer gold to be found in Alberta, but the particles recovered are invariably of "flour" size. Flour, float, or skim gold, to use synonymous terms, has been found in the sand and gravel bars of the Red, Deer, McLeod, Saskatchewan, Athabasca, and Peace rivers, as well as most of the other major watercourses of the province.

Under the provisions of the provincial Exploration Regulation of 1978, you, as a prospector, are permitted to conduct exploration without obtaining a permit, providing that you take surface samples only, do not occupy a site for more than 14 days, and use only hand tools (defined as tools that are easily transported by hand and operated by the physical effort of the operator).

Any gold you find is considered to be a geological sample. You may not sell it, give it away, or use it (as in jewelry manufacture). It belongs to the Crown. Legislation is under consideration, however, to revise the present regulations concerning the recovery and disposal of gold.

For more information, contact:

Director
Mineral Resources Division
9915 108th St.
Edmonton, AB
T5K 2C9

British Columbia

At one time, British Columbia held important ranking among Canada's provinces in the commercial production of gold. Not only were there lode gold deposits being worked, but the Williams, Grouse, Lowhee, and Lightning creek placers were attracting great numbers of eager miners. All that began to change late in the nineteenth century when the known deposits began to give out. The last commercial operation shut down in 1971. Activity has resumed in recent years, although on a much smaller scale.

What this activity should imply to the amateur gold seeker is that there is gold to be found in British Columbia. There are countless streams, bench and terrace deposits, and unprospected and unworked areas where colors, flakes, grains, and even nuggets are to be found. In addition, there are hundreds of old mine dumps that are waiting to be examined with sophisticated electronic equipment.

The Ministry of Energy, Mines and Petroleum Resources (address below) has available hundreds of maps, bulletins, and other publications regarding the geology of the province. Write and request a copy of the Ministry's publications list. *Notes on Placer-Mining in British Columbia,* Ministry of Energy, Mines and Petroleum Resources, 1946, $1, is another publication you'll want to obtain.

For more information, contact:

Ministry of Energy, Mines and Petroleum Resources
Parliament Buildings
Victoria, BC
V8V 1X4

Geological Survey of Canada
100 West Pender St.
Vancouver, BC
V6B 1R8

Manitoba

No significant deposits of gold were known in Manitoba until 1911, the year important discoveries were made at Rice Lake, about one hundred miles northeast of Winnipeg, leading to the opening of the Central Manitoba Mine, which became the province's biggest. In the years that followed, other discoveries were made and other mines began operation, but none survived beyond 1968. Currently, however, many of the old mine dumps are being worked. As all of this may suggest, the gold resources of Manitoba occur exclusively in bedrock—Precambrian bedrock. No placer deposits of any consequence are known to occur.

Gold Mines of Manitoba, by J. W. Stewart, Education Series ES 80-1, available from the Mineral Resources Division (address below), summarizes mining activity in the province and lists the principal geological references.

For more information, contact:

Mineral Resources Division
Department of Energy and Mines
969 Century St.
Winnipeg, MB
R3H OW4

New Brunswick

Gold occurrences have been reported at some eighty locations in New Brunswick, chiefly in the northeastern, northwestern, and central regions of the province. And, according to Dr. Richard Potter, director of the Mineral Resources Branch, there is also gold to be found in the southern regions of New Brunswick.

Before you begin prospecting, you must obtain a license from the Department of Natural Resources (address below). It costs $10. At the time you're handed your license, you'll also be given a copy of a booklet titled "Information Concerning the Mining Act." This tells you what you need to know about recording and staking claims and other such topics.

Gold Occurrences in New Brunswick, compiled by B.M.W. Carroll of the Department of Natural Resources, is "must" reading for anyone planning to prospect and mine in the province.

For more information, contact:

Department of Natural Resources
Mines Branch
P.O. Box 6000
Fredericton, NB
E3B 5H1

Newfoundland

Just about all of the gold production with which Newfoundland has been credited has been derived as a by-product of base-metal ores, chiefly zinc, lead, and copper. But a handful of gold mines once proved profitable. These included the Goldenville Mine in Mings Bight and the Browning Mine at Sops Arm. Other lode gold discoveries were made at Sims Ridge at Sops Arm, Little Bay, and White Bay. All of these occurrences are clustered pretty much in northeastern Newfoundland.

Placer deposits are rare. Those that have been found have invariably been associated with the lode discoveries mentioned above.

A good summation of the gold occurrences of Newfoundland is contained in the book *Mines and Mineral Resources of Newfoundland,* by A. K. Snelgrove, Department of Mines and Resources, 1953.

For more information, contact:

Government of Newfoundland and Labrador
Department of Mines and Energy
P.O. Box 4750
St. John's, NF
A1C 5T7

Northwest Territories

Despite its vastness (Texas could fit within its borders five times) and the fact that mining is its principal industry (but it's lead and zinc mining), there is only one active placer mine in the Northwest Territories. It's located on the Liard River, not far from the British Columbia border.

In spite of the lack of commercial activity, there is gold to be found in the Northwest Territories. Staff geologists have compiled a tall stack of information concerning the Territories' many widely distributed gold showings. Each report gives the location, history, and current state of development of the occurrence. Xerox copies of all of these reports cost about $65. If there's a specific area you wish to know about, however, or for any additional information, write:

Geology Office
Department of Indian Affairs and Northern Development
Box 1500
Yellowknife, NT
X1A 2R3

Nova Scotia

Gold was being panned from the black sands of Nova Scotia's beaches as early as 1849, and commercial mining began in 1860 when gold-bearing veins were discovered in the Mooseland District of the province. By the 1940s, more than one hundred gold districts had been established that had produced gold commercially at one time or another. (A map available from the Department of Mines and Energy [address below] shows the locations of most of these old mining districts.) So prevalent are the old mining areas that literature distributed to tourists warns of the perils of exploring the innumerable shafts, pits, trenches, and quarries that dot the Nova Scotia landscape.

A license must be obtained to prospect in Nova Scotia. In your license application, you must specify which claims or tracts you plan to investigate, describing them in accordance with reference maps issued by the Department of Mines and Energy. This is provided for under the terms of the Mineral Resources Act of Nova Scotia, a digest of which you can obtain from the Department of Mines and Energy.

For more information, contact:

Nova Scotia Department of Mines and Energy
P.O. Box 1087
Halifax, NS
B3J 2X1

Ontario

Ontario produces about 43 percent of Canada's gold, and there are more than 2,000 sites in the province where gold has been reported or geologically indicated. While the great bulk of these occurrences refer to lode gold, many streams in northern Ontario are known to have placer deposits, although no commercial placer operations have been established.

A good starting point for anyone planning to mine or prospect in Ontario is a publication titled *Rocks and Minerals Information,* issued annually by the provincial Ministry of Natural Resources (address below). This lists government maps, reports covering gold deposits in various parts of the province, and the names of publications of the Geological Survey. Be sure to obtain the *latest* issue. The Ministry of Natural Resources also has available mineral deposit maps that plot and describe the province's gold occurrences.

It's vital to know about the Mining Act of Ontario, the provisions of which are detailed in numerous publications issued by the Ministry of Natural Resources. The act mandates that you must obtain a license to prospect in Ontario. Once you've discovered an area that you feel warrants placer mining, you must stake a mining claim and acquire title to the land. In fact, no gold may be recovered from the site until actual title is issued.

For more information, contact:

Public Service Centre
Ministry of Natural Resources
Whitney Block
Queen's Park
Toronto, ON
M7A 1W3

Quebec

The early 1940s saw the heyday of gold mining activity in Quebec, with more than 25 mines in operation. About half that number are active today, with several others in one stage of planning or another.

The principal gold-bearing regions of the province are located in the Abitibi-Temiscamingue area of northwestern Quebec, an area adjacent to the gold producing regions of Ontario. A comprehensive report on the gold-mining prospects of this region is available from the Ministere de l'Energie et des Ressources (address below). The work of Maurice Latulippe and Maurice Rive, the report is titled *An Overview of the Geology of Gold Prospects and Developments in Northwestern Quebec.*

For more information, contact:

Government of Quebec
Ministere de l'Energie et des Ressources
1530 boul. Entente
Quebec, PQ
G1S 4N6

Saskatchewan

While there is little chance of finding valuable deposits of gold in the stream and river gravels of Saskatchewan, the province's low-grade placers have been attracting an increasing number of weekend prospectors and gold-seeking vacationers. Many concentrate their efforts on the North Saskatchewan River between Edmonton and Prince Albert, the Waterhen River, and other rivers in the north central region of the province.

A small gold mine once operated at Mallard Lake northeast of La Ronge, and there have been gold showings in the streams near La Ronge. Placer deposits have also been found in the Flin Flon area.

No license is required to prospect in Saskatchewan; however, prospecting and panning can be conducted only on "open ground," that is, land that is not staked or covered by mineral claim. In northern Saskatchewan, there is little open ground now. Claim maps, illustrating what open ground there is, can be examined at a Mining Recorder's Office. These are located in La Ronge, Creighton, Uranium City, and Regina.

A detailed report on gold occurrences in Saskatchewan can be derived from the publication, *Mineral Occurrences in the Precambrian of Northern Saskatchewan,* which can be obtained from the office of the Geological Survey (address below). The report costs $5.

For more information, contact:

Geological Survey
Saskatchewan Mineral Resources
1914 Hamilton St.
Regina, SK
S4P 4V4

Yukon Territory

The Yukon Territory was the scene of one of the world's great gold rushes. In 1897 and 1898, miners from every corner of the globe swarmed to the Klondike region of the Yukon, streaming from San Francisco and Seattle to Skagway, Alaska, then up the Dyea Trail to Whitehorse. Or they arrived by boats that carried them down the turbulent Yukon River to Dawson. By 1900, tens of thousands of miners were working Yukon streams, creeks, and river valleys.

It's not as frantic today. In one recent year, however, some 26,000 ounces of gold were derived from 75 placer operations involving about 200 people. This activity is closely regulated under the provisions of either the Yukon Placer Mining Act or the

Yukon Quartz Mining Act. Literature explaining these statutes is available from the Yukon Territory's Geology Section (address below). Note that suction dredging is not allowed in the Yukon, except on placer claims in good standing.

Gold mining in the Yukon often presents a unique problem. The ground must be thawed before it can be excavated. This is achieved by wood fires, by heating rocks and then covering them with moss or blankets to retain the heat, or by forcing hot or cold water or steam into pipes driven through the sand and gravel to the bedrock.

Two maps detailing the distribution of known placer occurrences in the Yukon area are available from the Geology Section. They cost $1 apiece.

For more information, contact:

Geology Section
Department of Indian Affairs and Northern Development
200 Range Rd., Takhini
Whitehorse, YT

Additional Reading

Books

THE BOOKS LISTED below, and scores of others as well, are available by mail from these three sources:

Exanimo Establishment
Box 448
Fremont, NE 68025

Keene Engineering
9330 Corbin Ave.
Northridge, CA 91324

Miners Incorporated
P.O. Box 1301
Riggins, ID 83549

Books published by Ram Publishing Company (P.O. Box 38649, Dallas, TX 75238) can be ordered directly from that firm.

Blasters' Handbook

Explosive Products Div., E. I. du Pont de Nemours & Co., Wilmington, Delaware; 1977; $18.

For lode gold miners, the definitive book on the use of explosives, with the latest information on the use of water gel.

The Complete VLF-TR Metal Detector Handbook

Roy Lagal and Charles Garrett; Ram Publishing Co., Dallas; 1980; $7.95.

A detailed explanation of ground-canceling metal/mineral detectors, plus information as to how to use these instruments in different types of treasure-hunting situations. Recommended.

Dredging for Gold: The Gold Divers' Handbook

Matt Thornton; Keene Industries, Northridge, California; 1979; $6.95.

Authoritative and comprehensive, this excellent book opens with an explanation of gold dredges and then instructs the reader in how to conduct dredging operations. There are also chapters on the basic elements of placer geology, panning, and equipment for underwater diving. Recommended.

Electronic Prospecting

Charles Garrett, Bob Grant, Roy Lagal; Ram Publishing Co., Dallas; 1980; $3.95.

The definitive book on the use of metal/mineral detectors in prospecting for gold and silver. Recommended.

The Gold Hunter's Field Book

Jay Ellis Ransom; Harper & Row, New York; 1975; $5.95.

Subtitled "How and Where to Find Gold in the United States," this book is divided into two sections. The first is an introduction to gold hunting, covering such topics as panning, placer mining, and staking a claim. In the second section, author Ransom gives detailed state-by-state (and, for Canada, province-by-province) reports on North American gold-producing districts, based on state and county surveys and material from the U.S. Geological Survey.

Gold in a Campground

Jerry Keene and Matt Thornton; Keene Engineering, Northridge, California; 1980; $1.50.

The fundamentals of panning, sluicing, and dredging.

Gold Panning is Easy
Roy Lagal; Ram Publishing Co., Dallas; 1980; $3.75.
A comprehensive instruction manual, with emphasis on the gravity-trap pan, an invention of the author's. Recommended.

Handbook for Prospectors
W. M. von Bernewitz, revised by Harry C. Chellson, fifth edition; McGraw Hill Book Co., New York; 1973; $21.95.
Originally published in 1926, this 426-page classic has been revised and updated. It deals with the techniques of prospecting and mining, covering the fundamentals of mineralogy, petrology, and geology as they apply to lode and placer gold. Mining and ore-processing methods are explained in depth. Recommended.

How and Where to Find Gold
Verne Ballantyne; Arco Publishing Co., New York; 1976; $3.95.
The basics of placer mining by a prospecting veteran.

How and Where to Pan Gold
Nugget Publishing Co., Tombstone, Arizona; $3.
An experienced prospector, Winters recommends equipment and sets down procedures useful in gold hunting, his text enlivened with scores of anecdotes. There's an excellent section on staking claims. A best-seller, the book has been updated several times, most recently in 1980. Recommended.

Looking for Gold
Bradford Angier; Stackpole Books, Harrisburg, Pennsylvania; 1975; $5.95.
Information on more than 500 gold-bearing regions in the United States, with material on grubstaking, staking claims on federal land, and leasing and selling claims. Also provides information on food and shelter.

The Modern Treasure Finder's Manual
George Sullivan; Chilton Book Co., Radnor, Pennsylvania; 1975; $6.95.
Information on how to find buried coins, historical artifacts, and gem minerals; plus chapters on gold prospecting and mining.

Pan for Gold on Your Next Vacation: A Guide to Low-Cost Unique Adventures
Janet Ruhe-Schoen; Pilot Books, New York; 1980; $2.50.

Professional Treasure Hunter
George Mrocskowski; Ram Publishing Co., Dallas; 1979; $6.95.
The author's adventures in searching for lost mission relics, Civil War artifacts,
   Confederate gold, and other treasures.

Prospecting and Operating Small-Gold Placers.
William F. Boericke; John Wiley & Sons, New York; 1936; $11.95.
A somewhat technical explanation of placer deposits and how they have been formed;
   includes sections on panning, sluicing, and sluice box construction.

Magazines and Periodicals

California Mining Journal
P.O. Drawer 628
Santa Cruz, CA 95061

Exanimo
Box 448
Fremont, NE 68025

Dig
Found Enterprises
133 Prospect St.
Auburn, MA 01501

Gold Prospectors News
            and Mining Journal
Gold Prospectors Association of America
P.O. Box 507
Bonsall, CA 92003

International Treasure Hunter
International Treasure Hunting Society
P.O. Box 3007
Garland, TX 75040

National Prospector's Gazette
Ames, NE 68621

Pay Dirt
P.O. Drawer 48
Bisbee, AZ 85603

The Treasure Hunter
P.O. Box 5
Mule Creek, NM 88051

Western & Eastern Treasures
P.O. Box 7030
1440 West Walnut St.
Compton, CA 90224

Western Prospector & Miner
P.O. Box 146
Tombstone, AZ 85638

Index

The futurist, John Naisbitt, introduced us to the idea of high tech and high touch. We are now introduced to a contemporary replacement for that old and once useful concept — digital dialogue — the meeting of the digital mind and the analogue heart. Let yourself be provoked by these ideas in order to come up with your own business solutions…or watch from the sidelines while others relegate you to the past.

— Dr Stephen Lundin, Author of *Fish!*

As a fighter in the digital revolution, I'll concede that *Digilogue* understands what few on my side are willing to acknowledge: there is beautiful, emotional value to old-school analogue ways. The answer for business owners and marketers is to understand both worlds and utilise the best of each to reach and engage your customers. *Digilogue* provides you with the perfect roadmap.

— Dina Kaplan, co-founder of Blip.tv, *Fortune* magazine's Most Powerful Women Entrepreneurs and Fast Company's Most Influential Women in Web 2.0

In a world where digital is disrupting everything from retail to media to travel at an accelerating rate, companies have to ask the tough question 'Are we facing our Kodak moment?' *Digilogue* is a valuable guide to anyone who wants to address this challenge, providing practical ideas on how to reinvent businesses of all shapes and sizes in the digital tsunami of change.

— Rachel Botsman, TEDster and 2013 Young Global Leader, World Economic Forum

As we make the shift to digital products and new distribution and service models, the friction of old high touch processes sometimes clash dramatically with new, more efficient high-tech engagement models. It's clear from Anders' latest book that this transition is not a simple on/off state, or a choice of one over the other. *Digilogue* is a great platform for nuancing the balance of this transition.

— Brett King, Banking Futurist and Founder of Moven

Futurist Anders Sörman-Nilsson's *Digilogue* offers excellent insights on how to deal with these forces, how to embrace change and how to exploit the opportunities it offers — a must read for every business leader!

— Jan Pacas, Managing Director, Hilti Australia

The digital evolution has transformed the way in which we bank — who would have thought 20 years ago, customers would be able to communicate with their bank online or more radically, purchase a banking product online? The traditional analogue branch experience is now a convergence of the customer's journey in the digilogue world. The best time to develop your strategic response to the digital disruption was yesterday. The next best time is today — ignore at your peril!

— Matt Janssen, Head of Retail Lending, Bank of SA, Westpac Group

Digilogue shows us that there is not a stark dichotomy when it comes to choosing either a digital (clicks) or analogue (bricks) strategy. Instead, those brands that are able to create an authentic and masterful blend of the two will be the ones to survive over the long term. Anders Sörman-Nilsson helps us upgrade our thinking to ride the wave of digital disruption while paying homage to the analogue past that continues to engage the customer's heart.

— Richard Ruth, Consultant — HR, Eli Lilly & Company

At Hong Kong Broadbank Network, when we look into the digital future, we turn to Anders as our guide. Anders' global perspectives, as articulated in this book, allow us to learn from other people's experiences without having to go through the ups and downs ourselves.

— Niq Lai, Head of Talent Engagement & CFO,
Hong Kong Broadband Network Limited

We all want to be part of the customer engagement revolution through the digital world but many of us still want to retain the relationships we have so carefully developed. So how do we integrate the heart and minds of the physical with the seemingly expedient digital world? *Digilogue* gives us a platform on how to consider this confronting and disrupting challenge. We at People's Choice Credit Union have embraced this thinking and are evolving our business to ensure we embrace both the past and the future into a seamless digital analogue relationship with our customers.

— Peter Evers, Managing Director, People's Choice Credit Union

Anders Sörman-Nilsson links the past with the future to show how business models continue to be disrupted. His magic is to personalise and embroider insights into the business recommendations using his own family history and personal stories. The reality of how technology, speed of interactions and relationships are rapidly evolving and driving innovation is softened and brought to life in an engaging way through the writer's genuine, personal and heartfelt touch.

— Petrina Conventry, Chief Human Resources Officer
and Director at Santos

In the great tradition of Swedish diplomats, Anders has shown us a middle path in which atoms and bits stand united and where we can compromise between the heritage of the past and the promises of the future. *Digilogue* is an inspiring, insightful manifest for everyone who claims to be a 21st-century business leader.

— Magnus Lindkvist, Futurologist

There is a palpable panic present in the business world as the digital revolution transforms how we connect, transact and share ideas and ideals. As a futurist, Anders Sörman-Nilsson understands the impact of this change and where it is taking us from and to, but more importantly, he understands the need for high touch in a high tech world. If you want to remain relevant and impactful in the economy of the future, Anders' thinking is priceless.

— Dan Gregory, CEO, The Impossible Institute

While new innovation drives the growth of our company, the way we communicate with our customers remains predominantly analogue...The *Digilogue* strategy map provides the catalyst to transform how we now operate in our everyday lives, into our everyday business. At the same time, Anders reminded me that we need to leverage our analogue touch-points to also forge emotional connections through these new digital channels.

— Marketing Executive, *Fortune* 500, Pharmaceutical Company

Digilogue neatly lays out the need for an enhanced sense of urgency amongst senior executive ranks when it comes to corporate strategy development...If you're a senior executive and haven't tooled up on disruption then you'd best get busy — start with *Digilogue*. Anders once again does a great job in bringing these critical corporate insights into focus in a manner which is both easy reading and full of practical roadmaps for the application of *Digilogue*'s key messages into the reader's organisation post haste.

— Peter Lang, Group Executive, McMillan Shakespeare Group

DIGILOGUE

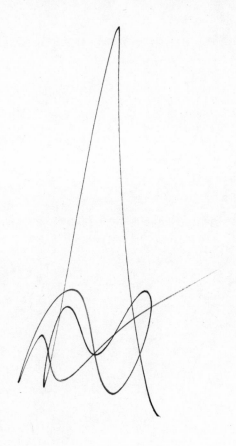

anders sörman-nilsson

DIGILOGUE

how to win the digital minds and
analogue hearts of tomorrow's customer

WILEY

First published in 2013 by John Wiley & Sons Australia, Ltd
42 McDougall St, Milton Qld 4064

Office also in Melbourne

Typeset in 10/13 pt DIN Light

© Thinque Pty Ltd

The moral rights of the author have been asserted

National Library of Australia Cataloguing-in-Publication data:

Author:	Sörman-Nilsson, Anders (author)
Title:	Digilogue : how to win the digital minds and analogue hearts of tomorrow's customers / Anders Sörman-Nilsson.
ISBN:	9781118647936 (hbk.)
	9781118641385 (pbk.)
	9781118641392 (ebook)
Subjects:	Product management.
	Business planning.
	Success in business.
	Thought and thinking.
	Digital communications.
Dewey Number:	658.827

Cover design by Hema Patel, Grey NYC

Figure 8.1: © Bettmann/CORBIS

Printed in Singapore by C.O.S Printers Pte Ltd

10 9 8 7 6 5 4 3 2 1

Disclaimer
The material in this publication is of the nature of general comment only, and does not represent professional advice. It is not intended to provide specific guidance for particular circumstances and it should not be relied on as the basis for any decision to take action or not take action on any matter which it covers. Readers should obtain professional advice where appropriate, before making any such decision. To the maximum extent permitted by law, the author and publisher disclaim all responsibility and liability to any person, arising directly or indirectly from any person taking or not taking action based on the information in this publication.

To mum, for your enduring energy, passion, dedication, motherly affection, sharing and chutzpah.

To dad, for your continuous encouragement, global inspiration, sense of history, taste, culture, humanity and gentlemanly nouse.

To my brother, for your presence, accountability, quirky challenges, groundedness, BS radar, patience and being my best friend since the day you were born.

CONTENTS

ABOUT THE AUTHOR

Anders Sörman-Nilsson (LLB MBA) is the founder of Thinque — a strategy think tank that helps executives and leaders convert disruptive questions into proactive, future strategies. As an Australian-Swedish futurist and innovation strategist, he has helped executives and leaders on four continents map, prepare for and strategise for  foreseeable and unpredictable futures. Since founding Thinque in 2005, he has worked with and spoken to clients like Apple, Johnson & Johnson, Cisco, Eli Lilly, SAP, IBM, Xerox, ABN AMRO Bank, Commonwealth Bank, McCann Erickson, BAE Systems and Young Presidents' Organisation, across diverse cultural and geographic contexts.

Anders is an active member of TEDGlobal (Oxford 2009 and Edinburgh 2011), has keynoted at TEDx, and has spoken at the prestigious Million Dollar Round Table. He has a Global Executive MBA from the University of Sydney Business School and has completed executive education at the Indian Institute of Management, Bangalore, London School of Economics and Stanford Graduate School of Business.

Anders divides his time between Sydney, Stockholm and New York City.

ACKNOWLEDGEMENTS

This book is the result of a family affair. I really appreciate the input of Britt Sörman, Karin Fimmerstad and Lucie Fairweather for linking me back to facts, photos and figures. Thank you to Dad for objectively remembering the story of your adopted family, and your sense of historical dates and events. Mum, I appreciate you sharing the good, the bad, and the ugly of the family's skeletons in the closet, and the transparency you provided me in my business intelligence gathering, and putting your store and the brand on the line in this book — for the world to see. Gustaf, bro, big shout out for proofreading and challenging the contents of this book, and for spurring me to have some much-required writing breaks during our holidays together. To Emma Harpham, yes I know we are not technically family, but you are adopted family, and your energy, passion, centredness, cool, and incredible engagement at Thinque enabled me enough space to bring this book to reality during a time of intense work pressures (and by extension to Aaron, Max and India for lending your wife and mum to represent Thinque around the world). Ware, bro, thanks for your research insights, and spot-on futurist mind — we love having you at Thinque.

To the team at Wiley. Kristen for seeing the merit in this book, for partnering in a strategic and humane way, and for believing in both the concept and in me as a human being. Charlotte, for incredible editing, and proactively optimising the content and ideas. Elizabeth, for keeping up with my precise nature, managing my time on time and packaging the book to something I am truly proud of. Alice, for polishing and tweaking everything from big concepts to Swedish typos and Keira for her attention to legal detail, scouring the world for permissions and triple checking our sources. To Gretta and Katie for communicating the essence of the book and representing its digilogue soul in an age of flux. To Hema Patel at Grey NYC for conceiving an amazing cover and executing an awesome design and for proving, yet again, that a great book deserves to be judged by a great cover (and for being a continued, and treasured, part of my life).

To the Global Executive MBA cohort at the Sydney University Business School. Big shout outs for challenging, supporting, and confirming the commercial need for this book, and helping me shape it globally with your input over 18 months. Particular thanks go out to Jan Pacas, Adam Holloway, Matt Tapper, Peter Lang, Dan Beecham, Petrina Coventry, Ron McCalman, Michael McGee and Charlotte Park — you all know why in your individual ways.

To my agency, Ode Management — for representing me globally, and sending me on speaking assignments in every nook and cranny around the world. To Jules for seeing the alignment, partnering, and growing together, to Leanne for your directness and mentoring, to Heidi for starting the journey with me and being an unsung hero, to Reggie and Lauren for helping me partner with the best bureaux around the world, to Rachel and Steph for keeping my travel i's and t's dotted and crossed, to Tanja for your gentle gung-ho mindset and peak performance, and to Becs for being the boomerang marketing guru who keeps me strategic. Finally, last but certainly not least — to Jay, for believing in me 100%, your authenticity, your dedication, and advice, and bringing Anders to the USA in a big way.

To my mentors: Jonas Ridderstrale, Matt Church, Mark Veyret — kudos and respect. To John Naisbitt, thank you for providing me the shoulders to stand on, and revolve and evolve your futurist insights. To Mike Walsh, thank you for setting a new standard for a new generation of futurists and for challenging my thinking around digital.

On a personal front — thanks for the stories, laughs and inspiration: Aaron and Kaitlin, Luke, Joel. For sharing my challenge of the status quo and engaging in risky global travel:

Mattias. For challenging absolutely everything: Simon. The depth and longevity: Fredrik, Therese, Lars, Julia, Shire, Carl-Patrick and Ben G. The friendship, travel, emotional advice and design execution: Miki. The constant challenge, check-in, and being there when it matters: Dave and Megan. For allowing me to care, step in, and take a stand: Noddy. The brand focus and fascination with food and great conversations: Daniel. The global appreciation and cognitive exchange: Dave Cummings. The man-dates: Rasmus. Learning about connection: Nicole. Totally grounded: Julia. Being there and always picking up where we left off: David Brightling. Adrian for being Adrian. To the world's best flatmate, an inspiration, and a genuinely lovely human being, and a person I miss deeply from a distance: Hala Razian. To authenticity, quirkiness and genuinely brilliant eccentricity: Charlie H. And finally, Zhenya, for connecting with me before, during and after my writing of this book — a wonderful surprise, and a digilogue adventure. To my cat Finnegan and my dog Cicero — thank you for being in the moment, and reminding me to connect with sleep, fun, food and snuggles.

Finally to my clients who make all of this possible, and who have shared generously for the inspiration of this book. Richard R, Matt J, Melanie M, Jennie L, John A, Libby D, Jan P, Sam M, Angie T — you mean more than you know and I truly appreciate our friendships and your support.

PROLOGUE

The fight is on, and it's personal. It's about family. It's about history. Legacy. Survival. It's about psychological warfare. It's about fighting digital disruption.

In the red corner: old-school analogue. High-touch, face to face, personal and human. In the blue corner: new-school digital. High-tech, interface to interface, public and computer. At stake — the future of how business connects with clients. This is timeless wisdom versus youthful fitness. This is experience versus fresh perspectives. This is street-smarts versus school-smarts.

The pre-match talk has been all about technical knockouts. In its last bouts, Digital Disruption knocked out Kodak, Blockbuster and Borders. On form and match fitness, Analogue Anachronism is the underdog. But never underestimate a wounded animal, backed up in a corner. It's picking this fight for a reason. This fight is personal. More than prize money is at stake. This time Analogue Anachronism represents the displaced — the outsourced, the computerised, the disenfranchised, the laid-off, the defunct, the bankrupt, the unemployed. An old, faded, almost indecipherable tortoise tattoo marks Analogue Anachronism's proud, leathery chest, wisely contemplating its opponent's fresh and fit hare-emblazoned biceps, bursting with energy and flexing for the paparazzi. This is a gladiatorial combat, and at the end, one will get the thumbs up, and one will get the thumbs down. And not in the Facebook way. This fight, it seems, is about life and death. Who is your money on?

This battle is also about my family, which makes this scene deeply personal. In the red corner, Analogue Anachronism is my mum's old-school champion. If it wins, her third-generation family business might have a chance. If not, the writing is on the wall. As a digital native — born, bred and inspired by the fit fighter in the blue — I feel torn, and somewhat guilty. Does my mum's storefront even stand a chance? Is death by digital a fait accompli — a fate she must endure? Will clicks kill bricks and mortar? My strategic, pro bono advice to my mum is on the line, and the stakes are high. Her life's legacy and work is at risk.

Will digital or analogue get up? Who will win the hearts and minds of the audience? Does the underdog even stand a chance against Digital Disruption? What is the future of business? How will we interact with our clients tomorrow? How do we protect historical brand equity? Can a computer interface really replace a human face? Should we focus on digital bling, or remember the analogue thing? Is there scope to reconnect with humans by disconnecting from technology? How can analogue businesses, large and small, stay relevant in the digital future? And how can digital businesses, large and small, create a human, personal, analogue brand?

Before I start answering these questions, have a look at figure 1, where I outline the contrastive definitions of analogue and digital. These definitions form the basis of this book, as I use the broader terms (*analogue* and *digital*) to describe technologies, brands, marketing strategies and customer interactions.

Figure 1: analogue versus digital

Analogue	Digital
Hearts	Minds
Face to face	Interface to interface
Facial recognition	Facebook recognition
High-touch	High-tech
Old school	New school
Personal touch	Digital connect
Nine to five	24/7/365
Slow	Fast
Word-of-mouth	Word-of-mouse
Offline	Online

This is what the *Oxford Dictionary* has to say about our fighters:

analogue |ˈanlˌôg, -ˌäg| (also analog)

adjective

relating to or using signals or information represented by a continuously variable physical quantity such as spatial position or voltage. Often contrasted with digital (sense 1)—analogue signals: "the information on a gramophone record is analogue"

(of a clock or watch) showing the time by means of hands rather than displayed digits.

Origin: early 19th century: from French, from Greek analogon, neuter of analogos 'proportionate'

digital |ˈdijitl|

adjective

1 (of signals or data) expressed as series of the digits 0 and 1, typically represented by values of a physical quantity such as voltage or magnetic polarization. Often contrasted with analogue.

relating to, using, or storing data or information in the form of digital signals: digital TV, "a digital recording"

involving or relating to the use of computer technology: the digital revolution.

2 (of a clock or watch) showing the time by means of displayed digits rather than hands or a pointer.

Origin: late 15th century: from Latin digitalis, from digitus 'finger, toe'

So, using the *Oxford Dictionary*'s definitions, the analogue–digital divide could be portrayed as shown in figure 2.

Figure 2: analogue versus digital clocks

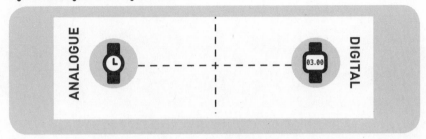

So, who will be defeated (or throw in the towel)? And who will win the hearts and minds of tomorrow's customer? Or, can the analogue and digital somehow co-exist?

The answers may not lie in either the digital or the analogue. At a time when we are LinkedIn, hyperconnected and online 24/7/365, our minds have become largely digital. But our hearts have remained analogue. Digital minds. Analogue hearts. Thus, the answer spells *digilogue* (see figure 3). Digilogue is the translational sweet spot, the convergence of the digital and the analogue.

Figure 3: digilogue — the convergence of the digital and the analogue

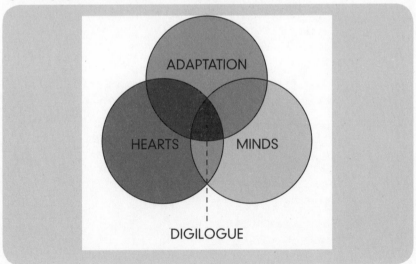

Digilogue is what enables your business to provide value to digital minds, and connect with analogue hearts. Let the fight begin.

INTRODUCTION

Think about the future of your business and ask yourself the following:

- Is your business being digitally disrupted?

- Is your analogue business model fit enough for a digitised future?

- How do you plan to adapt in a way that is mindful of your brand's history, yet focused on its continued, profitable future?

- Which communication touch points in your business can be digitised, and which ones must never be digitised?

- How do you stay smart and timely in a way that simultaneously oozes timeless wisdom?

- How do you ensure that you win the analogue hearts and digital minds of tomorrow's customer?

As a global futurist and strategist at the think tank Thinque, I get to tackle these disruptive questions with executives and leaders every day and turn them into proactive, bespoke, future strategies. While my work in Sydney, Shanghai, Stockholm, San Francisco and Singapore is diverse, it always focuses on co-creating better futures for brands and the organisations they symbolise.

We're asked to provide positional advice and coordinates for a constantly shifting digital landscape, yet frequently are also asked to disregard the historical journeys that led the brands to the present. But this is a mistake. We cannot disregard the strategic moves, the organisational culture, the personal stories and the evolution that got us to now. Timeliness must co-exist with timeless wisdom. High-tech must be balanced by high-touch. Interface to interface must be offset by face to face. The new school must collaborate with the old school. Innovation must recognise heritage. And the digital must blend with the analogue. This book recognises that we cannot throw the analogue baby out with the bathwater.

To this effect, *Digilogue: How to win the digital minds and analogue hearts of tomorrow's customer* will provide you with:

- a one-page strategy map so you can chart your brand's future course

- concrete tips to win digital minds and analogue hearts today and tomorrow

- global case studies that showcase how to go digilogue in your business.

PART I
The fighters size each other up

The gloves are (nearly) off. Part I is a veritable street fight between two amped fighters: Digital Disruption and Analogue Anachronism. Both tap into their core strengths, dig deep for control and come out swinging. This part of the book analyses the tensions, dynamics and backstage antics of both sides, and explores their comparative strengths. It also highlights and investigates the seeming evolution of sophistication in the fighters' behaviours.

Chapter 1 takes a global view of digital disruption, and how it is decimating businesses, big and small, around the world. It zooms in on a menswear retailer in Stockholm, Sweden, and does a deep dive on what survival lessons business leaders more broadly can extrapolate from this family business's battle with the digitisation of everything. The chapter tracks and decodes digital as the key wave of change affecting business today and for the foreseeable future, and provides you with a way out if you are under the false impression that you can stem the tide of change that digital disruption personifies.

Chapter 2 bears surprising news as it investigates what the customer of tomorrow is looking for. Curiously, we'll find out that decision-making and consumer patterns are shifting in a way that is as focused on the past as it is on the future. We'll reach an understanding that to win the hearts and minds of tomorrow's savvy customer, business leaders need to provide value to the rational, digital minds, and connect profoundly with the emotional, analogue hearts of those customers.

After two rounds of fighting and counter-punches by the contestants, chapter 3 explores the convergence of the digital and the analogue — that translational sweet spot where tomorrow's customers expect you to be. The chapter strategically examines how the lines between the digital and the analogue are increasingly blurring, as customers are taking technology and their futures into their own hands — literally. The chapter takes a macro view of how tradition and technology can combine to make both lederhosen and laptops sexy and economically viable at the same time, and we take a trip down under to New Zealand and look at how a forward-looking sports apparel company managed to make supply chain management attractive by converging the digital with the analogue, and more. It's an ambitious chapter, and hopefully it will hold up a nice self-reflective mirror for you as you contemplate both your personal and professional life in an era of a digital paradigm shift.

This leads to chapter 4, the final chapter in the part, which will prepare you to get down and dirty with the strategic message in this book. It provides you with a one-page Digilogue Strategy Map, which you can use to find the convergence between the digital and the analogue — the *digilogue* — in your business. The chapter examines how digital businesses that at face value seem to lack an analogue heart are able to reach out through their screens and connect deeply with customers' analogue hearts, and how analogue businesses, at face value ill-suited to the digital world, are able to provide value to customers' digital minds. The chapter also provides you with the first part of your strategic mapping exercise, and will (kindly) provoke you to think and write deeply about the future of your own business.

1: Digital disruption

Round 1. The crowd goes berserk. The skimpily clad Las Vegas glamour girl is strutting the perimeter of the ring holding a big #1 above her head. The announcer amplifies the electric atmosphere. Motivational words by coaches blended with saliva get spat in the faces of the combatants. It's a prologue of punches to come. The boxers size each other up. Total focus. The battle is about to begin.

In this round you will learn what an entrepreneurial Swedish farm boy born in 1885 in West Gothland can teach us about today's changing retail landscape, how a Chicago-designed barber's chair from the 1920s ended up as an antique in my Sydney living room, and why disruption isn't a new phenomenon. I will illustrate the difference between the old, linear paradigm of disruption and the new, exponential paradigm of *digital disruption*, and the impact this increase in the speed of disruption is having on business today and into the future. Through this analysis you'll see that digital disruption is a disintermediation force that threatens businesses, big and small, and that it is causing a wave of change throughout the entire global economy. We will discuss whether the coveted customer has reached a digital point of no return, before looking at what Alcoholics Anonymous can teach us about self-reflection, insight and adaptation to the external environment. Go.

Everything that can be digitised will be digitised

For many, 1902 was a year of opportunity, modernisation and a break with the past. In the United States, Electric Theatre — the first ever cinema — opened its doors in Los Angeles on 2 April, and President Roosevelt became the first President to be seen riding in an automobile on 22 August. In Australia, British female subjects won the vote on 16 June. Over in Europe, the Carnegie Institute for Science launched on 28 January, on 15 February the Berlin U-Bahn (underground) was opened, on 29 May Lord Rosebery inaugurated the London School of Economics, King Edward VII was crowned in London on 9 August, and in Egypt on 10 December the first Aswan Dam on the Nile was completed.

To a 17-year-old Swedish boy on the farm Sörboda in the West Gothland countryside by the name of Georg Johansson, none of this mattered greatly. For him, 1902 was the year when he decided to break his ties with his rural past, and made entrepreneurial plans to take the canal boat to Stockholm to embrace opportunity, modernisation and an urban future for himself. On 20 October, Georg and his older brother Gottfrid, in a gesture to provenance, changed their surname to Sörman, indicating their origins on the farm of Sörboda. Young Georg's compass pointed east to Stockholm while his elder Gottfrid's pointed west toward America.

In 1903, the two brothers departed in separate directions. A few days after the first west–east Atlantic radio broadcast connected America with Europe, Gottfrid followed in the footsteps of one million Swedish farmers who had left the Old World for the New World during a century of emigration, embarking on a ship to the Atlantic, while Georg followed in his father's footsteps and got on a canal boat to Stockholm. Georg arrived at the quay on the island of Riddarholmen in a city still dominated by the horse and buggy, and untouched by Henry Ford's 1908 Model T. To a place where a camera lens attracted child-like curiosity and spontaneous poses, and where Charlie Chaplin ('The Tramp') would feature in the short film *Kid Auto Races at Venice* in cinemas a decade of modernisation later. Georg's arrival happened two years before the birth of the eventual Swedish femme fatale Greta Garbo, and nine years prior to Stockholm coming of age in its successful curation of the 1912 Olympic Games. This was a Stockholm still shaped by the late-industrial era, a place where urban and rural commingled and collided, and a city on the verge of a more international, more sophisticated, more elegant future. Georg was there to seize this opportunity.

The east–west connection shaped Georg from an early age. Sörboda, Georg's place of birth, was near the Swedish engineering miracle Göta Canal (*Göta kanal* in Swedish), constructed in the early 19th century. Göta Canal was built to connect the capital Stockholm in the east on the Baltic Sea with the harbour city of Gothenburg in the west on the Atlantic. (The 614-kilometre canal stretches from an inlet of the Baltic near Söderköping to Gothenburg.) When the canal was completed, it became the central artery for east–west commerce and trade, because of the way it connected Stockholm to Gothenburg — and from Gothenburg the rest of the world. Like the Caledonian Canal in Scotland, also designed by Thomas Telford, Göta Canal consists of a series of locks, and uses natural waterways as strategic connectors, and Georg grew up near one of these locks. The canal was a cornerstone in the modernisation efforts by the new Swedish king, King Karl the 13th, with the chief project manager, Baltzar von Platen, proclaiming the virtues of the innovation by saying that mining, agriculture and other industries would benefit from 'a navigation way through the country'.

During young Georg's childhood and adolescent summers, he would watch the goods from Europe, Scandinavia and the United Kingdom on their way up from the Atlantic to Stockholm, and see the Stockholm-designed products on their way back the other way into global commerce. Georg's dad was a captain on one of the canal boats, and young Georg was mesmerised by the goods, products, exotic wares and textures being transported between east and west, and the things he would see when he ran down to watch his father tie the marine knots at the local lock. Inspired by what he witnessed during his adolescence, Georg had an entrepreneurial epiphany and it was decided that his and Gottfrid's sister Anna would take over Sörboda, while Georg took the canal boat east to Stockholm. Georg's father, the canal boat captain, had facilitated an introduction in Stockholm for Georg, and according to the county records of Ekeskog in West Gothland, the 17 year old assumed the title of *handelsbiträde* — retail assistant — at a shop in Stockholm. As he transitioned from his boyhood agricultural roots to the urban landscape of his manly future, Georg Johansson, the rural Johan's son, became the urban man from Sör, Georg Sörman.

Georg's first business visit to Stockholm was in the good old days, when a blackberry was just a fruit, a tweet was a sound a bird made, and a phone still had a rotary dial — if you even had a phone. It was a time of community and of high-touch connection. A time when the word *disruption* referred not to your smart phone but to supply lines in the Boer War. From 1903 to 1917, young Georg was biding his time, learning the urban ropes of retail and observing the changing tastes of Stockholmers as the city came of age. During this time, he started courting his future wife, Ethel, a woman who was to prove critical to Georg's eventual business success. And over these years, Georg spawned the entrepreneurial dream to dress the men and women of Stockholm at his future fashion boutique on Stora Nygatan, in historic *Gamla Stan* (the Old City).

Doors to Georg's bricks and mortar boutique opened on 25 October 1917 in the midst of a tough economic climate during First World War, only a few months after the Battle of Passchendaele on the Western Front, where the horrors of war were playing out for many young men in Europe and around the globe. Thankfully for Georg and a generation of Swedish men (and women) Sweden was outside the conflict because of its neutrality. Sweden tried as best it could to get on with business, despite the economic impacts from the south. Always mindful of the hardships others were enduring, Stockholmers got on with their affairs as best they could. The start-up enabled the men and women of Stockholm to browse his collection of clothes by analogue window-shopping or physically talking about their needs to a human sales professional.

Unlike business today, Georg did not have to think about digital bling — he could focus purely on the analogue thing in his business, and the analogue thing was great customer service, attention to detail, artisanship and knowledgeable staff. The sound *kaching* was associated with the cash register and not a mobile banking app. And the sound echoed enough for Georg, who diligently went about his business and eventually established himself as part of Stockholm's business guild, and could afford a family expansion. In 1919 he had married Ethel, and in 1920, 1924 and 1925, young Per, Sven and Karin were respectively born into a retailer's family. Business in the Old City was good, partly as a result of the boat traffic and commerce from Lake Mälaren and from the Stockholm archipelago, which tended to congregate and be transacted on the island where the store was located. Georg had his driver's licence, but after an early encounter with a lilac hedge, he decided that a driver would be responsible for him getting from A to B. The company driver, immaculately dressed in a uniform, became the store's delivery man and paid personal visits to the store's remote customers with deliveries.

But nothing stays the same. Ever. For good and for bad. Because of technological change, demographic shifts and evolution in consumer behaviour, commercial boat traffic on Stockholm harbour slumped, in favour of rail and cars, and, combined with the impact of the Great Depression, business in the Old City suffered at the beginning of the 1930s. The whole family had to dig in and ensure ends were met, meaning Per, Sven and Karin had to help out in the store. Meanwhile, Ethel showed particular creative flair by taking the tram and looking around Stockholm through 1935 and 1936 for a new locale that would be buoyed by the redirection of traffic flows and demography, finding that locale on the island of Kungsholmen, near the transportation hub of Fridhemsplan. Here the business might take advantage of the convergence of rail and car traffic, while benefiting from the westward expansion of central Stockholm.

By 1937 Georg had invested in a new location on the island of Kungsholmen in Stockholm. Twenty years after its establishment, it was Ethel who had become the driving force for rejuvenation, reinvention and rebirth of the brand Georg Sörman. To make a mark on the new locale at 41 St Eriksgatan, together Ethel and Georg changed the facade of Stockholm by commissioning the design of a neon logo by Graham Brothers Neon Co for the store's signage (see figure 1.1, overleaf). The logo — including a gentleman with a walking stick and a top hat — communicated stature, sophistication, class and style, and had good equity with the people of Stockholm. The logo design would later became the statuette — the Oscar-equivalent — for the retail signage awards in Stockholm, and in 1998, during Stockholm's year as the cultural capital of Europe, was named as the best sign in all of Stockholm. In the judging panel's published motivation behind its decision, they argued, 'The Sörman man — for generations of Stockholmers — has for 60 years enriched the nocturnal urban space. In his austere elegance he appears as the Oscar of signs'. There was analogue trust that the shop and its professionals would deliver on the promise that the brand communicated.

Figure 1.1: Georg's neon logo

Georg's interest in the retail clothing business proved genetic, and the family legacy of commerce, professionalism and connecting people with goods continued with Georg and Ethel's son Per (born 1920), the oldest of their three children. Per had grown up in the business and worked alongside his father from a young age, and so it was he who was slated to take over custody of the family brand, and to guard its legacy. Per was dyslexic but had learnt from practical experience the tools of the trade, and made up for his impairment with street-smart entrepreneurialism and an attention to financial affairs learnt during the Great Depression of the 1930s.

One year prior to Georg's death in 1968, Per graduated from his apprenticeship and stepped up as the managing director of the company. Change had become a part of the business ethos at the store and the brand had evolved from a fashion retailer, to a family clothing shop, to having a sole focus on gentlemen's clothing by the time Per took custody of the brand. Per, like his father, was a hard worker but also an entrepreneur, and was constantly on the lookout for new niches and growth opportunities. He realised that the tailors in the shop presented a latent opportunity for more bespoke services, including customised service for the 'large and tall' client segment, who weren't catered for in Stockholm, or around Sweden for that matter. Catering for these people could open up the brand's allure and customer-centricity to a new client segment. By offering unique sizes and bespoke alterations, the brand built up a reputation around Sweden, and ensured that it could service a broader market, including using postal delivery to reach a geographically diverse clientele. Business was good, and client service ensured continued profitability, and positive analogue good will.

Per, like his father, was determined to pass on his legacy, and to see the brand live on for another generation. Client service, value for money, generous client size options, and a personal and familial touch were business lessons that he passed on to his daughter, Birgitta (born 1954), the youngest of his three children, (Margareta, Katarina and Birgitta). Like her father before her, Birgitta had grown up in the business and, when Per passed away at age 83 in 2003, she graduated from her 30-year apprenticeship and took over the reins from her father. By 2003, the business landscape had changed. While the business had reached the maturity stage of its life cycle (see chapter 9 for more on this) and had been coasting comfortably for years, the world around it had changed. And the people (and their offspring) who used to frequent the family-owned store during the time of Birgitta's father, and his father before that, had changed with it. Fashions had changed, retail had shifted, the demographics of the neighbourhood were different, and clients' analogue shopping habits had been upended with the rise of the internet. The only thing that wasn't different was the store.

In the blue corner: Digital Disruption. In the red corner: Birgitta. A lot can change in a century—or between three generations of retailers. Birgitta faces a very different set of challenges from Georg back in 1917.

But Birgitta loves history and the customer service lessons she inherited by cultural osmosis. The shop, every antique lover's dream, is testament to this respect for the family business's history (see figure 1.2, overleaf). But today, her prospective clients enter the analogue store after they have digitally window-shopped her offering, and they share their needs with her personnel—who unsuspectingly measure them up for a potential sale of tailored services. The supposed clients pay attention to these measurements, carefully memorising the digits of their sleeve lengths before exiting the store to the sound of a lonely door bell. If they have the human courtesy to Birgitta, they then sneak around the corner and place a digital order on their smart phones with their digital dealer. Birgitta's store has become a showroom for eBay, and she is doing eBay's front-end analogue marketing for them. Hence, cornered like a wounded animal, Birgitta's business/life-defining moment lies in how she fights on. Or, more accurately, where she should position her business on the continuum between the analogue, high-touch, tradition at one end — the thing she has grown up with — and the digital, high-tech at the other end — the thing that is threatening her very presence. Birgitta — my mother — and her store Georg Sörman have become digitally disrupted. Is it time to throw in the towel?

Figure 1.2: Georg Sörman storefront

Digital is an omnipresent force for change

Birgitta is not alone. Retailers around the Western world are struggling to cope with digital disruption. They need a support group. Digital natives and, increasingly, digital immigrants[1] have questioned the very need for analogue physicality—in more ways than one. (A *digital native* is a person who was born during or after the general introduction of digital technologies and, through interacting with digital technology from an early age, has a greater understanding of its concepts. A *digital immigrant* is an individual who was born before the existence of digital technology and adopted it to some extent later in life. Natives speak technology 'without an accent', whereas digital immigrants speak technology with one.) If you're dealing with digital disruption, you can cynically challenge it as an affront to human analogue connectivity, and you can passively aggressively lament the digital due diligence and price comparison research your customers now do, or romanticise about the good old days of margins. None of it will help you though. The internet is here to stay.

Digital disruption goes hand in hand with the internet. And the attack on bricks and mortar is deliberate. In 2011, eBay—the bane of my mum—(ironically) staged a traditional print campaign enticing consumers to 'Browse it at Westfield [Shopping Centre]; buy it on eBay'. Their 2012 campaign? 'Dress like a Model. Shop like an Accountant'. Ask travel agents, real estate professionals, car rental operators with shopfronts, mortgage brokers, car salespeople, owners of shoe stores, hairdressers, insurance professionals or financial advisers about digital disruption. These are people we used to do business with. Many of us don't anymore. While the services of some of these people might be more protected than others, the way we engage with them has fundamentally shifted—be it the backyard hairdresser who you found on Facebook, the online-only mortgage broker who rates highly on your comparison portal, or the real estate agents you digitally connected with because they were ranked highly on Rate-My-Agent.com. If a service can be digitised, it will be digitised.

And retailers and service-providers are not the only ones who are struggling. Book pages that used to be turned by hand are now turned by a swipe, newspapers delivered to a doorstep are now delivered to a browser, and cash that used to live in a physical wallet now gets transacted via a digital wallet. All for free, or virtually for free. Banking used to be a physical activity we planned for; now it's a digital bump, a mobile login or a Mint.com visit away. Music encoded on vinyl and amplified on decks can now be accessed—as opposed to owned—via digital streaming. Drama series once aired on television get file-shared as BitTorrents, and films once collected by enthusiasts quickly get digitally discarded to make more space on a digital hard drive. Memory, once a profoundly analogue thing, is also starting to become a digital thing.

This all presents a massively disruptive challenge. How do you ensure that your business doesn't end up trading real, analogue dollars for digital pennies?[2] In other words, how do you ensure that you don't follow in the footsteps of the music industry, book industry and film industry, which have all been decimated as they have digitised their analogue products, ending up with commodities that customers weren't willing to pay as much for? In a future world where cash may no longer be king, where our perception of the real and the virtual, the analogue and digital is increasingly blurring, how will your, let alone my mother's, business cope?

Consider the following digitisation statistics, which highlight that we have passed the digital point of no return:

- By November 2012, Google's digital advertisement revenue alone surpassed the analogue advertising revenue of *all* American newspapers.[3]

- US total analogue newspaper advertising revenue is lower today than it was in 1950, the dawn of print advertisement on a large-scale.[4]

- By August 2012, Amazon.co.uk was selling 114 digital eBooks for every 100 analogue books.[5]

- In 2012 in the United States (digital) Cyber Monday online sales reached $1.47 billion, up 17 per cent from 2011, representing the heaviest online spending day in history, according to digital tracking firm comScore. This firm also reported that sales for (analogue) Black Friday (the day following Thanksgiving) reached $1.04 billion.

- Forrester estimated a ratio of roughly 5:1 of web-influenced bricks and mortar retail sales to digital online sales. With direct digital online sales running at the time of writing at approximately US$200 billion annually, that means roughly US$1 trillion offline retail sales are influenced by digital marketing.[6]

- Aside from digital retail, Amazon generates over US$1 billion in digital advertisement revenue — from its suppliers.[7]

- Mobile sales of Angry Birds outnumbered the total analogue console games sales by Nintendo 6:1 by December 2011.[8]

- Digital accounted for 55.5 per cent of the £155.8 million spent on all music in the United Kingdom in the first three months in 2012.[9] These were only the legal downloads.

- The Top 10 brands in mobile payments do not include a single analogue bank brand.[10]

- An equal number of American heterosexual couples now meet online as meet in bars or restaurants; American homosexual couples are six times as likely to have met online as they are to have met through friends.[11]

But, are you truly at risk of digital disruption? Neophiles like me would, of course, always say that you are, and you will perhaps counter that people will always need cradle-to-grave services like nappies, tampons and deodorant. And I would agree. There is still no app controlling body odour. But I would add that how people get access to these fast-moving consumer goods will increasingly be via digital distribution, not by physical store visits. Think Amway home delivery, not Walmart SUV visit. *Well, Anders,* you may say, *there are things like taxes, right? Cannot escape them, so this means accountants and bureaucrats will always have jobs, and we are forced to see accountants every year.* True, but already by 2006, six million American tax returns were done in India — digitally — via outsourced, online services. *Aha,* you may counter, *but, Anders, that's because numbers can be automated and digitised and that is why we should focus on being creative professionals, and be part of the knowledge work economy like lawyers and designers, because that strategic/creative side of the brain can never be computerised, right?* Unfortunately, 99designs, Guru and MichiganOnlineLawyer changed that, too. *Well, if I have to endure that divorce I got on MichiganOnlineLawyer, at least I can rest assured that my extensive network of friends will find me a new partner, Anders.* Well, yes, if someone looking for a new partner will trust that analogue word-of-mouth. You may actually be better off building up your digital brand reputation,[12] and keeping an updated profile across multiple digital dating networks to ensure your channel marketing reaps rewards. (Now possible both via desktop and mobile devices, FYI.) The client/partner/date/customer is in power, and empowered to dis you at any moment.

Everything that can be digitised will be digitised

Some things sooner than others. If you're a recruiter, a mortgage broker, a travel agent, a real estate agent, an insurance broker, a stock broker, a financial adviser — be afraid. Be very afraid. The internet is a digital *disintermediator.* This means that intermediators — middle men and women — are being cut out faster than you can say '56K modem'. The internet shines a light on complex value chains, highlighting inefficiencies and so removing the need for many types of analogue connectors.

Specialist knowledge and the ability to find something or someone, and to lock in a deal between two parties, used to come at a premium. Google (and their digital predecessors) changed that. I remember visiting Phi Phi Island in the Thai archipelago in 1998 with my family, on a vacation we had booked via a family friend of ours who ran a travel agency. This was still rational behaviour back before the irrationally feared Y2K bug. We had briefed our friend (let's call her Sarah), on what we wanted and, given that no-one in my family knew anything about Thailand prior to this visit, we gave her creative freedom, and a generous commission, to book our travel, connections and transfers — something she executed with love and care. We arrived and were soon having a good time. I cannot deny it.

My analogue bubble, however, burst on that island. Within the arrangements that Sarah had organised for us, we were only able to visit the island — a magical plot of land in the Andaman Sea — during the day. We weren't allowed to stay overnight and we had to leave this paradise in the afternoon. My bubble burst during a lunch-time conversation with another Swedish family on the island. This family proudly (a bit too proudly in my view) told us how much money they had saved because they had booked the entire family holiday 'på nätet' (Swedish for 'on the internet'), and how (great) 'nätet' (it) was for creating tailored holiday experiences off the beaten track. This family were not only saving lots of Swedish Krona by booking directly via Altavista (the premium search engine in 1998) with the airlines, tour operators and boat operators, but were also bespoke tailoring their own adventures. To make our analogue matters worse, they asked us whether we too would be spending the night on the island. First-world problem, I know, but I felt somewhat ripped off. We realised that Sarah had essentially private-labelled a package holiday, and received a commission on an itinerary that was stock standard and mass produced for people who didn't know how to use the internet. Ignorance on the part of clients such as us had kept Sarah in business.

So we felt ripped off but some in my family were still uncertain. I remember the philosophical debates on the boat back to the mainland (as we were being waved goodbye to by the Robinson Crusoe family). Could you trust the internet? What would happen to Sarah's business? How could you pay for something online? How would you receive your physical ticket? Could you call Altavista if something went wrong? I didn't really care. I had discovered off-the-beaten-track travel, and I decided that I didn't need a middle man or woman to create it for me. But was there something bigger at play here? Was there a general trend away from custodians of knowledge, from multiple layers of distribution and from shadowy supply chains that would affect stock brokers, recruiters, real estate agents, financial advisers, mortgage brokers and insurances salespeople over the coming years? You bet.

Disruption isn't new: why a discarded Emil Paidar barber chair is now a collectable

Disruption is nothing new. The digital story of the demise of travel agents may be the modern version of the Luddites, the proverbial buggy whip designers or the barber stool manufacturers, who irrationally clung to their wares in the face of massive external change. The Luddites were members of the bands of English workers who, in 1811 to 1816, destroyed machinery, especially in cotton and woollen mills, that they believed was threatening their jobs. Since then, the term *Luddite* has become a descriptor for people who vehemently resist change, who want to maintain the status quo at all costs and who go militantly backward in the face of disruption. At the beginning of the 20th century, buggy whip designers believed they would always be needed and that their industry would be buoyed as long as the horse and buggy was the primary mode of transport. Soon they were replaced by Ford's new automobiles.

Disruption is nothing new, but while disruption, outsourcing, lay-offs and obsolescence might have taken years or even decades in the past, digital disruption can happen within months now. Let us contrast old-school and new-school disruption for a moment.

Thanks to disruption, my home in Surry Hills, Sydney, is adorned by an antique Emil Paidar barber chair (see figure 1.3, overleaf). I love this chair. (Perhaps too much, according to the four friends who helped me carry this 150-kilogram piece of equipment up several sets of stairs.) This 1920s chair would fit right into a scene from an American old-school mafia movie or the set of *Boardwalk Empire*. It's an incredible piece of engineering, and sitting in it to watch a movie on the big-screen projector is a great experience. Sturdy comfort. Great headrest. Nostalgia. The Bordeaux-wine coloured refurbished leather cushion is ageing beautifully and is starting to build up a nice patina. The leather provides a nice colour contrast against the enamel and steel hydraulic joystick lever, which a Chicago barber once would have used to swivel his esteemed clients around, in a display of the latest in barber chair design nearly a century ago. But, the chair has ended up as a collector's item in my living room. Why?

Figure 1.3: Emil Paidar barber chair in Surry Hills, Sydney

From the early 20th century until the late 1950s, Emil Paidar and its main competitor (and archrival), Koken, enjoyed a cushy existence in the American marketplace. Fashion dictated short, neat cuts, and the barbershop represented a type of man cave for business deals, providing rest and relaxation for the gentlemen of the time. Together with Koken, Emil Paidar sold about 10 000 chairs a year to 100 000 American barbershops, which provided it with steady profits. It's easy to think that this would be an industry that was recession-proof, right? I mean people will always need to get their hair cut. Yet fashions shifted, and with the 1960s demanding longer, wilder hair, the neatness and service associated with barbershops were no longer seen as cool. Consequently, some bricks and mortar barbershops closed. But even more disruptive to Emil Paidar was the market entry by Japanese barber chair manufacturer Takara.

Takara disrupted Emil Paidar's world. Takara's parts were directly interchangeable with Emil Paidar's parts (deliberately so, as part of the go-to-market strategy by the new entrant). They were also 20 to 30 per cent cheaper. Moving toward Takara products became a no-brainer for America's barbers, whose margins were being pressed and whose volumes of clients were shrinking. Hidenobu Yoshikawa, the founder of the Osaka-based Takara company, wasn't just about price, though — he enjoyed an upmarket, regal following. Literally. Japan's Emperor Hirohito and the Thai King Bhumibol had Takara barber chairs in their palaces at the time, and enjoyed their weekly man-pampering.[13] By 1970, Emil Paidar had dropped from 70 per cent market share and was on Nixon-initiated government support as the only remaining US-owned barber chair manufacturer.[14] Takara had acquired Emil Paidar's former competitor, Koken, to command 70 per cent of the US market. Emil Paidar made a last-ditch protectionist stand, and unsuccessfully filed under section 301(a)(2) of the *Trade Expansion Act (1962)* against the disruption.

This excerpt from the Tarriff Commission's report is illustrative of Emil Paidar's failure to react appropriately to disruption. The views of Commissioner Thunberg sounded the death knell for the viability of Emil Paidar's business model in the face of this disruptive competitor:

Paidar's competitive efforts since 1967 have seriously depleted its resources; its working capital is exhausted; its marketing efforts have been contracted; its sales volume is declining. Under these circumstances, I find that without assistance Paidar must cease to exist very shortly.

The writing was on the wall. This view is a sad indictment on the failure of Paidar to adapt, and to keep up with its disruptive competitor. But notice how long the disruption took. Takara Co had been around since the Second World War and entered the US market in 1957. Emil Paidar had 10 years to respond competitively, but chose to ask for government assistance instead. As a result of their failure to deal with disruption, I, as an antique's collector, am now the proud owner of a collector's item from the heyday of Emil Paidar design. The question remains: could Paidar have responded differently to the disruption? Is there a way for old-school companies to survive new digital threats? Yes, there are — and this book will show you how.

Exponential thinking: from good to great to obsolete

Unlike the Emil Paidar example in the preceding section, digital disruption doesn't take 10 years. In fact, even the digital disruptors are being disrupted. Facebook's initial problems post-IPO have been caused by its failure to adapt its advertisement sales model to mobile browsing patterns, Apple's iTunes distribution of ownership is being digitally disrupted by the rise of digital access to music via channels like Spotify and Rdio, and Nokia's market share of the smart phone operating systems market diminished from 44 per cent to 2 per cent in 24 months. Because of digital disruption, you can go from being good to great to obsolete within a matter of months. Digital disruption is a paradigm shift and it will be your downfall unless you're connected to its dynamics.

Let's take a look at the dynamics underpinning digital disruption. Exponential dynamics. Technology grows in an exponential, rather than a linear pattern, as shown by the popularised law of technology — Moore's Law. This law states that computing power doubles every 18 to 24 months — in other words, over the history of computing hardware, the number of transistors on integrated circuits doubles approximately every two years. The capabilities of many digital electronic devices are also strongly linked to Moore's Law: processing speed, memory capacity, sensors, and even the number and size of pixels in digital cameras. As those of you who are mathematically inclined are probably already aware, exponential equations have exponents on at least one variable. Exponential growth, when graphed, looks like a ski jump, whereas a graphed linear function is a straight line. This means that we see massive productivity gains in price performance, capacity and bandwidth each year and, as such, an acceleration of digital technology.

Progress in digital technology is exponential, not linear, so in order to keep up we need to start thinking exponentially, not linearly. What is fascinating about this exponential force is that, while it's disruptive and increasing the speed of change for old business models, it's also highly predictable. We can extrapolate how powerful digital technology will be in one, two, five or even ten years, based on this exponentiality.

Mobile e-commerce uses technology that follows this disruptive pattern. While online digital retail has grown at a steady linear pace (a compound annual growth rate (CAGR) of around 9 per cent), it will take mobile e-commerce (which is on an exponential growth curve at a 300 per cent greater pace) much less time than it took desktop- or laptop-based e-commerce to reach the 10 per cent market share penetration it now enjoys.[15] The future is exponential, not linear. And it seems the future will be coming at you at an ever faster pace. Digital disruption begs the question of whether not being part of digital technology and its concomitant exponential growth curves is such a smart move.

Table 1.1 illustrates the disruptive force of exponential growth.

Table 1.1: linear growth versus exponential growth

Linear growth	Exponential growth
$1 + 1 = 2$	$1 \times 2 = 2$
$2 + 1 = 3$	$2 \times 2 = 4$
$3 + 1 = 4$	$4 \times 2 = 8$
$4 + 1 = 5$	$8 \times 2 = 16$

The example shown in table 1.1 translates into a chart that looks like the one shown in figure 1.4. This is digital disruption in a graph. How fast is your thinking improving?

Figure 1.4: exponential increase of computing power versus linear growth

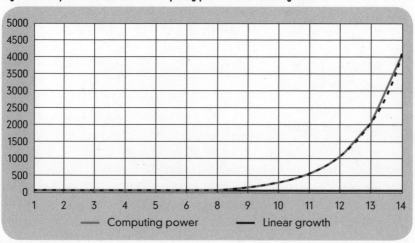

A digital wave of change

Based on the disruptive power of exponential growth, is it any wonder that analogue business models don't keep up? If digital disruptors are having to take their own medicine, where does this leave the rest of us, who are about as nimble as a 150-kilogram barber's chair? Kodak's moment came and went as quickly as we decided that photography was not a double-handed viewfinder creation but a single-handed arm's length digital transaction. Snap. Borders' bricks and mortar physical presence couldn't protect it from Amazon's one-click upsells, social cross-sells and kindling of eBook consumption. Despite its best analogue attempts, Blockbuster (US) couldn't build a block strong enough to resist the BitTorrent of disruption headed in its direction. Everything that can be digitised will be digitised. Kodak. Borders. Blockbuster. Out of business. Why?

Because they failed to do three things. These companies, and the leading voices within them, failed to:

- accurately spot the oncoming trend
- get a feel for the underlying currents driving the disruption
- position themselves in a way that took advantage of the tidal shift.

Digital disruption is a wave of change that is pushing companies and brands into an economic riptide, and good men and women around the world are drowning in their efforts to find a firm footing. I will revisit the concept of digital disruption throughout the book with concrete case studies and, ultimately, use it to demonstrate how you can position yourself profitably using the Digilogue Strategy Map. For now, let me provide some illustrations.

Caught in a riptide

I know what the feeling of being caught in a riptide is like. Not metaphorically, but literally. In February 2009, I was staying with a group of colleagues at a beach house on the New South Wales south coast at Culburra Beach. We were at a strategic retreat and enjoying stimulating conversations focused on the future of business. On day two of the retreat, we decided to have a mid-morning break and go for a swim. Having grown up in Sweden, I was a good swimmer — in Swedish lakes. In the Australian surf, I was less able, but I had never had a problem in the water on the guarded beaches adorned with red and yellow flags (one of the symbols of Australian summers).

As you may already know, there are two key rules when you visit an Australian beach. One is to always 'swim between the flags', which means that bronzed Aussie surf lifesavers can spot you, and save you should you find yourself in the middle of a disconcerting current or look like shark bait. (The other rule is to 'slip, slop, slap' — slip on a shirt, slop on some sunscreen and slap on a hat — which has nothing to do with swimming safety, but the legitimate and more long-term concern about skin cancer.[16]) Unfortunately, both these rules failed to help me on that fateful day on Culburra Beach.

Culburra Beach is an unguarded beach, so it has no flags to swim between. It is isolated. It is a rugged, long stretch of sand, and the next major landmass to the east is Chile, several thousand kilometres away. Nonetheless, my four male colleagues and I decided that a dip in the ocean seemed like a good idea. The race to throw on some 'boardies' (board shorts) and run down past the sand dune to the water's edge was on. None of us slipped, slopped or slapped. While I may not have been a great swimmer, I was a decent runner, and I managed to outrun my colleagues. On seeing the water's edge and the cooling waves lapping the shoreline, my Swedish instincts kicked in. I dove straight into the water. Headfirst. Without thinking. Without spotting what was going on around me. Luckily for me there was plenty of testosterone in the air, and in their competitive urge to not be the last guy into the water, my colleagues also dived in. Quickly, more quickly than me, they recognised something wasn't right. They felt that the underlying current was gripping all of us, and dragging us out to sea. Dragging us toward the break. And the surf was also feeling particularly macho today, and had decided to knock us about. Very soon, my colleagues realised we had to get out of the water, or we'd all run the risk of being dragged into the riptide and out into an aquatic equivalent of a washing machine. While their freestyle strokes enabled them to awkwardly body surf and (just) make it back to shore, my breaststroke, eminently suited to quiet Swedish lakes, did little to comfort my increasingly panicking and oxygen-starved body. I felt like a mitten being hurled around in the wash cycle.

Similar to the failings of companies such as Kodak, Borders and Blockbuster, I had failed to spot the riptide we had all jumped into, I had failed to activate my early warning system by not feeling the underlying current, and I had lacked the skill required to position myself with the wave to get back to shore. Three potentially devastating mistakes.

Luckily, men have now reached an evolutionary stage where we can ask for directions. We can also comfortably ask for help. My colleagues quickly noticed my upstretched arm clamouring for attention. One of them, Nils, who is an extremely able surfer, was given the task of swimming out and dragging the breaststroker back to shore, while the others were co-ordinating a land-based rescue as a Plan B. Nils called out instructions as he approached, which had the effect of calming me slightly, and as he reached out to me in the middle of the foamy wash, he told me to relax, to tread water and lean back while he commenced his life-saving exercise. It felt extremely coordinated and well executed. (Hopefully, in retrospect, Nils agrees.) Even so, it took us 10 minutes for him to drag and throw me in the direction of the beach each time a wave was starting to break, which felt like an eternity. He was using his body surfing skills to spot waves, feel underlying currents and position us on the waves to safely get us back to shore. A massive feat, especially given Nils is about 15 kilograms lighter than me and I wasn't much help in the rescue.

The panic only really subsided once I felt solid sand under my feet. Then the fear set in. I could see on the faces of my colleagues that they, too, were exhausted and emotionally affected by the severity of the event. We had all gotten back to shore safely, by the seat of our board shorts. Scenarios were playing out in my head. Thoughts of loved ones were vividly projected. Guilt trips over unfinished footnotes. But I was safe.

And as I revisited this experience over the course of months and years with my colleagues and friends, I realised that there were some massive lessons to be learnt here. Primarily, I needed to learn how to swim properly. While it took me a while to get out there and have a proper crack at freestyle swimming, friends and my brother encouraged me, advising me to start ocean swimming and to take total immersion swimming lessons. So, in the four years since, I have had to admit that I wasn't a skilled swimmer, take adult swim classes, and get back to basics and face my aquatic fears. Literally a deep dive into the deep end of the Pacific swimming pool. But, with increasing confidence, and a couple of 1-kilometre races under my belt, I also developed an interest in triathlons, and have since competed in two Olympic-length triathlons involving longer ocean swimming, too. And just to clarify the critical 'slip, slop, slap' version of ocean swimming and survival in the water? Stop to spot what's going on around you, relax and get a feel for the underlying currents around you and, finally, ensure that you position yourself to take advantage of the ocean's waves and movements to safely get you back to shore.

The survival lessons are not dissimilar in business. During a time when many businesses are finding themselves in an economic riptide, when there are no flags to swim between, and when there are few, if any, life guards around, you must become adept at spotting waves of change, getting a feel for the underlying currents of change, and positioning yourself to take advantage of the disruptive waves around you. This is why it's critical that you:

- stop to spot the digital disruption that is occurring right now
- get a feel for the underlying currents of technological and customer behaviour driving this disruption
- position your business to take advantage of the trend and safely get you back to shore.

This must become your survival technique. Before it's too late. Don't venture anywhere without it. (Check out chapters 4 and 10 for help with workshopping your future positioning.)

Surfing Swedish waves of change

While my ocean story may be a visceral representation of what can happen when you don't heed survival advice, as a business owner I can empathise with business leaders who feel the same panic in the face of disruption — digital or otherwise. This is why I started a conversation about digital disruption with my mum, Birgitta.

In May 2011, I was sitting down in Stockholm with my mum, my toughest pro bono client, and talking to her about the future of the business. She said to me, 'Anders, the internet is killing our business! What will happen to St Eriksgatan (the equivalent of Main St), when everyone shops online, buys brands from outlets and bargain hunts via eBay?' She continued, 'Certainly, there must be some things people will never buy online, like shoes?' I had to use the tough-love approach and break it to her — Zappos had digitally disrupted shoe retail, too. At this point, my mum's eyes glazed over, and she refused to continue the conversation about her business's response to digital disruption. It was as if she spoke analogue and I spoke digital, and we were failing to translate our respective languages. We needed a third party, a neutral Swede, to translate for us!

I continued to prod, and asked Mum what would happen if she told her analogue story of the 96-year-old family business via digital media, if she could show the analogue craftsmanship and expertise that her personnel possess, if she could let people properly digitally window-shop her beautiful, antique-filled store, to entice them to enter the analogue store. She responded, defending her ageing demographic of clientele, 'But, Anders, don't be silly, our clients don't know how to use the internet!' Perhaps with a tone of passive aggressiveness, and with a certain lack of respect for my elder, I countered with a 'So, they only know how *not* to buy from you online . . .'. On that occasion, this is where the conversation ended.

What is critical here is that my mum has at least spotted digital disruption. She cognitively understands, unwillingly at times, that everything that can be digitised will be digitised. For better and for worse. This spotting is the first step. It's essential that you too spot this wave of change in order to make sense of it, and then to flexibly position yourself to take advantage of it.

Awareness of yourself and your surroundings is the first step

Awareness of yourself and your surroundings and observational clarity are a must for continued growth and sustainability. A close friend of mine has been battling an alcohol addiction for the past three years. While his story is not a perfect case study, or a miraculous turnaround yet either, I have learnt something profound from him, and the community he is seeking help from — Alcoholics Anonymous (AA). AA is the most famous self-help organisation in the world and, according to the psychiatrists, psychologists and doctors that our friendship group consulted with, it's still the most effective way of getting someone out of old behaviours and helping them turn their lives around. According to the research, it's not always clear why AA is so effective.[17] But the organisation attributes a lot of its power to the 12 steps outlined in the Big Book. When I helped my friend along to a meeting, and listened to the stories of the recovering alcoholics as they shared their journeys, everyone came back to the steps.

The first step in the 12-step process of AA is this: 'We admitted we were powerless over alcohol — that our lives had become unmanageable'.

My friend has tried drug and alcohol counselling, compulsive exercise, music, and even turned to other drugs but, in my conversations with him, he's acknowledged the power of the steps — and particularly the first step, which he says was the most provocative and impactful admission he's had to make. He has admitted to himself that he is powerless over alcohol — and that his life had become unmanageable. Since our friends came together as one to support him and accompany him to AA meetings, he has reported improvements. We are all aware that his journey is a marathon and not a sprint, and that any joyous declaration of victory would be premature. However, it is heartening to see his insight, and that the wakening up to a discovery based on the facts and an observation of self and the external environment are helping our friend to get better and to live a sustainable life. I believe we can all learn from this — both personally and professionally. Growth flows from insight, observation and honesty, and taking the first step.

AA tells another story, too. The analogue community of AA — the personal connections, the touch, the face-to-face conversations, the authenticity, the raw emotion, the vulnerability, the heart, the feeling, the tears, the inspiration — can never be replaced by a computer interface. Everything that can be digitised will be digitised. But some things cannot be digitised. AA belongs in that category. Its analogue brand equity will remain rich. Its story will continue living on. The service it provides is mostly analogue, high-touch, enduring.

This is heartening. Not just for AA members and their families and loved ones. But also for the rest of us. For those of us who crave the analogue, the heart to heart, belly to belly, the genuine, the human, the raw, the attention to detail. People like you. People like me. People like my mum.

However. Round 1 goes to Digital Disruption. 'Will this be an easy win for the new-school champ?' the commentators and analysts ask. Digital Disruption came out swinging and through several body blows inflicted some serious pain. We learnt in this round that everything that can be digitised will be digitised, that digital disruption is happening at an exponential rate, and that this exponential rate trumps the linear thinking of Analogue Anachronism when it comes to speed. Customers are increasingly doing their due diligence digitally and brands, like my mum's, that are failing to provide value in the digital world are failing to entice customers in the analogue world. This is looking like a lay-down misère. What would Georg say?

Ask yourself...

Use the following to get yourself thinking:

1 What digitisation disruption is affecting your industry right now? Write it down.

2 What challenges and/or opportunities does the digital disruption pose for you? Add the challenges and opportunities in two separate columns.

3 What is the underlying meaning of this disruption? List the key drivers of the disruption—for example, convenience, speed, empowered consumers or globalisation.

4 What key uncertainties does this create for the future viability of your business? List the top five uncertainties—for example, commodification, margin squeeze or sustainability.

5 What does this mean for your future positioning? How do you want to position yourself in the minds of tomorrow's customers? Premium, fast, cheap, high-quality, green? For which key target segment? Conservative investors, wealthy stay-at-home mums, sporty millennials/gen Ys? Research five keywords that appeal to this key segment and focus your digital marketing on them.

Apply it...

Now use your thoughts and strategic insights:

* Imagine for a moment that you are a futures trader. You're trading 2020 futures based on the perceived underlying value of your business. In its current shape, and given digital disruption, would you back your own business?

* Do a pre-mortem. We often do post-mortems on projects and businesses. But, before it's too late, instead do a pre-mortem and look back on your business with the benefit of hindsight in 2020. What are the things you missed? Which

trends did you spot today and do nothing about? What were the fundamental shifts in business that you ignored? What did it cost in numbers and emotions not to pursue these opportunities?

- In 35 words or less, write down exactly what services and products you want to provide and to whom, and outline why these people should choose you over everyone else in 2020.

Endnotes

1 Definition of digital native from the *Oxford Dictionary*, oxforddictionaries.com/definition/english/digital%2Bnative, last accessed 8 April 2013.

2 H. Blodget, 'Jeff Zucker to CNBC: Online ads much slower than anyone thought', *Business Insider*, www.businessinsider.com/2008/12/jeff-zucker-online-ads-much-slower-than-anyone-thought#ixzz2PqQNcnQ1, last accessed 8 April 2013.

3 S. Tibken, 'Google makes more money from ads than print media combined', Cnet, news.cnet.com/8301-1023_3-57548432-93/google-makes-more-money-from-ads-than-print-media-combined, last accessed 8 April 2013.

4 H. Blodget, A Cocotas, 'The future of digital', *Business Insider*, au.businessinsider.com/future-of-digital-slides-2012-11?#-1, last accessed 8 April 2013.

5 C. Williams, 'Kindle books "outselling print" on Amazon.co.uk', *The Bookseller*, www.thebookseller.com/news/kindle-books-outselling-print-amazoncouk.html, last accessed 8 April 2013.

6 M. Griffin, 'Companies are gearing up to track every retail transaction through your smartphone', *Business Insider*, au.businessinsider.com/using-mobile-to-bridge-the-gap-between-online-and-offline-marketing-2011-9, last accessed 8 April 2013.

7 H. Blodget, A. Cocotas, 'The future of digital', *Business Insider*, au.businessinsider.com/future-of-digital-slides-2012-11?#-1, last accessed 8 April 2013.

8 H. Blodget, A. Cocotas, 'The future of digital', *Business Insider*, au.businessinsider.com/future-of-digital-slides-2012-11?#-1, last accessed 8 April 2013.

9 M. Sweney, 'Digital music spending greater than sales of CDs and records for first time', *The Guardian*, www.guardian.co.uk/media/2012/may/31/digital-music-spending-bpi, last accessed 8 April 2013.

10 K. Patel, 'Survey: Consumers don't trust Google or Apple with mobile payments', *AdAge*, adage.com/article/digital/consumers-trust-google-apple-mobile-payments/229163/, last accessed 8 April 2013.

11 M.J. Rosenfeld, RJ Thomas, 'Searching for a mate: The rise of the internet as a social intermediary', *American Sociological Review*, 77(4), 2012: pp. 523–547. (www.stanford.edu/~mrosenfe/Rosenfeld_How_Couples_Meet_Working_Paper.pdf)

12 R. Botsman, 'Welcome to the new reputation economy', *Wired*, www.wired.co.uk/magazine/archive/2012/09/features/welcome-to-the-new-reputation-economy?page=all, last accessed 8 April 2013.

13 Unknown, 'Japan: The great barber-chair coup', *Time Magazine*, www.time.com/time/magazine/article/0,9171,876764,00.html, last accessed 8 April 2013.

14 Unknown, 'Japan: The great barber-chair coup', *Time Magazine*, www.time.com/time/magazine/article/0,9171,876764,00.html, last accessed 8 April 2013.

15 J. Yang, 'US e-commerce landscape and trends', Harvard Business School, Independent Student Research, www.slideshare.net/joshyang/ecommerce-landscape-201216, last accessed 8 April 2013.

16 Unknown, 'Key statistics — skin cancer', Australian Government Department of Health and Ageing, www.skincancer.gov.au/internet/skincancer/publishing.nsf/content/fact-2, last accessed 8 April 2013.

17 B.I. Koerner, 'Secret of AA: After 75 years, we don't know how it works', *Wired*, www.wired.com/magazine/2010/06/ff_alcoholics_anonymous, last accessed 11 April, 2013.

2: Analogue versus digital

Digital Disruption came out swinging. And it's pounded Analogue Anachronism. Analogue Anachronism looks like a wounded animal, and only when backed into the corner in the final few seconds of the round did it lash back at its nimbler opponent. Is this the end of the story? Is it towel time?

In this chapter, I will illustrate why the analogue can be resilient. Here you'll find out why analogue emotion plays an equally important part as digital reason in our brand preferences and decision-making. I will bring you to New York City and show you how a barber's shop repackaged the story of the barber's chair in a bout of retail renaissance, we will uncover why microbreweries are trending up while macrobreweries are trending down, and we will explore why we keep insisting on wearing analogue watches when their digital counterparts are so much more precise. Stories of legacy, culture, identity, provenance and craftsmanship speak directly to our analogue hearts, and, as you will see, are the reasons we buy luxury priced tap water, why ritual animal slaughter is big business, and why only Champagne from Champagne can be called Champagne. And you will find out why Skype dating can never ever replace holding hands in analogue reality. Finally, we will link these lessons back to your business, and show you why there is plenty of fight left in Analogue Anachronism.

Hearts versus minds?

I am about to enjoy a home-cooked dinner with my friends Kaitlin and Aaron. Just 150 metres from us, the Pacific Ocean is lapping one of the most famous beaches in the world — Bondi Beach. It's April 2012, and I am about to embark on a speaking and executive-consulting trip to Europe and the United States. While Aaron is mixing up a pre-dinner cocktail and Kaitlin is putting the finishing touches on a vegetarian lasagne, I grab the latest edition of my favourite magazine, *Monocle*. Flicking through the pages of this magazine is a corporal pleasure. Every time. It's a visual explosion of beautiful photography, a kaleidoscope of engaging artwork. Communication goosebumps. Aaron disrupts my love affair with this inanimate object to start telling me about his and Kaitlin's latest trip to New York, one of the destinations on my upcoming trip. He tells me of the places they visited, of oysters in Williamsburg, the mixologist at The Standard on the High Line, the speakeasy bars in Chinatown. All quite interesting, I am vaguely listening and taking notes. Then. All of a sudden, he mentions the magic word. Barbershop. Barber chair. He elaborates — FSC Barber, Freemans Sporting Club, Horatio St. In the West Village. He had me at hello. I am suddenly intrigued.

I ask Aaron whether he really means there is an old-school, analogue barbershop in the West Village. In 2012? *Of course*, Aaron responds. And I shouldn't have been surprised. During a wintery consulting trip to Chicago in January 2012, I had visited what I imagined was a similar establishment because of my irrational obsession with Emil Paidar chairs (see chapter 1 for more on this). Unfortunately, the former had been a 'faux old' experience. But I wanted to learn more from Aaron about this one. He told me it was walk-in only. Old-school barber chairs, with private label and upmarket shaving kits. Antique colognes. Cool 1950s haircuts. Lots of tattoos. Think Jitterbug meets *Boardwalk Empire* — yes, decades colliding, but apparently it worked. Aaron, ex-navy, and I both enjoy a bit of man pampering. He talked of the hot towel and of the Truefitt & Hill shaving kits. Then he started describing the trust for the man with the single blade; the experience of putting your life in the hands of another. And when he started mentioning this, he wasn't talking of his time in the Persian Gulf. He was referring to a barbershop experience, which sounded more like a five-star spa retreat experience.

Let me show you, he announced. I asked him whether he YouTubed this experience. *Nah, but The Sartorialist did*, Aaron responded. The

Sartorialist. A reformed fashion sales guy, who had turned fashion photographer and blogger in 2005. A man of wisdom, style and sophistication. A man accustomed to the best, to the inner circle. A man of black and white intrigue. Aaron did a quick online search using the terms 'FSC Barbers' and 'Sartorialist', and up popped Scott Schuman, aka The Sartorialist (his *nom de guerre*). What was even more interesting was that Intel had collaborated with The Sartorialist on a video series called 'A Visual Life' and, in one of the videos, The Sartorialist visits the FSC Barber in West Village. I started watching the video, and the craftsmanship of the barber was immediately on display. All the visceral highlights that Aaron had recounted of his visits had been beautifully captured. Analogue. I had to go there. (For a video blog including The Sartorialist video, visit www.thinquetank.com.)

Why? Story. Emotion. Artisan. Intimacy. Craftsmanship. Old school. Relaxation. Reconnection with self. De-stress. Tune out. Inspiration. Oasis. Local. Of the place. Bricks and mortar. This all spoke to the heart. I was hooked — irrationally. My next visit to NYC was going to last 43 hours. Going to FSC Barber was going to be one of my top priorities. And so it was. On Saturday, 28 April 2012, I went there — sleep-deprived, jet-lagged, and with a maladjusted hydration level. I asked for the 90-minute session, the whole kit and caboodle, relaxing in an Emil Paidar, just like home. Out from FSC Barber came a new human being. Hyperbole? Perhaps. But I felt wonderful.

As you can tell, I am a big, irrational fan of this retro revival — this analogue, old-school establishment. Just before leaving, I had a brief encounter with the manager. With my mum's store in mind, I asked him how they had nurtured the history, how they'd survived all these years, and how they had built the brand and stayed relevant. The manager told me relevance wasn't an issue because they were only nine years old. Faux old. *Not again*, I thought. But then I did a double-take. Here is a new brand, wanting to 'seem' old. Wanting to pass on a legacy. As founder Sam Buffa says, '[In Europe] the barbershop was where men would discuss sports, politics, and culture while getting a haircut or shave...and we want each of our locations to represent that same kind of engaging atmosphere'.[1] The location had been the site for a barbershop for generations according to the manager. There is massive brand equity in this story. In the analogue. Biding its time in the red corner.

Rational versus emotional

Our emotional hearts are in a constant battle with our rational minds. We buy on emotion. And then we post-rationalise the decision, to make ourselves right. In other words, if we connect to something with our hearts, we want it and, often, we buy it. Occasionally (sometimes by procurement's force), we also run the decision past rational filters in our minds. Successful sales and marketing professionals, therefore, know that they must connect with emotional hearts and provide value to rational minds. This is why love brands, or 'lovemarks',[2] both business-to-business (B2B) and business-to-customer (B2C), have to win both the hearts and minds of their clients, so that when we are about to make a decision to part with dollars, we both feel good about it and can intelligently rationalise our decision to ourselves and others.

Increasingly, the latter process is one that takes place digitally. We browse, compare, download, flick through charts, get digitally educated about specs, and we punch data into Excel spreadsheets — often before we decide to have a face-to-face conversation with an analogue human being. Our rational minds have gone digital, but our emotional hearts have remained analogue. It's absolutely critical for business survival today to win the analogue, emotional hearts and the digital, rational minds of your clients and prospects.

What could this mean for my mum's business? A business that actually is old. Could the analogue highlights be recounted differently? How would the importance of analogue affect brands and businesses across the board in their efforts to stand out in a digitally disrupted world?

This got me thinking. I thought of the 135-year Eli Lilly pharmaceutical company that I consult with, and how the story of health through the ages gets told. I thought of my dealings with the French beverage company Pernod-Ricard — established in 1797 — and how they communicate the analogue in their provenance marketing. My mind flashed to the beef technology company that engages me as a futurist to talk of traceability and story. I remember the breakfast conversation with the winner of the Cannes advertising Lion for their communication of Tasmanian pure, analogue waters. How Johnson & Johnson hold up their family values in their marketing. And how when I spoke for Cisco in Barcelona, the reference to San Francisco was critical to their brand equity. Is the analogue, the locality, the sense of grounding, the communication of time, depth and artisanship something we should do more with? Or perhaps package in an even more engaging way? Does analogue history have a place in our fast-moving world?

Digital rationale and analogue emotion: music, clocks and cinema

Think about the analogue environment for a moment. Do you remember your first digital download? Let me remind you, it was a probably an illegal act. But there's no shame if you don't remember it. Most likely, it was a fleeting moment in time. Now contrast that with this. Do you remember the first record, tape or CD you ever got given or bought? I can vividly recall running around at our modest summer house on Ingarö Island outside Stockholm for an entire summer in 1987 singing 'It's the final countdown' by Europe. And I didn't speak a single word of English. Nobody really remembers their first digital download, but vinyl is forever. I still have the record. The computer on which I first downloaded from Napster has long been digitally discarded. No wonder analogue vinyl sales are on the rise around the world.

According to statistics released by Nielson, vinyl sales were at 0.3 million in 1993, rising slightly but staying steady until they shot to 1.9 million in 2008 and then dramatically soared to 3.6 million in 2011.

Take a look at your wrist. The one where you might wear a watch. Do you? Is it an analogue or a digital watch? In other words, does it display arms, mini-engineering movements and Roman numerals, for example? Or does it display digits? I remember the Casio digital innovations of the 1980s being really cool, especially next to the analogue old-school watches my parents used to make me wear. For us, being given a watch at age seven was a sign of maturity, and of being ready for school. And my parents refused to provide me with a digital Casio watch. To me, digital was really cool, and it made sense to use a digital watch because I could count to 12, but I didn't necessarily know how to interpret the analogue arms of an analogue watch. I learnt with time.

Today, many men (and let me focus on men for a moment) still wear watches — analogue watches. Why? We don't actually need them. We are surrounded by digital time. My iPhone, iPad and Macbook Air — even my kitchen stove — give me a more accurate digital indication of true time than my analogue Swiss Movado watch that I inherited from my grandfather Per. I regularly have to take this watch in to get it fixed, and have to manually wind it up so that it will keep going. I have to 'analogue calibrate' this watch on a daily basis. But I do this gladly, and with love. Perhaps not surprisingly, this 70-year-old piece was designed by a company with the motto of 'The art of time'. Notice it is not the *science* of time. The language is precise, the watch is not. My analogue watch contrasts sharply with digital clocks that actually know the accurate time. But I keep wearing Per's Movado. For now — and hopefully for a while longer (but I can never be sure, judging by its analogue arms...). Why do I keep wearing it? Is it just an expensive piece of man jewellery? No. We all want a little piece of magic ticking away on our arm.

Precision is not what we are after in an analogue watch. (Unless you're actually a pilot or a diver, which, let's face it, very few men of the modern era are. Despite our man-crushes on John Travolta, Leonardo Di Caprio and James Bond.) Accuracy is cheap because of the technological Quartz innovation in the 1970s that commoditised precision and accuracy. But our hearts are moved by pictograms and approximations — which is why we meet someone for coffee when both arms of the watch are most erect, at 12 o'clock, rather than at the digitally meticulous 11.57.

The analogue watch is also about story. The marketing for these luxury brands picks this up, and encapsulates the communication of timeless (excuse the pun) wisdom. In Patek Philippe's words, 'You never truly own a Patek Philippe, you merely look after it for the next generation'. This simple statement, set for an advertisement against a black and white family/son scene in, what looks like a wood-panelled office located in a Swiss chateau, tells the story of elegance, genes, heritage, timeless quality, curation, sophistication and priceless taste. It sets the watch apart from the idea of ownership and moves it into the space of legacy and guardianship. Smart move. Other watch brands communicate the story of the uniqueness of the movement, highlighting when the brand was established, its Swiss origins, and the blend of engineering and design. The ads highlighting these aspects tend to occupy full pages in upmarket magazines. That same story then gets transmitted to the in-store experience, which oozes analogue sophistication, and is centred on finding your 'identity fit'. Fascinating. We wear analogue watches because they connect with our hearts, not our minds. A digital Casio or my iPad could never speak to my heart like my grandfather's Movado does. There is a need for things we don't necessarily and rationally need. Accessory brands thrive on this.

Now, we too must learn from them. Analogue wins hearts. They speak the same language. Digital may rationally be the way to go, but analogue stays in the fight. Digital may democratise, but the analogue still intrigues. Digital is fast, analogue is slow. Digital gives you a snapshot preview, analogue is the film. Tension. Digital enables instant access, analogue requires physical effort. Check out this creative tension in figure 2.1. And remember — our rational minds may now be digital, but our somewhat irrational hearts are still analogue.

Figure 2.1: analogue versus digital

Analogue	Digital
Hearts	Minds
Face to face	Interface to interface
Facial recognition	Facebook recognition
High-touch	High-tech
Old school	New school
Personal touch	Digital connect
Nine to five	24/7/365
Slow	Fast
Word-of-mouth	Word-of-mouse
Offline	Online

Let's look at film and movies in a little more detail. Despite the instant access to digital downloads, we still crave the cinematic, analogue experience. While we as consumers still go to cinemas in solid numbers — attendance rose during the financial meltdown in Western economies — what is happening in the analogue setting of the cinema, which requires your physical attendance, and in movie production takes us back to the battle between the analogue and the digital. This fight is about how the movies get shot and, ultimately, about how they get delivered. During a consulting project in 2011 with the Independent Cinema Association of Australia, I learnt that there are two schools of thought. Of course. Digital (in the blue corner) and Analogue (in the red).

This battle is borne out in the 2012 documentary *Side by Side*, produced by Keanu Reeves, which juxtaposes the views of directors like Martin Scorsese with those of James Cameron and David Fincher. On the one hand, we hear that digital democratises and enables new forms of creativity and delivery. On the other, we hear that analogue film is the true art form; that it passes on the legacy of a profession in jeopardy of bastardisation, and a deeper connection to the audience. The latter sounds somewhat snobbish, with a total disregard for innovation and a blanket insouciance for the digital newcomer, while the former sounds slightly brash and disrespectful. While Reeves's documentary, incidentally, was shot in digital, it premiered at the analogue Berlin International Film Festival and featured at the 2012 Tribeca Film Festival. Interestingly, independent cinemas (which you'd think would ordinarily be on the side of analogue film) are able to compete more quickly, faster and more cheaply on digital terms with the big commercial chains. Why? Because film can be expensive, prohibitively challenging to access, and requires lots of physical space for storage. Digital film sits on a USB — at most. It levels the playing field. But the deeper point remains — we still crave connection, community, face-to-face meetings, the smell of the popcorn, the oily, salty fingers, the fading lights, the booming sounds, and the overall experience of cinema. Even when we could rationally enjoy the same content digitally via our iPads. Sometimes we just want to be analogue, and evidently some filmmakers use the analogue medium to tune into this analogue craving.

Digital minds; analogue hearts

We tweet, status update, and check-in in virtual life, but sometimes we forget about real life. Being present comes at a premium today. We are distracted from distraction by distraction. An intellectual diet of 140-character statements is the equivalent to a diet of fast food. Quick sugar rush, yes, but a creeping sense of void, of a lack of substance, depth and true sustenance. Despite the best efforts of the Twitterati to summarise our greatest works of fiction in tweets, I still much prefer to read *The Great Gatsby* in an old paperback. Why? Because it speaks to the heart.

Home delivery and fast food didn't kill fine dining and incredible restaurant experiences. They just meant people demanded more of the analogue, in-store experiences. There is massive opportunity for the analogue. Remember — our minds may have become digital, but our hearts are still analogue. This is critical. Digital minds; analogue hearts. Sometimes brands and leaders focus too much on the digital bling and forget about the analogue thing. In your rush to innovate, to be on the leading edge, to be connected, you must remember that you cannot throw out everything that's analogue. If you want to speak to the heart, retain the essence of the analogue. If you want to connect profoundly with your clients, and keep them loyal, speak to the heart. Digital minds, yes — but analogue hearts.

Smart digital brands get this. If I said 'Google' to you, would you think digital or analogue? Most likely, you'd think digital. Google is one of the great digital disruptors today. If you are forced into competition with Google, you may soon be out of business. Think of struggling or digitally disrupted brands like Yellow Pages, Nokia, *News of the World*, Borders, Kodak, Netscape, NAVTEQ and MapQuest. These have all been in competition with Google, and the company is such a disruptive force that there is constant debate about whether its success as a disruptor should be stymied because it's on the verge of engaging in anti-competitive behaviour — in other words, being overly competitive. While search engines were its original domain, Google has since expanded into advertising (Google AdWords), media (YouTube), books (Google Books), news (Google News), maps (Google Maps), data storage (Google Drive), digital photography (Picassa), social networking (Google+), operating systems (Android), computers (Chromebook), browsers (Chrome), mobile banking (Google Wallet), phones (Nexus) and telecommunications (GoogleVoice). What am I missing? Yet, no contact phone number is listed on Google's central home page, dedicated to organising the world's information. Google provides value to our digital minds, yet how does it connect with our analogue hearts? Maybe it's nailed the digital value piece to such an extent that it doesn't need to?

No, Google needs to. Remember — digital minds, yes; but analogue hearts. This is why I was amazed at what I found in the snail mail one day. The innovation was very thin. Thinner than the latest iPhone. It was revolutionary. It was a flash of innovation genius. You could hold it in your hand, between two fingers. Amazing. Insert whatever hyperbole you like. It was a printed, analogue voucher. An analogue gift voucher for $75 from Google — to spend with Google, digitally. An analogue voucher espousing the virtues of digital marketing. An analogue trial to get digitally hooked. A way for Google to be human.

If you're a consumer dealing with Google, when did you last connect with an analogue human being? If you're a small-business owner, when did you last connect with your analogue account manager? When did you last call Google because you couldn't find something on their website? When did you last contact them to complain because YouTube didn't stock your favourite movie? When did you last let them know that an AdWord advertisement took you to a store you hadn't really anticipated (which happens to me all the time)? In all likelihood, unless you have met them socially, you have never, ever, talked with human Googlers. They can be lovely people, and they even include an ex-flatmate among their number (and yes, we're still very civil). I am a big fan of Google and its employees. But they are very elusive. In some ways, the organisation is devoid of humanity on the surface. Digitised to the max. In a modern marketing move, reminiscent of the Tin Woodman's search for a heart in *The Wizard of Oz*, the digerati at Google stretched out and went analogue. Facebook soon followed suit. Analogue voucher. Digitised can be dehumanised. To connect with analogue hearts, even the digital disruptor Google had to inject some analogue into their client dialogue.

Apple's differentiated analogue play in a digital world

Now let's look at Apple. (Not the fruit.) Digital, right? Well, yes and no. All the other computer companies that Apple competed with at the turn of the millennium went digital in the way they connected with clients. Think of the brands you no longer use (unless you're forced to at work). These brands went digital in a big way. I remember ordering one of those brands' latest innovations digitally. It was amazing. No human beings were involved — at least none that I could perceive. And for the occasional introvert like myself, that was kind of nice. The website was reasonably user-friendly. It told me about the specs I needed, and I got to mass-customise my very own computer. It arrived from Asia a week later. Excited I opened the box, and found an instruction manual the size of the computer. An instruction manual I was going to have to look through to operate this piece of engineering ingenuity that had been designed just for me. I used the laptop for a couple of years, and even finished my law thesis on it. But I had zero connection to the brand. Digital convenience, yes; brand loyalty, no. Around the same time, Apple realised that, to compete, it couldn't be a heartless, pure-digital company. It needed to re-engage the hearts and minds of its consumers.

Apple's leaders asked themselves an important question. What is the best in analogue, high-touch, personal customer care? They did not see it in their industry. None of their competitors was a beacon of heartfelt interactions. In a brainstorming session, several members of the Apple leadership team referred to the Ritz-Carlton in New York. They described the client journey at this luxury hotel — the personal greeting at the door and the caring concierge who helped you through the entry. The escort to the check-in counter and the human, handwritten note welcoming you to your room. Perhaps the bottle of Champagne Taittinger waiting for you because they remembered you loved it on your previous visit. The great conversation over a glass of 18-year-old Bunnahabhain with the bartender, who could magically solve all of your life's problems.

What if a visit to an Apple store could mirror this experience? This is exactly the analogue standard they decided to set. This analogue representation of a digital brand that packages 1s and 0s has been critical in Apple's renaissance. Now, you enter the Apple store, and you're greeted by a concierge. This person escorts you to where you need to be and introduces you to the person who can help you check in. If you've already made an appointment to see the technology genius at the genius bar, this person will have been pre-briefed about your technological problems and, over a casual conversation over the 'bar', will solve them for you.

On 25 August 2012, Apple became the world's most valuable company at $621 billion. Of its total profit, 12 per cent is derived directly from this channel.[3] As of August 2012, Apple has 395 stores worldwide, with global sales of US$16 billion in merchandise in 2011. And they lead the United States retail market in terms of sales per unit area, outperforming Best Buy 4:1 and Tiffany's & Co 5.6:3.[4] In 2011, Apple Stores in the United States had revenue of $473 000 for each employee.[5] Analogue is working for Apple.

This has triggered a veritable analogue retail renaissance, with the usual digital suspects all making forays into analogue bricks and mortar: eBay in London and NYC, Amazon in 7–11 stores across the United States, Google with its Android Shop-in-Shop in Melbourne, and PayPal with its pop-up merchant meeting space in NYC. They have all been intent on bridging the online–offline divide, and to ensure a consistent brand feel. While these efforts have not necessarily been about opening up another direct sales channel, the brand presence and the signals these send are important. Banks and credit unions are also inspired by Apple's retail success. As you can imagine, this industry is one that has shifted from physical branch banking to desktop and mobile banking. In the process, banks and credit unions fear that we will lose our loyalty to their brands. Rightly so, because the connection with the analogue heart is critical to brand loyalty. Thus, if you visit an Umpqua Bank outlet or a People's Choice Credit Union, you will find that they feel a lot more Apple than old-school Bank of America or Westpac do. So for the nostalgics, the news is good. Analogue still has a role to play. When digital disruptors go analogue, and when banks aspire to be like Apple, there is a future for bricks and mortar. But the future may not look quite like the past. The future will look more like clicks *and* bricks.

Think about it this way. I love Skype. But it has limits. Ask anyone who has been in a long-distance relationship in the last 10 years. This digitised voice and video over internet protocol interface innovation has probably been both the saviour and disruptor of many romantic relationships in the past decade. It can serve to sustain the flame. It can conveniently enable you to have a live conversation that is surprisingly intimate. But a computer interface can never replace a human face. At some stage, you're going to want pillow-talk, or to hold hands, depending on your world view. You're going to want to smell pheromones, you're going to want to taste your lover's lips, you're going to want to find their long hair in your shower. Okay, too far — but you get my drift.

If you have never had a romantic Skype affair, replace the above example with your family members and grandparents who live overseas — perhaps Greece — and think about your Skype conversations with them. How they show you your distant pimply cousins via an old-school webcam, the way they hold up a plate of today's souvlaki, and their featuring of your vain aunt who keeps staring at herself rather than you, when you're speaking to her. (If you're not Greek, insert other nationality and cultural foods here.) At the end of the day, nothing beats an analogue big fat Greek wedding, a tipsy Swedish Midsummer's dance around a phallic symbol, a Chinese New Year or a stinking hot Australia Day. By cultural extrapolation, you get my drift. Skype, good; but analogue, critical. Full stop. A computer interface still cannot replace a human analogue face.

Analogue culture — analogue opportunity

This cultural piece is imperative to the discussion of the analogue — because culture, identity and community speak to the analogue heart. As I am writing this, I have been spending the last few weeks living in Williamsburg, Brooklyn. My Jewish cousins in New Jersey recommended that I check out some different parts of Williamsburg and Brooklyn, which contrast sharply with the hipster-esque part of the area where I have been staying. So I went for a jog. While my lycra-clad legs didn't raise an eyebrow in Bedford Avenue near the L train subway station, as soon as I passed by Marlow & Sons restaurant and went under the Williamsburg Bridge to the southern side of the steel construction, the demographics shifted radically. I stood out. My cousins, the Braunsteins, had told me that this neighbourhood would remind me of my visit to the ultra-Orthodox Mea Sharim area in Jerusalem in 2010. Indeed it did. While there are many differences, what I found intriguing were the commonalities, and the sense of identity and community in this area. The sense of dress was religious, even pious. The Hebrew language and its various fonts adorned shop windows, and occasionally Yiddish words sprung across deli windows. Food was branded kosher. Wherever I looked I was reminded of the analogue, of aspects of culture and identity that can never be digitised such as proximity, physical location, history, differentiation, belief and provenance. These all shaped what I experienced on my 6-kilometre route through Williamsburg on the southern end of the bridge.

This is not a story unique to Judaism. (My cousins would vigorously disagree — love you, Moshe and Hannah.) My jog could have been on a US marine base in Kaneohe, Hawaii, or through the Armenian Quarter in Old Jerusalem, in Harlem, NYC, on the streets of Palermo, Buenos Aires or through the playa at Burning Man. All of these neighbourhoods have their own sense of analogue identity — for religious, secular, spiritual, sexual, tribal and anthropological reasons. These things run deep. And the sense of coming together in a tribal fashion communicates one thing — analogue importance. Heart. Yes, you can maintain a sense of digital diaspora. But it's harder. We crave the sense of intimacy that the analogue brings. This provides immense analogue opportunity for business. (For a different spin on this, check out chapter 8 for a discussion on how analogue location drives digital innovation.)

Even what seem to be unspoken cultural traditions can be branded with the help of the analogue. Think of kosher and halal. Both involve the ceremonial slaughter processes of animals. Both have roots in religious texts. But you'd think in an age when few us are connected to agriculture, farms and meatpacking that we wouldn't want to know how an animal is 'harvested' — the industry's term for slaughtered. When you buy milk, you might opt for the organic full-cream milk with a picture of 'Daisy' on the front, and with a charming stamp from the local farmer who personally milked Daisy for you. But when it comes to your steak, some in the meat industry assume you don't want to know about the steak's life cycle, and how it was ended. Yet many do. Because we buy kosher and halal.

These two types of slaughter brands communicate that the slaughter took place in a certain fashion, with detailed associations of exactly how Daisy's throat was cut, what was said prior to the incision, and which direction she faced. Too much information? Well, this analogue information doesn't just have religious value, it also has commercial value — the US kosher market alone is worth US$12.3 billion. And not just for religious reasons, it turns out. As you can see in figure 2.2, quality, food safety and healthfulness surpass religious diet restrictions as the reasons given by consumers for choosing kosher. We buy the analogue story and the brand associations that go with it, even when it comes to a process that many of us find grotesque to think about (despite the process being one any omnivore is reliant on for protein intake).

Figure 2.2: US reasons for choosing kosher as percentage of total responses

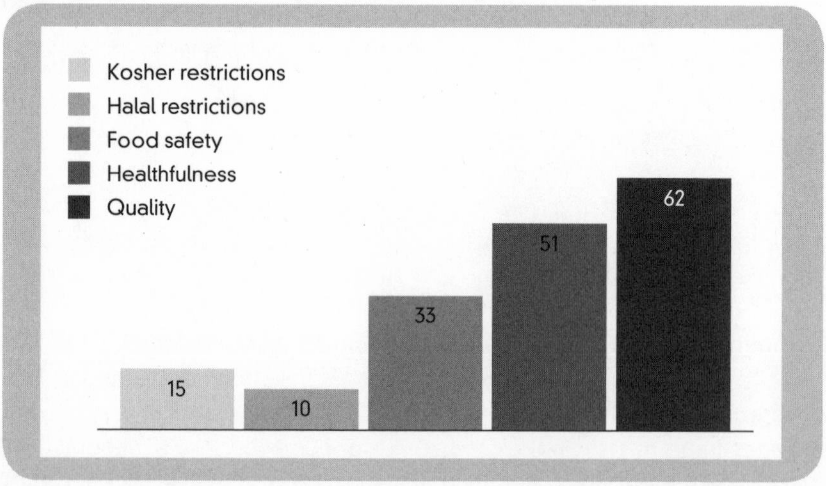

Source: Mintel Oxygen Report, February 2009.

Analogue provenance marketing

Think of these various types of cultural (ir)rationale (such as halal and kosher) as analogue provenance. Provenance is the place of origin or earliest known history of something — for example, *an orange rug of Iranian provenance.* It connotes the beginning of something's existence, and can also refer to a record of ownership of a work of art or an antique, and be used as a guide to authenticity or quality. The word comes from late 18th century French — from the verb *provenir* (meaning 'come or stem from'), from the Latin *provenire,* from *pro-* ('forth') and *venire* ('come').

Provenance marketing is the key reason only Champagne from Champagne can be called Champagne. It's why Parmigiano-Reggiano cheese can be branded as such only if it comes from the geographic provenance of Parmigiano-Reggiano in Italy. Eating Chicken Parmigiana sounds a lot more exotic than eating generic Chicken Swiss cheese or Chicken Bega. Analogue provenance — local origin creates a massive branding asset. Think of Swiss watches. German engineering. Turkish delight. Danishes. Great Dane. Colombian coffee. Texan Longhorn. King Island Beef. New Zealand lamb. Lebanese labneh. Brooklyn beer. Swedish massage. The examples are endless (even though the last one is of dubious authenticity).

So what industries are tuned into this analogue marketing secret? Generally industries and brands that command a premium — the brands that vigorously fight generics and commoditisation. *What do you mean?* You may be asking. *The more analogue, local and historical we get, the more we can charge?* Absolutely. Have a look at these examples:

- Think back to the analogue examples of watches for a moment, and about the connotations associated with them. Swiss-made. The Rolex movement. Established in 18*XY.* Patented engineering. Handmade. Limited edition. Artisan. Guildship. Craftsmanship. Precision. Attention to detail. Alps. Rich bankers. Heritage. Legacy. Provenance. Want to buy one?

- Let's switch to clothes — specifically, Icebreaker Outdoor Clothing. I provide more in-depth analysis of this brand in chapter 3, but for now focus on the key points — 100 per cent New Zealand sheep's wool. Transparent supply chain. Know the farmer who curated your Icebreaker top. Get to know the sheep who gave its winter wool. Scan the baaah-code. Okay, can I pay $200 for an undergarment please?

- Now look at premium beers. James Boag. Tasmanian handcrafted. From the pure waters of Tasmania. Natural. Against a silhouette backdrop of the Tasmanian wilderness. Feeling thirsty? Go ahead, you deserve one!

- How about premium meats? Tottori Wagyu Beef. Happy cows. Kobe, Japan. Massaged. Beer- and sake-fed. Want to pay 400 per cent more for your steak? I'd love to. New Zealand lamb. The grass is greener 365 days a year in New Zealand. Grass-fed. Organic. From the lands of 'Middle-earth'. Carbon-offset. Herb-crusted. Yes, please.

- And whisky. Scotch, of course. Bunnahabhain. Weird Scottish Gaelic name. Intriguing. From the Hebrides. James Bond. Since 1881. Islay Single Malt. Yes, on the rocks. And can you please throw a couple more logs on the open fire?

All of the examples in the preceding list represent analogue stories of provenance — and they're powerful stuff.

But does the savvy, digitally rational consumer of today really buy this? Yep — we swallow it hook, line and sinker. Think of water for a moment. The stuff you have on tap, sure. But, even more importantly, the stuff that's packaged in a branded bottle next to your hotel bed. Voss, Fiji, San Pellegrino, Perrier, Aquafina. This is analogue provenance marketing in a bottle — literally. And the fascinating thing is that many of the packaged water bottle brands are simply packaging tap water. Sometimes at a 10 000 per cent mark-up per unit of water. Some bottled water is more expensive than petrol. And, as shown in figure 2.3, we gladly pay for it. In Australia, sales of bottled water have been relatively stagnant over the last five years (in part due to the global financial crisis). However, this flat period came off a period of very strong growth from 2000 onwards, and IBISWorld predicts sales will begin to grow again over the next five years.

Why? Well, usually because of the provenance story. Look at the case of Fiji water, and their claims of being located thousands of kilometres away from the nearest industrialised country, the purity of the rain and being sourced from tropical aquifers — this all plays on our analogue heartstrings. Oh, and Brangelina happen to drink it. Lucky they didn't tell us about the bat that peed in the water. Despite our environmental misgivings, 9000 millions of gallons of the stuff were consumed in 2008.[6] Smart marketing, and highly irrational consumer decisions. If you package and brand something effectively, you can turn even a commodity like water into a premium brand. This is what can happen when you appeal to analogue hearts. Big business.

Figure 2.3: US bottled water sales, 2008

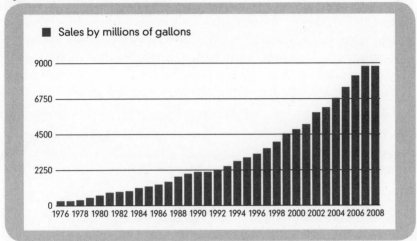

Source: USDA Economic Research Service and BMC, 2008.

What about flavoured water? Like beer. Once again, analogue story is on the rise. In the US, Prohibition limited the diversity of beer brands and, instead of destroying production of alcohol, concentrated power with big brewers, at the expense of quality, variety and, ultimately, consumer experience. As you can see in figure 2.4 (overleaf), prior to Prohibition there was a plethora of US microbreweries but, by 1930, the industry virtually didn't exist, and brewing power rested with the commoditised mass brewers. Since 1990, microbreweries have mushroomed, and they have surpassed 1900 levels to reach a 125-year high. Why? Microbreweries tell a local, analogue story of provenance. It's about craft, artisanship, family — above all, it's heartfelt. We perceive the quality to be higher and we imagine the taste to be superior to the big boys' variety. Again, we love to listen to the story of how the hop for the Lagunitas IPA grows and under what conditions, and we emotionally tune into the story of the guild of brewing professionals who care for each handmade bottle of beer that you consume. We even get excited about the yeast. I mean, who usually gets excited by yeast? Turns out, in this context, loads of people do. The microbrewery sales share in 2011 was 5.7 per cent by volume and 9.1 per cent by dollars, which means we like to pay more for micro than macro. Microbrewery retail dollar value in the United States in 2011 was an estimated $8.7 billion, up from $7.6 billion in 2010.[7] In Australia, the craft segment is growing at 42 per cent. In 2012, it constituted a 2.6 per cent by volume share of beer, but 3.7 per cent of its value. In the same year, 17 new microbreweries were added to the total of over 200 craft breweries in Australia. Analogue pulls both on our heartstrings and on our purse strings it seems.

Figure 2.4: decline and growth of US microbreweries

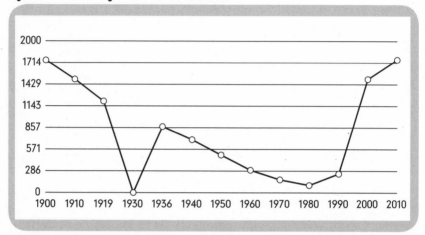

Source: Brewers Association, 2010, Boulder, CO.

What compounds this counterattack by the analogue is the idea of the *locavore*. Locavores are people interested in eating food that is locally produced, and that hasn't been moved long distances to market. Think farmers' markets. Farm-to-table restaurants. Apples with a couple of bumps. Direct trade. Single origin. Economies of small scale. Genuine. Personal. High-touch. Rural meets urban. This fascination with the analogue local is best encapsulated in a great sketch called 'Is the chicken local?' from the TV satire on hipster culture, *Portlandia*.[8] In this scene, Carrie Brownstein and Fred Armisen fret over the menu choices at a cafe, and ask the extremely well-informed waitress if she can tell them a little about the provenance of the chicken on the menu. Being a hipster, farm-to-table establishment in Portland, of course she can. She informs the couple that the place serves only 'local, free range, heritage-breed, woodland-raised chicken that's been fed a diet of sheep's milk, soy and hazelnuts'. Most people would be happy with that. The characters go further, however, and ask the waitress whether the hazelnuts, too, are local. How big is the parcel of land on which the chicken lived? The waitress comes back with a certificate from the farm and a photo of the chicken, Colin, who they'd be enjoying on this special occasion. After enquiring about whether Colin had lots of friends at the farm, the couple eventually stand up, without ordering, after deciding that they are going to visit this local farm.

Beyond the satire, and the commentary on this provenance trend, what is also ironically interesting is that what enables much of this traceability and transparency of origin is digitally enabled supply chains. Digital can enhance the analogue storytelling, as I illustrate in chapter 3. But the analogue, authentic story of provenance comes first.

Ten years ago, if I had said: Orange, Apple, Blackberry, you would have thought of fruits. In an ironic twist of digital branding, these three technology brands refer to something analogue, tangible, tasty and fresh. Remember — we mustn't throw everything analogue out when we go digital. The analogue speaks to our hearts. It's story, touch, identity, community and tribe. It goes deep. It's intimate. It connects with our analogue hearts. You must guard this brand equity and learn to package it better. Whether you're a 'faux old' analogue brand like FSC Barbers, a digital disruptor like Google, or provenance old like my mum's business, Georg Sörman, analogue is crucial. We wear our hearts (and our analogue watches) on our sleeves. Your analogue watch is a symbol of the continued importance of the analogue — an expressive finger to digital disruption. While the analogue heart may seem irrational, this irrationality commands an emotional premium. A premium brands around the world seek to command. That is why analogue matters.

Round 2 — Analogue Anachronism. We buy on emotion and then we post-rationalise the decision. Our rational minds may have become digitised, but our emotional hearts have stayed analogue. Both analogue brands and digital brands must connect emotionally with our analogue hearts, and this chapter has illustrated that there is plenty of brand equity and public support for the analogue underdog in this fight. The underdog plays it smart. He realises that he doesn't move as fast, as nimbly or as flexibly as his opponent but when he delivers a blow, it goes deep, real deep.

Ask yourself...

Think about the following with regard to your business:

1   What aspects of your communications can be digitised and which aspects must never be digitised? (List these in two columns.)

2   How can you differentiate your brand from the digital disruptors by telling the analogue story of your brand in an engaging fashion? (Create a one-page landscape storyboard of the interesting bits of your history and provenance.)

3   Which bits of your provenance can you use in your marketing that are aligned to your brand's positioning aspirations?

4   How do you best charge a premium for an analogue brand experience with you?

5   Which aspects of the artisan service or product you provide do you take for granted, and how could you better tell a story about it to educate your clients and prospects?

Apply it...

Use the following practical strategies within your business:

•   Think about how you'd digitise the analogue stories of professionalism, history, expertise and thought-leadership of your brand ambassadors—for example, how can you best use a one-minute video series to amplify your analogue story digitally?

•   Blog about your brand's expertise and the origins of it. Make sure you include key search terms, and optimise your blog with professional images that highlight the analogue sides of your business.

•   Engage a digital community via analogue means—on Facebook. Show retro images of your business (if it truly is old), or share interesting trivia about the profession to engage enthusiasts.

Endnotes

1 Unknown, 'Freemans family', FSC Barber, www.fscbarber.com/fscbarber/freemans-family, last accessed 8 April 2013.

2 K. Roberts, *Lovemarks: The Future Beyond Brands*, powerHouse Books, New York, 2006.

3 Unknown, 'Yearly and quarterly financial results', IFO Apple Store, www.ifoapplestore.com/stores/charts_graphs.html, last accessed 8 April 2013.

4 D. Segal, 'Apple's retail army, long on loyalty but short on pay', *NY Times*, www.nytimes.com/2012/06/24/business/apple-store-workers-loyal-but-short-on-pay.html?_r=0, last accessed 8 April 2013.

5 D. Segal, 'Apple's retail army, long on loyalty but short on pay', *NY Times*, www.nytimes.com/2012/06/24/business/apple-store-workers-loyal-but-short-on-pay.html?_r=0, last accessed 8 April 2013.

6 P. Gleick, 'The war on tap water: An exclusive excerpt from Peter Gleick's *Bottled and Sold*', Circle of Blue, www.circleofblue.org/waternews/2010/world/the-war-on-tap-water-an-exclusive-excerpt-from-peter-gleicks-bottled-and-sold/, last accessed 8 April 2013.

7 Unknown, 'Brewing industry facts', Brewers Association, www.brewersassociation.org/pages/business-tools/craft-brewing-statistics/facts, last accessed 8 April 2013.

8 'Is the chicken local?', *Portlandia*, www.youtube.com/watch?v=ErRHJlE4PGI, last accessed 8 April 2013.

3: Bringing the old and the new together

Two rounds down. Digital Disruption came out swinging in Round 1 of this bout. Analogue Anachronism looked like a wounded animal, but showed in Round 2 that this wasn't going to be a lay-down misère. Its hits connected deeply. To the surprise of the audience, and to the pleasure of those vying for the underdog, it looks like this bout may last longer than a few rounds.

In this round, the contenders will learn some new moves and see their roles in the ring changing slightly. They have been sizing each other up, tapping into their core capabilities and exposing their weaknesses. In this round, both tune into Muhammed Ali's mantra 'dance like a butterfly, sting like a bee'. We will investigate how to run an analogue marathon with a digital running mate with only 16 hours' preparation, why your digital mirror can modify analogue behaviours, and how Bavaria trumped all other German states by combining tradition and technology in its economic strategy. Furthermore, we will see how a New Zealand sports apparel brand made supply chain management sexy by merging the digital and the analogue, how beef branding is experiencing a resurgence because of RFID tagging, and why cyborg anthropologists are really interested in the behaviours of Swedish toddlers.

The convergence of the digital and the analogue

I am standing on the starting line of the 2011 NYC Marathon on Staten Island. I am unprepared. I'm nervous. In fact, 17 hours ago I didn't know I was running this race. Oops. How did this come about? A convergence of factors — including technology, phone calls, client engagements, mobile communications and a fortuitous travel schedule. If you were a little New Age, you might call it synchronicity; if you like Gwyneth Paltrow, maybe you'd call it 'Sliding Doors'. If you didn't like running, you might call it idiocy. It is a beautiful day in NYC, and I am off for a 26.2-mile jog. And I know the last 26 miles will be a challenge.

Okay. Let me backtrack slightly. What happened and how did I find myself on the starting line, together with 50000 jogging enthusiasts? I had left Sydney, Australia, for the United States to speak, facilitate and consult in Charlotte, North Carolina, and Indianapolis, Indiana. Luckily, as it turns out, I had brought my running gear. (I'd done so because I was preparing for what was going to be my first-ever Half Iron Man in Lake Taupo, New Zealand, which was due to be held on 8 December 2011, a month after my US visit. In other words, I needed to work out while in the United States.)

My US visit began in Charlotte, where I gave a presentation called 'Future Thinking' — without thinking about my own future 65 hours later. The conference was great, and I had a chance to connect with and share the stage with great thinkers like John Sculley, ex-Apple CEO, Randi Zuckerberg, ex-CMO of Facebook, and Vivek Kundra, President Obama's CIO. Afterwards, on Thursday afternoon, I'm at the airport on the way to NYC to see my friend Simon, and I receive a text from my Australian girlfriend at the time, who asks me whether I am going to be running the NYC Marathon this Sunday. Thanks to mobile communications, at least I am aware that the spectacle is taking place. For a moment, I think about participating and acknowledge it's a great idea, but quickly realise I have not been training for that distance and anyway — registrations probably closed a year ago.

Pling plong. Another text message. This time from Sweden. My friend CP — Carl-Patrick — is pregnant. Well, technically, his girlfriend, Therese, is pregnant. (In the egalitarian Sweden, with six months of paternity leave, Swedish men sometimes describe themselves as pregnant, which, of course, is a gross misrepresentation of reality, and says something about our sense of humour.) Because I'm just about to board the NYC-bound plane, I text him back to congratulate him, telling him I will call him in the morning. A short flight later, I am in Nolita at the Nolitan Hotel and enjoying a great late-Thursday dinner with Simon. I have completely let go of the idea of running the NYC Marathon. (At check-in the staff asked me whether I was in town for the marathon, and I responded no. I told them I was there for RnR, hanging out with mates and enjoying microbrewery beers. At most, I'd enjoy a Central Park Saturday jog.)

The following morning, Friday, I am in executive consulting calls the whole morning, but I manage to sneak in a quick Skype call to Stockholm. The bliss of VoIP. Carl-Patrick picks up. 'Carl-Patrick Höglund'. So formal — but it's good to hear his voice. He asks me what the weird number I'm using is and where I am. When I tell him I'm in NYC, he quickly asks whether I'm running the marathon this weekend. I am starting to get angry with this question. 'Nah,' I say, 'didn't even know it was on. Sorry.' CP tells me I should connect with his dad, aptly named Carl-Greger, or CG for short. CP and CG — like father, like son. CG is in NYC with a group of marathon veterans. Normally they run wine marathons in France (where they enjoy alternate drink stations, some with Bordeaux, some with Gatorade — very French), but they have got a crew together for this weekend. Sometimes, one of the old leathery guys pulls out. 'That's how I got in a few years ago,' CP tells me. Fantastic. I promise CP I'll call CG, give him a big congrats on 'being pregnant' and tell him to give Therese a hug for me. Skype disconnected, 1s and 0s and feelings and ideas exchanged.

Later that afternoon, just to have a stab at it (like the Aussies like to say), I call CG. He tells me I have no chance — everyone is fit for fight and, by the way, you can get expelled from future events if you sell your bib. Ah, OK. 'Well, CG,' I say, 'please pop me on stand-by consideration.' It's Friday night and my thoughts are more on NYC night life anyway, and which Village we're going to enjoy a splendid dinner at, than lycra and whether I need a new pair of Asics. *If it happens it happens*, I think. Before dinner I'm able to access the peer-to-peer network StubHub, and I digitally pick up Saturday night tickets for Simon and me to go to Madison Square Garden to see the New York Rangers play Montreal Canadiens. I also make a reservation for Saturday night at the Swedish Restaurant Aquavit in Midtown, so we can enjoy a nicely deserved degustation menu. I know — an elaborate and very NYC weekend. And clearly no plans to go running on Sunday morning.

Saturday begins rather late. Simon and I catch up downtown, go for a splendid walk around Battery Park and check out the building platforms on the site of the former World Trade Center. As the sun is starting to set at around 4 pm, we're chilling out with a couple of Brooklyn Lagers when my phone rings. A hidden number in the iPhone. It's CG, asking me if I'm still keen to run — Lars has a sore knee and won't be running. 'Can you be at our hotel in 45 minutes?'

'CG,' I say, 'can you give me five minutes to think about this?' Holy moly. Quick — honest — assessment of my fitness level. It's going to be a struggle. But, hey — how often do you get a last-minute chance to run the NYC Marathon? The distance I'd been training for is 13.1 miles (okay, some swimming and biking was involved too, but nonetheless...). The race is 26.2 miles. I check Google Maps. Madison Square Garden is 10 minutes from the pick-up location. *Bugger it*, I think. I am in. No more beers, lots of carbs at Aquavit. I have an early night while Simon and his mate Olof continue on. As I drift off to sleep, my last thought (just after *What am I doing??*) is *Well, at least I will have Nike+ by my side*. With this, I'll get encouragement, tracking, updates, and even cheers and matching soundtrack. This could work.

My digital self

What is Nike+? Why is it important in the context of my marathon? And why does it represent a convergence of the digital and the analogue? In 2011, I had begun mapping my runs via this iPhone application. So it can tell me that in 2011, I had run 178 kilometres, spread over 16 runs, that I had burnt 16 402 calories, and that my average speed had been 4:59 minutes per kilometre. I didn't grow up numeric. I do love analysis, patterns and trends though. Nike+ had become my digitised self. A digital mirror on my analogue life. Every step you take, every move you make, every breath you take, the digital world will be watching you, Anders. I know — this sounds very Sting (or even *1984* Orwellian), but to me, Nike+ was one of the earliest examples of the convergence of the digital and the analogue. And it was going to help me to run the NYC Marathon.

Digital and analogue working together in tandem. Analogue sweat, digital analysis. Digilogue. A combination of military GPS technology,

Google Maps, iPhone sensors, and Nike digilogue ingenuity enabled Nike+, not another human being, to be my greatest supporter on this run. Yes, of course, there was the atmosphere — the four rows of humans screaming, shouting and playing music all through Brooklyn's diverse neighbourhoods. The 'New York, New York' by Sinatra at Staten Island. The steel drums in Harlem. The excitement of the Central Park finish. But it was the pre-recorded cheers from Dirk Nowitzki, the thumbs-up encouragement via Facebook, and the timing and analysis from that little iPhone app that really got me through. It provided me with hindsight, insight and foresight of how I'd done, was doing and should do, what I needed to eat during the race, and whether I was meeting my targets.

This digilogue convergence contrasted with my previous, analogue marathon experience. In 2005, my brother Gustaf and I ran the Sydney Marathon. Yes, we wore iPods and had music. But the main analogue push of encouragement came from running together, side by side. We had set a target and we missed it — partly because we didn't really know how we were tracking. In six fast years, the technology had changed, and digital equipment previously reserved for elite sports teams and individuals had become an affordable consumer technology, enabling an amateur runner like me to compete in the NYC Marathon — and finish the race eight minutes faster, and with less training, than a race over the same distance six years prior. While I wish I could have experienced the marathon with my brother, this digital and analogue convergence signified something even bigger. The digitised self — the way we are now quantifying and digitising our analogue behaviours.

Digital insight affects analogue behaviour

This convergence of the digital and the analogue is a paradigm shift. And it has implications across the business world. Think about it. In a world where you can track, for example, exactly what you eat, whether you slept well, how many drinks you had last night, what your risk profile as a driver looks like, what your heart rate is looking like and what your blood sugar levels look like, this digital transparency affects how you do business and what decisions you make. This digital information can lead to analogue transformation. If you can see exactly how you behave, this tool can become the proverbial man or woman in the mirror — the person Michael Jackson told us we should look at.

Digital information can affect personal, analogue transformation. This is powerful on a personal level, because these tools can send you real-time data — which without tracking making it so easily available, you might ignore. It's also powerful because, just as an elite sports person has to have fitness tests before changing clubs, and before and after competition, your state of fitness can be constantly monitored by outside bodies. This digital and analogue convergence means insurance companies, financial advisers, medical doctors, the government, your employers and your family might become increasingly interested in your digitised self — as much as they are in your analogue self (or more). Management literature includes a saying that what can be measured can be managed. Self-management is now possible to a whole new degree. It helped me run a marathon. What could it do for you?

A lot it seems. According to the Australian Bureau of Statistics (ABS), if we listen to our analogue gut feel too often, we might blow our diets. In other words, our 'gut feel' or our analogue approximations are often wrong. Significantly wrong. Think about the last time you went to the doctor and she asked you about how many standard drinks you have per week. Or maybe that awkward conversation with a personal trainer on your first appointment when he asked you how often you work out. Or perhaps the tense conversation with your financial planner about your spending habits. Do you tell the truth, the whole truth and nothing but the truth?

Well, you might convince yourself that your facts are truly facts, but most likely they are not. Most likely they are a combination of white lies and optimism bias. Together with the Australian Productivity Commission, the ABS surveyed Australians and our estimations of spend in various areas of our lives. As you can see in figure 3.1, we grossly misrepresent to ourselves — and others — what we spend. When researching how much we actually spend when we say that we spend $100, the ABS and Productivity Commission found some large differences. When we say we spend $100 on transport, for example, we actually spend $104; when we say that we spend $100 on alcohol, we actually spend $158. And when we say we spend $100 on gambling, we actually spend a whopping $735. The convergence of digital digits and the mapping of analogue behaviours have massive potential to make us smarter, and the convergent sweet spot holds opportunity for business.

Figure 3.1: Australian perceived spend versus actual spend, 2012

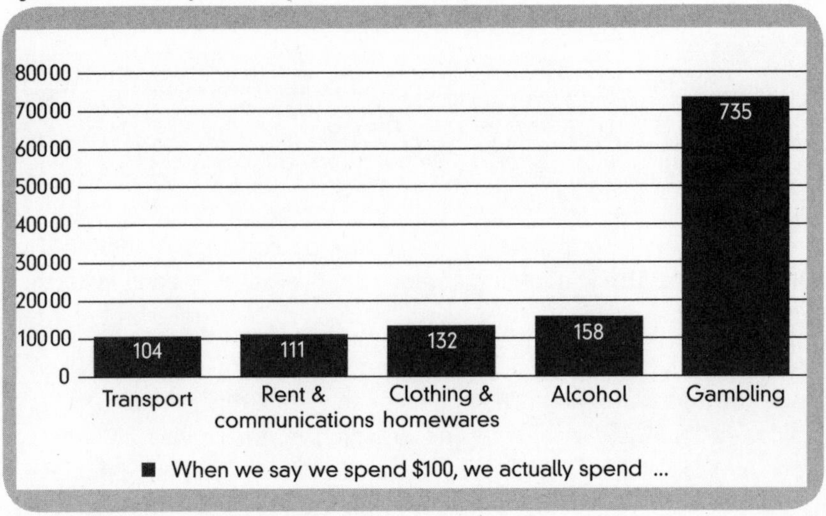

Source: Australian Productivity Commission, 2012.

Your digitised self is transforming analogue behaviours at a global level. On web platforms like PatientsLikeMe, patients are exchanging their clinical data, and sharing advice and experience in a peer-to-peer fashion. This digital information leads to analogue transformation. In surveys on PatientsLikeMe, for example, users of this peer-to-peer network report changing analogue doctors and adapting their analogue risk behaviours as a result of sharing digitised data.[1] The survey revealed the following:

- 41 per cent of HIV patients reported reduced engagement in risk behaviours

- 12 per cent of patients changed their physician as a result of using the site

- 22 per cent of mood disorder patients agreed they needed less inpatient care as a result of using the site.[2]

Imagine what this digital mirror could do for preventable diseases like obesity, for heart conditions and for certain degenerative diseases.

And it's not just health behaviours and data. Websites like Mint or Movenbank provide you with a digital mirror on your analogue financial behaviour. This provides you with financial hindsight, insight and foresight by visualising your financial data — credit card spend, mortgage repayments, savings rates, kids' college education savings, pensions, investments — all digitised in one place. Mint enables you to transparently view your analogue financial behaviours and how these affect you in the short and long term. For a person like me who is not naturally gifted numerically, the visualisation of data and financial numbers is critically important. This convergence between the digital and the analogue enables us to open our eyes to reality. It's digital information leading to analogue, personal transformation.

Digital smarts

At school, digits weren't my forte. I went to a German School in Stockholm and my inability to make sense of Herr Vanderleeden's Year 9 maths made me pretty unpopular. We had all been sent to this school to become logical, rational and practical, and grasping numbers was part of that story. A grade of 4 indicated a 'Genuegend' (sufficient) ranking but it was very close to a 5 — or a 'Mangelhaft' (deficient) — and I was nervously treading a fine line between the two. The fact that my German was not great in Year 9 might not have helped my understanding of digits either. Swedish person, speaking Swedish, but learning maths in German.

Shifting to Canberra, Australia, and entering the education system there helped slightly, as digits down under, in English, seemed to make a little more sense to me. However, I also had to sit back after school with a special maths tutor, Mr Still, and under his watchful eye I managed to achieve a D+. (The 'D' denoted capability; the '+' indicated effort.) Of course, A was the grade everyone strived for, but it seemed my report card always read something like 'Anders — sufficient capability, big effort; trying hard, but still not getting it'. Tough news, but I've accepted it.

So rather than educate myself about digits, I have tried to make the digit-savvy part of the population talk to me in my language — visuals, metaphor, trends, patterns, graphs and diagrams. To this effect, I have fired three accountants in six years, and let go off four different bookkeepers. Now, my finance team — a financial adviser, two accountants and a bookkeeper — all know how to translate digits into a language I can understand. A language that is more analogue. Excel doesn't compute with my brain. I apologise.

Digital, however, provides digits. It is based on data. Numbers — 1s and 0s. Ouch. And I love the digital age. I am fascinated by Digital Disruption. The great thing, though, is that data can be beautiful — only when we turn it into information, into knowledge and into insight, however. The translational sweet spot between digital information and analogue transformation and understanding happens on a computer interface, through programs like Roambi and Mint, and Amex financial overviews. These translational sweet spots and graphical user interfaces are seriously important. Without them, no convergence would be taking place. Apple's success largely lies in their ability to take something rational, logical, numerical and practical, and make it beautiful, easy, empathic, intuitive and fun. They essentially package 1s and 0s, and it just works. No instruction manual. Just have a crack. Nike+ takes the millions of bits of information that your running, your sweat and tears generate and they turn it into heat patterns, GPS coordinates and Dirk Nowitzki cheers. Powerful. I wonder what would happen to education if more teachers had this translational capacity. How much education wouldn't get lost in translation? Intelligence only flows when we can decipher patterns, see trends and view the digital forest that the digital data trees construct. The convergence is only powerful to the extent that it translates understanding, and two different — digital and analogue — modes of thought. The profound opportunity here is that digital provides analogue insight, and when that insight is powerful enough, we change and adapt our behaviours.

Laptops and lederhosen

After reading the preceding section, you may think it perhaps ironic that I now refer to German convergence as an example of this translational sweet spot. Specifically, I want to chat about the state of Bavaria — the fiercely independent, Catholic southern province of Germany, just north of Austria. The Texas of Germany. A place that you may associate more with tradition than technology. In Year 4 with Frau Eriksson, I was asked to pick a German Bundesland (state) to give a presentation on. I picked Bavaria, or Bayern as it's called in German. I can't remember exactly why I picked it, but I think my dad might have had something to do with it. I do remember Dad helping me out with the presentation and, as I recall it, the cover of my binder was Blue and White, the colours of the state — we even pimped the presentation booklet with some blue and white ribbon. It was a geography assignment, and I remember scoring well on it. I referred to aspects of Bavaria's geography, but also its culture and history — lederhosen, Oktoberfest, the Alps, strange hats, awkward foot-slapping dance moves. Old-school companies like BMW, Audi, Munich RE, Siemens and Adidas. Tradition. Old-school manufacturing. Heavy industrial base.

My assignment didn't once refer to technology. But today, Bavaria personifies the convergence of tradition and technology. The analogue and the digital. And it's not just because a retailer launched a pair of lederhosen with in-built MP3 technology at the Consumer Electronics Conference in 2007.[3] The state has had an outspoken policy of 'laptops and lederhosen' since the late 1990s. Technology and tradition. This policy of combining the old and the new, the analogue with the digital, is credited with the re-emergence and relevance of Bavaria as a German economic powerhouse in the digitised 21st century.[4] The state is renowned as having the highest proportion of its workforce employed in the high-tech sector in all of Europe.[5] In nine years, under the governance of Edmund Stoiber, the state grew its GDP by 18.5 per cent, which was the biggest increase of any German state in the same time period.[6] It is now the fourth largest biotech centre in the world, after Silicon Valley, Boston and London–Oxford.[7] Laptops and lederhosen, eh. Convergence and combination. Tradition and technology. A translational sweet spot that goes hand in hand with the digital disruption we are currently experiencing.

Digital and analogue convergence

The translational sweet spot in this convergence between the analogue and the digital, in the connection between analogue hearts and digital minds is where digilogue comes in. As you can see in figure 3.2, the analogue circle overlaps the digital circle and they prop up the adaptation circle, which overlays the bottom two circles. Adaptive companies get digilogue in their ambition to provide value to digital minds and to connect with analogue hearts. Digital versus analogue doesn't have to be an either/or proposition — it can be an and/also proposition.

Adaptive companies understand this translational sweet spot between the digital and the analogue. And increasingly, customers are demanding this convergence of the digital and the analogue. It's about doing what's right for them, not necessarily what's most convenient or easiest for you as a businessman or businesswoman. The challenge for you is to become versed in a new language, and translate digital into analogue, and analogue into digital. Adaptation and true connection with both digital minds and analogue hearts happens at the digilogue sweet spot. Adapt or fade into irrelevance.

Figure 3.2: digilogue convergence

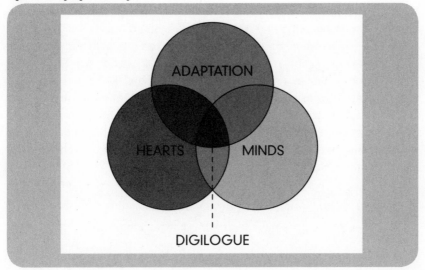

There was a beautiful ad on television a few years ago for Berlitz language education. It featured 'Hans and Frank' in the German coastguard, both sitting in a bunker on the North Atlantic German coast. It is Hans's first day of work and he is a little nervous, chewing his nails and biting his lips. Frank is the old sturdy German guy who is instructing his younger apprentice in coastguard technology. 'Das ist das wichtigste Gerät, das ist das wichtigste Gerät, das ist das wichtigste Gerät', and so on, meaning that is the most important piece of technology you need to be aware of today (so as not to stuff up on your first day of work). Hans is staring overwhelmed at the beaming lights and the flashing radar, and trying to make sense out of all of these instructions.

Feeling like he has completed the briefing, Frank gives Hans a slap on the shoulder and is off to get a soy latte (or something). So, there's Hans by himself, first day of work, German coastguard, looking rather uneasy and praying that Frank will get back soon. All of a sudden a crackly message comes through on the transmission: 'Mayday, mayday, mayday, mayday. This is the British fleet commander' (in thick Essex accent). Hans responds nervously (in a thick German accent): 'Hallo, zis is ze Tcherman Kostkard speaking'. The British commander screams, 'Mayday, mayday, we are sinking, we are sinking!' Hans, looking slightly more relaxed, says, 'Vot are you zinking *about*?' Occasionally messages get lost in translation. In the convergence between the digital and analogue languages, it's crucial that your message translates seamlessly across the digilogue sweet spot.

There is a deep story of transformation in this convergence. In my strategy consulting, and the work we are doing at Thinque with organisations on change leadership, what we find across the board — from FMCGs (fast-moving consumer goods) to retail, from pharmaceutical to high-tech — is that only when organisations recognise the old school, the traditional, the analogue, the timeless wisdom of its people, can transformation happen. You need to acknowledge the hard work that got your business to where it is today. Pay respects to the old guard whose sweat led to historical successes. Give credence to the culture and brand equity built up over generations. This is imperative. (I provide more case studies on this throughout the book.)

Retaining the analogue is a critical component in getting people to buy into and accept new behaviours as a way of meeting Digital Disruption head on. Could Bavaria have gone high-tech without remaining high-touch? Maybe. But the combination of 'laptops and lederhosen' attracts both hearts and minds. It provides value to digital minds, and it connects with analogue hearts. Gen Ys and Xers can get behind the idea. It makes sense and is inspirational. Boomers and veterans feel respected, and are re-energised. The policy becomes an identity for the emerging Gen Zs. The old guard don't feel like someone is about to throw out the baby with the bathwater. Analogue tradition can be told via digital media. The policy of laptops and lederhosen recognised that the more some things change, the more other things stay exactly the same. Adaptation can begin when it sits on a solid footing of hearts and minds, and at the convergence of hearts, minds and adaptation lies digilogue. Digital minds, analogue hearts. Helping Bavaria to adapt to changing times and positioning it for the future. Inspiring.

Icebreaker: laptops and lederhosen in a New Zealand context

How can a company apply this convergence of the digital and the analogue? On that late autumnal marathon day in NYC in 2011, a New Zealand–made Icebreaker hoodie was keeping me warm. Natural, high-tech fibres. What? Yep — merino wool from the Southern Alps of New Zealand; 'Middle-earth' country. *You wear wool when you run?* you may ask. Yep, of course. I received this amazing sporting hoodie at TEDGlobal 2011 in Edinburgh, and I love it. It keeps me warm in winter, and cool in summer. It's made out of all-natural fibre, yet it's still high-tech — it has to be, because Icebreaker competes with the best synthetic outdoor sports apparel and clothing companies in the world. It's a competitor of

Patagonia and North Face. The company is truly digilogue. It is channelling the energy from the convergence of the digital and the analogue. How? Its founder, Jeremy Moon, a cultural anthropologist, was given a pair of pure merino jockeys in 1994. At the time, merino wool was considered such a commodity that it was frequently blended with traditional wool. Sound itchy? Well, they were not, and Jeremy's business idea was hatched, revolving around nicely designed undergarments and performance garments for tough environments. Designed from nature. Utilising the adaptation of the merino sheep to the New Zealand alpine environment. Taking a thread that had been used by Italian suit makers for decades, and applying it in a different context. Genius. As an anthropologist, Moon was tuned into the consumer insight that outdoor people and nature fanatics want natural materials, preferably sustainable eco-materials. But they also want high performance and, up until the rise of Icebreaker, synthetics had had some advantages on the performance side. Moon realised that a natural material that had evolved in cold winters and hot summers would be ideally suited to outdoor apparel. He would, however, have to take a high-tech approach to this high-touch Merino wool.

This digilogue approach enabled Icebreaker to double in size during the global financial meltdown. It has won both hearts and minds around the world. The company, privately held, sells its outfits in 3000 stores across more than 30 countries. At a time when retail across the Western world has been digitally decimated, Icebreaker has gone from strength to strength. The marketing focuses both on how the clothing is worn, but also how it's born. This enabled the company to turn a feature with Sarah Palin — sporting one of its jumpers — on the front cover of *Newsweek* to its advantage. The title of the article: 'How do you solve a problem like Sarah? She is bad news for the GOP — and for everyone else, too'.[8] Moon's reaction: Palin looked both 'smart and hot'.

The company is extremely focused and strong at each stage of its supply chain. Its brand storytelling begins with its supply chain. That's right — it uses the unsexy area of supply chain management in its branding. If you, like me, sport an Icebreaker jumper, you can turn the garment inside out, and find a barcode or, as the company calls it, a baa-code. This code enables you to track your garment back to the farm from which the merino came. On the website, you can digitally access a YouTube video featuring the farmer, his family and their sheep. You can view maps of the location, and its history and relevant statistics. What makes the story and brand so engaging is that the story is traceable and not mere

_____

_____

spin. Icebreaker prides itself on having long-term partnerships through the supply chain, working with merino wool farmers on a long-term contractual basis. This gives the farmers certainty and market access into the future, and ensures supply and high quality for Icebreaker. This also enables Icebreaker to track batches of wool in a digital way, so that it can tell its analogue marketing story. Story through supply chain — high-tech, high-touch. Phenomenal stuff. This has won Icebreaker hearts and minds throughout the world.

This story — how an animal that grows in rugged, natural conditions can help humans escape into rugged, natural conditions — involves something pastoral. A sense of analogue human and animal interaction. Of being at one, and of disconnecting from the high-tech urban world by reconnecting with nature. It's also about form and function; design aesthetic and functional ruggedness. This digilogue branding has helped turn a commodity, merino wool, into a premium product. And the product doesn't come cheap. But it engages digital minds and analogue hearts, and, of course, we want to pay extra for the jumper brought to you by farmer Ray Anderson from Branch Creek and his sheep.

One of the things I love about the Icebreaker story is that the new enables the old to live on. History gets to live on because the company is future-minded. It uses technology to tell the story. It uses technology to transparently market its supply chain. It uses technology to shine a torch on the otherwise media-shy farmers they collaborate with. It is not surprising that the company is led by an anthropologist like Jeremy Moon. He thinks about anthropology — the study of humankind and, in particular, the comparative study of human societies and cultures and their development[9] — as much as he thinks about technology. Anthropologists like Jeremy Moon, and my friend Michael Henderson in New Zealand, are by their own admission obsessive about observation.[10] Through observing consumer patterns, the value chain and partnership dynamics, and the tribal needs of farmers, Moon was able to spot an opportunity. As my futurist colleague Mike Walsh says, in our rush to innovate and think about the future, we mustn't just think technology — we must also think anthropology.[11] In this instance, the anthropological observation of the old, of culture, of story, of heritage, of place and of provenance has been amplified by technological savvy. Digital and analogue converging. This is critical not just from a business perspective but also from a cultural perspective, because what Moon enables here is culture, tradition and a particular language to be handed down generationally. The new school enables the old school.

The importance of translation in the continuation of the analogue story

Around the world, old analogue languages are at threat. I do not mean this metaphorically. Ancient cultures and languages are becoming extinct every day. Out of the world's 7000 languages, nearly half are considered at risk of extinction in the next century. Enduring Voices, which is backed by the National Geographic Society, highlights that if most of the 2800 to 3500 endangered languages do die in the next century, this works out to about one language lost every two weeks. This means that the wisdom of elders will never be passed on to younger generations. The heritage bloodline will be cut. Ways of being will become a thing of history. This is a sad loss of diversity, and an indictment of development. But it needn't be this way.

We've all heard of languages that have died out because they only lived in oral tradition. When the language isn't spoken anymore, the language and the tradition dies with it. Symbolism, written alphabets and records of language are critical for their ongoing sustainability and survival. Technology enables this transfer of history into the future. If only a small percentage of the 24000 hours of YouTube videos uploaded every day[12] were about the passing on of tradition, culture and language — the analogue — these aspects would live on. The challenge is that analogue and digital speak different tongues, and often seem diametrically opposed. Until elders and leaders start following the lead of high-tech anthropologists like Moon, we will see more and more cultures and analogue stories dying. Let us use the new to tell the stories of old. Just like Icebreaker, just like Bavaria.

Back to the future: farmers' futures

So, the lesson is to use the convergence of the digital and the analogue to your advantage. So that tradition, culture and the analogue can live on. This is applicable cross-industry — and to prove it, let's look at one of the prima facie unlikeliest of places where the digital might support the analogue. Agriculture — even more old school than textile manufacturing from merino wool.

However, while agriculture sounds very analogue at face value, it has gone very digital over the last decades. The industry has implemented strategies such as precision farming, GPS tagging, RFID, robotics, digital organics, futures trading and traceability. (And we've also seen the growth of FarmVille.) So let's consider a challenge that one of my clients' core clientele has been having. This company is a technology vendor to the beef industry, and its technology assists consolidated farms and family farms track, measure and analyse their herds of beef, spanning the entire supply chain. The farmers, specifically feedlot owners and managers, are facing a tough environment. Pressures include increased volatility in beef commodity prices, the concomitant commoditisation of brands, increased input costs as a result of droughts and shifting productions habits, consolidation of buyer power, fragmented supply chains, social media scrutiny, green trends and increasing consumer demands on product quality. It's a tough world out there for them. Yet, there are solutions, and technology is one. Going back to their roots is another. In fact, the solutions lie in the convergence of the digital and the analogue.

It is perhaps ironic (or perhaps fitting) that beef has become commoditised — since branding began with beef. The word *brand* comes from the Old Norse *brandr*, which means to burn. The custom of branding cattle that originated from a certain farm is the origin of modern-day branding. Farmers used to burn their farm trademark into the hide of the cattle to ensure neighbouring farmers and buyers would know the provenance of the cattle. Maverick branding — which Sam Augustus Maverick lent his name to — was the opposite of this, and involved not branding one's cattle. Maverick was considered an independent-minded Texan, and his cattle, which were unbranded and thus seen as generic, would roam the plains. History books have two explanations for Maverick's reasoning. One is that Maverick, a lawyer by training and a businessman by choice, had little interest in farming his inherited cattle, and thus couldn't be bothered branding them and instead let them roam freely. Sceptics argue that Maverick didn't brand his cattle because of the fact that most everybody else did. Thus, Maverick's folks could collect all accidentally unbranded cattle and claim them as their own, even if they strictly may not have had the same provenance. On balance, the history books lean toward the former explanation. Either way, the word *maverick* is now used to connote both unbranded calves and people who challenge the status quo.

So branding began with beef, and now beef has become commoditised. This makes it challenging for farmers to extract a premium for quality. It also makes them very susceptible to market fluctuations and a volatile commodities market in beef. With corporatisation of farming, hyper-efficient productivity, and a relentless focus on yield and weight gain, my client's core clientele finds itself with little wriggle room to extract better margins. At the same time, buyers like Walmart and McDonald's demand more high-tech traceability and carbon footprint reports from the farmers, and consumers want locally produced beef and organic solutions. Some want local farmers' markets, where they can meet the person who looked after Daisy the Cow and Oscar the Ox. Rarely, however, do either the corporate buyers or the end consumers want to pay a premium that matches the added costs involved with these value-adds in the supply chain.

Technology offers a potential solution for these farming folks. Think digital and analogue convergence for a moment. Think Icebreaker for beef farmers. Imagine that they used GPS and RFID tagging to ensure total transparency and traceability. This would enable more efficient administration of feed and animal health products, and provide analytics of weight gain and yield. It would also enable the farmers to tell a story of provenance. Technology would enable them to seem small-scale again. To be 'local' but potentially act 'global' in scale. To tell the story of Oscar from Amarillo, Texas. To show the farm on Google Maps. To tell the world via YouTube about Sam who reared and looked after Oscar. Meanwhile, Walmart and McDonald's could verify the information, and feel certain that the supply of meat would be high quality, from animals in premium health, and see a reduction in social media risk. This would bring branding back to its origins. Technology enables differentiation and de-commoditisation. What is required is a mindset that converges the digital and the analogue.

Large multinationals use social media so that they can seem local, personal and grassroots. Technology enables them to do this. Similarly, technology now enables tradition and the analogue to be personal on a much larger scale. It used to be that you went and bought your meat, eggs and chicken at the local farmers' market. You used to know the farm they came from, and perhaps your kids went to the same school as the farmer's kids. On a small scale, historically, the farmer might have been able to tell you where the meat came from, which chicken's eggs would make it to your plate, and which batch had made the greatest sacrifice of them all to end up as your chicken parmigiana. But on a larger scale, this gut instinct became impossible and inaccurate. Because of urbanisation and the corporatisation of farming, few urbanites come into direct contact with the person who makes their food.

However, knowing your farmer, being a locavore and trusting the origin of your food supply is still critical. The more some things change for real, the more other things stay exactly the same. In a digitally disrupted world, we still crave the story, the local and the traditions of animal husbandry. Technology enables the return to tradition. It enables farmers large and small to tell a better story of the analogue history and origin of your foods. The new school enables the continuation of the old school.

Anthropology and technology

When we observe human behaviour and our convergence with technology, what becomes evident is that technology has enabled a mental extension of ourselves. While humans have always used tools like hammers, nails and boomerangs to extend our physical selves, mobile phones, laptops and smart phones now extend our mental selves and leverage us in news ways. What this means is that we are becoming cyborgs. A *cyborg* (cyber-organism) is not just the Terminator or Robocop of Hollywood fame that you might think of, but rather is 'an organism to which exogenous components have been added for the purpose of adjusting to new environments'.[13] Cyborg anthropologist Amber Case argues that every time we use a mobile device or computer, we are behaving like cyborgs because we are using an exogenous component for the purpose of adjusting to new environments. These devices don't enable us to hit something with greater physical force, propel a strangely shaped wooden tool through the air in a strange arc or hold heavy pieces of wood together, but they do enable us to beam ourselves onto Skype windows or Cisco Telepresence screens, and into a digital Second Life, where we might be represented even while we sleep. While digital immigrants are still entering this cyborg world, digital natives have grown up cyborg.

Digital natives — this group includes millennials (born during the 1980s and early 1990s) and generation Z (born 1994 to now) — inherently represent the tension between old school and new school. In the social media game FarmVille, 500 million acres of land are under management in the virtual world at the time of writing. Compare this to the 950 million acres of land that are under real management by real farmers in the United States. There are more virtual farmers in FarmVille than there are real farmers in the Western world. Digital natives access virtual rodeo games on Xbox Kinect, while rodeo television numbers are experiencing a

19.7 per cent increase year on year, as the style of farming of yesteryear (represented in some way by rodeos) is growing increasingly attractive to a younger generation of viewers.[14]

Digital natives speak digital without an accent,[15] as you might recall from chapter 1, in contrast to a generation of digital immigrants who speak digital with a heavy accent. In this context, author and humourist Douglas Adams made the observation that:

> Anything that is in the world when you're born is normal, and ordinary and is just a natural part of the way the world works. Anything that's invented between when you're fifteen and thirty-five is new and exciting and revolutionary, and you can probably get a career in it. Anything invented after you're thirty-five is against the natural order of things.

If you are part of generation Z, born after 1994, you do not know a world without the internet. (Or maybe your kids are part of this group.) According to AVG IT Security more two to five year olds in the Western world now know how to play Angry Birds on their iPhones than know how to tie their shoelaces.[16] Contrast this with the millennials or Gen Y who grew up hybrid. The internet hit these guys in a big way at the time of puberty, so it has shaped them in a way that is consistent with Douglas Adams's observation. The future of farming may well be a case of returning to the analogue past, and using digitisation to better tell a story of how our food makes its way from farm to fork.

The convergence of the digital and the analogue, of technology and anthropology, plays out in provocative ways. Let me share an example with you. On a December 2012 visit to Whatif Innovation on 2nd Avenue in NYC, I met with Katie Hillier, who is a digital anthropologist. Yes, this is her job description. Katie is 29 years old, which places her squarely in the millennial generation (born between 1977 and 1994). She is an anthropologist and sociologist by training, but digital has become her career — she applies her powers of observation digitally. This means that Katie represents brands on behalf of Whatif Innovation, and her job is to digitally observe consumer behaviour and see patterns, both quantitative and qualitative, in the way consumers interact with brands through technology. The old-school anthropologist used to wear khakis and, perhaps armed with a Moleskine diary, travel deep into the jungles of South America, say, and take Ayahuasca with native tribes. These days, Katie hangs out in digital jungles. And respect to her. Because of her, big multinational brands are gaining new insights, digitally, about consumer behaviours. It's anthropology and technology.

At the older end of the spectrum in Adams's observation is my mum. Let me share two stories. Remember — anything invented after you're 35 is against the natural order of things. At the time of writing this chapter, I got to hang out with my mum during a family holiday. As I wrote this line, I had a cold and, because mum was trying to fulfil her maternal duties, technology met tradition in a way that, let me just say, didn't quite work out. The house at Margaret River, Western Australia, that I had been renting had a tea kettle, which sat on a base that is attached electrically to the wall. You probably know how to use such a kettle — whenever we needed to refill the kettle, we just lifted it off the basin, popped the lid and filled it up with water from the rainwater tank. Then we popped the kettle back onto its base, flicked a switch and the water heated to its boiling point. Of course, with kettles like these you don't need to place the kettle on a gas stove over a naked flame like you may have needed to do in the past. But this is exactly what my mum did — bless her heart. She removed the kettle from the base, filled it up with water, and promptly proceeded to place it above a naked gas flame to the point when the whole contraption was on fire and all the plastic had melted. Tradition and technology didn't blend in this instance. In the end, she poured water into a saucepan, used that to boil the water and (finally) made me a ginger and lemon tea. I don't want to seem unappreciative here, but I think you get my point.

At other times, she is extremely digital. When consulting with her face to face in November 2012, I shared my laptop screen with her, because she was sitting next to me in Stockholm. She has grown very accustomed to the squeeze and pinch on her iPhone (and for this I am very proud). However, on this instance, she was having trouble seeing what was on my screen, so she kept taking her thumb and forefinger and trying to squeeze and pinch the screen to make it larger and more visible. Now

I am sure this touch screen capability is being developed in a computer lab somewhere in Silicon Valley or Foxconn, so my mum is being a bit futurist in this regard. But it does also show you that she speaks technology with an accent. She is trying to speak digital, but with some grammatical mistakes.

Digital natives and cyborg anthropologists

At the younger end of the generational spectrum, we are seeing a true convergence of the digital and the analogue. Take a look at one of the leading-edge tech countries in the world, Sweden, to get a sense of what the kids of tomorrow, born digital, may grow up to be. Here are some statistics about this new digital generation:

- 93 per cent of children and adolescents between 9 and 16 years of age are online at least one hour per day; among 2 to 9 year olds, 51 per cent are online at least one hour a day.

- 84 per cent of the 9 to 16 year olds play online computer games at least one hour a day; among 2 to 9 year olds, 65 per cent play online computer games at least an hour a day.

- 87 per cent of the 9 to 16 year olds use their mobile devices at least one hour a day; amongst 2 to 9 year olds, 16 per cent use their mobile at least one hour a day.

- Between 2009 and 2010, the use of the internet in this generational cohort mushroomed by 40 per cent (because of smart phone mobilisation).

- Children between 12 and 15 years of age utilise the most internet access points — stationary computers, laptops, mobile devices, TV Wifi and gaming consoles — of any age group in the Swedish populace.

- 96 per cent of all 12 to 15 year olds have access to a laptop, compared to 74 per cent in the population at large.

- 85 per cent of 12 to 15 year olds have access to the mobile internet, compared with 55 per cent of the population at large.

- 50 per cent of all three year olds, and 40 per cent of two year old Swedes, use the internet.[17]

How will their cyborg childhoods impact them and us mentally? This remains to be seen, but it is illustrative of the convergence of the digital and the analogue worlds.

Of course, this is not just a Swedish story. One of Apple's most successful advertisement campaigns globally in the last five years was inadvertent — it wasn't even generated by Apple. If you do a YouTube search for 'baby' and 'apple ipad', you will find a bunch of user-generated videos of babies and toddlers interacting — meaningfully — with the iPad. They are born digital. Remember, digital minds, analogue hearts. The irony here is that today's infants may be born digital, but they may grow up playing FarmVille, rodeo and Lord of the Rings digitally to connect with analogue tradition.

Round 3. A draw. Seemingly the two can dance and spar together. The boxers embrace towards the end of a round that is more like a dance than an assault, more capoeira than boxing. Their sweat and saliva intermingle, and neither is able to deliver a meaningful blow. There is a moment of recognition that they are both in this together. The referee has to separate the two boxers who have been dancing like butterflies more than stinging like bees in this round.

We've seen in this round that digital and analogue combine, co-conspire and intermingle in successful business models — from Nike Plus's contribution to unprepared marathon running, to Bavarian futurism inspiring biotech and cultural sustainability; from New Zealand supply chain storytelling, to a re-engineering of maverick beef branding through beef tracking technology. In the analysis it's becoming evident that the more some things change, the more other things stay exactly the same. We are connected to old-school stories, languages and plotlines via our digital devices. Laptops and lederhosen. Digital mirrors on analogue behaviours. Analogue marathons, with digital heat maps. Analogue animal husbandry converged with technology traceability. Anthropology combined with technology. Digital natives versus digital immigrants. And, of course, baa-codes. Digital and analogue are converging, and often one enables the other. Sometimes, the digital helps to modify analogue behaviours. But, most importantly, we need to remember to find the correct balance between the digital and the analogue. Just like Bavaria — think laptops and lederhosen.

Ask yourself...

Use the following to get yourself thinking:

1   Where does the digital and analogue converge in your business?

2   What do you see adaptive companies in other industries doing to monetise this convergence of the digital and the analogue?

3   What concrete opportunities does the digilogue present for your business to connect with new customer segments and their various communication preferences?

4   If you were selling change and adaptation internally, how could you use convergence as an argument for internal change?

5   How can you offer hindsight, insight and foresight to your clients based on the digital digits that they provide you with?

Apply it...

Here's how to put your thoughts into practice:

•   What would be your brand's adaptive rallying cry—your 'laptops and lederhosen'? What would culturally gel internally? Write it down.

•   Map out your key digital and analogue touch points with your clients—listing analogue on the left and digital on the right. Work out which ones would benefit from convergence, or being translated, so that you become easier to work with.

•   What are three examples of great businesses you enjoy being a client of? What specific digilogue customer service do they provide you with that makes you truly connect with the brands. Apply these ideas in your own business.

(continued)

Apply it... *(cont'd)*

Digilogue convergence

Now take your strategies further with the following:

*   Use this digilogue Venn diagram and, in each circle, start mapping down how you are currently communicating with both the hearts and minds of your customers.

    As a way to get started, here's what you could list under each area for the Icebreaker brand. **Hearts:** story of farmers, story of the origin of the wool. **Minds:** supply chain transparency, natural fibres for natural climates. **Adaptation:** turning New Zealand merino wool from a commodity into a premium brand on the global market.

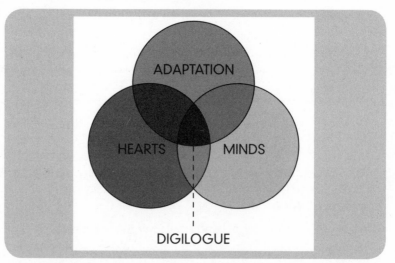

*   List the initiatives and strategies that are more heart-focused, as well as the ones that are more mind-focused.

*   Where does the convergence lie? Are there translational sweet spots between your analogue modes of communication and your digital modes of communication?

*   What would be some truly adaptive ways of connecting with your clients in the future?

*   List the best examples of adaptive digilogue strategies you have come across so far, highlighting the ones that might be applicable in your business.

Endnotes

1 'About PatientsLikeMe', PatientsLikeMe, www.patientslikeme.com/about, last accessed 8 April 2013.

2 P. Wicks, M. Massagli, J. Frost, C. Brownstein, S. Okun, T. Vaughan, R. Bradley, J. Heywood, 'Sharing health data for better outcomes on PatientsLikeMe', *PubMed*, US National Library of Medicine, www.ncbi.nlm.nih.gov/pubmed/20542858, last accessed 8 April 2013.

3 Unknown, 'High-tech Lederhosen: Is that an MP3 player in your pants...?', *Der Spiegel* Online, www.spiegel.de/international/zeitgeist/high-tech-lederhosen-is-that-an-mp3-player-in-your-pants-a-473317.html, last accessed 8 April 2013.

4 C.P. Wallace, 'Land of laptops and Lederhosen', *Time Magazine*, www.time.com/time/magazine/article/0,9171,351188,00.html, last accessed 8 April 2013.

5 J. Hoper, 'The laptop and lederhosen formula', *The Guardian*, www.guardian.co.uk/world/2002/sep/02/germany.eu, last accessed 8 April 2013.

6 J. Hooper, 'The laptop and lederhosen formula', *The Guardian*, www.guardian.co.uk/world/2002/sep/02/germany.eu, last accessed 8 April 2013.

7 C.P. Wallace, 'Land of laptops and Lederhosen', *Time Magazine*, www.time.com/time/magazine/article/0,9171,351188,00.html, last accessed 8 April 2013.

8 Unknown, 'I can see New Zealand!', IceBreaker, blog.icebreaker.com/tag/newsweek/, last accessed 8 April 2013.

9 Definition of anthropology from Wikipedia, en.wikipedia.org/wiki/Anthropology, last accessed 8 April 2013.

10 Unknown, 'Michael Henderson (Author)', Wikipedia, en.wikipedia.org/wiki/Michael_Henderson_(author), last accessed 8 April 2013.

11 Unknown, 'Mike Walsh', www.mike-walsh.com/mikewalsh, last accessed 8 April 2013.

12 D. Goodwin, 'New YouTube statistics: 48 hours of video uploaded per minute; 3 billion views per day', searchenginewatch.com/article/2073962/New-YouTube-Statistics-48-Hours-of-Video-Uploaded-Per-Minute-3-Billion-Views-Per-Day, last accessed 8 April 2013.

13 Definition of cyborg from Cyborg Anthropology, cyborganthropology. com/Book, last accessed 9 April 2013.

14 A. Rimas & E.D.G. Fraser, *Beef: The Untold Story of How Milk, Meat, and Muscle Shaped the World*, William Morrow Paperbacks, US, 2009, p. 168.

15 Definition of digital native from the *Oxford Dictionary*, oxforddictionaries.com/definition/english/digital%2Bnative, last accessed 8 April 2013.

16 J. Valentino-DeVries, 'Learning to play 'Angry Birds' before you can tie your shoes', blogs.wsj.com/digits/2011/01/19/learning-to-play-angry-birds-before-you-can-tie-your-shoes, last accessed 8 April 2013.

17 M. Olsson, S. Dagbladet, 'Tablet a threat against children's learning', 18 Dec 2012 olssonbetalar.blogspot.com.au, last accessed 11 April 2013.

4: Digilogue

Less boxing, more capoeira. Coordinated moves choreographed into a temporary standstill. Even the two combatants are surprised by the curious turn of events. What will unfold next? The commentators are dumbfounded.

In this round, we are forced to ask the question: can we have our cake and eat it too? Can we have a polyamorous love affair with both the digital and the analogue? Could it be that the two don't have to be mutually exclusive? What can analogue brands learn from digital brands, and vice versa? We will learn from brands like TED, GoGet Carshare, Zappos and Toni & Guy how clicks and bricks can combine into wonderful Digilogue Strategy Maps that enable brands to win both digital minds and analogue hearts of tomorrow's customers. Let us get down to business.

Providing value to digital minds and connecting with analogue hearts

I am a bit torn. I'm in the midst of a complicated love affair — with two lovers. One is the Penultimate application on my iPad and the other is my companion of 10 years, the classic Moleskine diary.

Penultimate is great. She is digitally savvy, fit for purpose and tuned into the latest trends. She is just an upgrade away from becoming even better. Moore's Law is on her side (see chapter 1). Penultimate contains thousands of pages of my hand-drawn notes, models and client briefs. In one place. (Actually kind of in two places — Penultimate syncs my digital notes to the cloud via Dropbox.) However it's stored, what it ultimately means is that I will never lose my notes (at least hypothetically). I can mark up documents, provide feedback on designs, and email my latest notes from meetings to colleagues, clients, staff and contractors. I also use it instead of a flipchart because Penultimate can project wirelessly at conference venues. Digital, geeky cool. I love Penultimate in so many ways, and I enjoy showing her off. But sometimes (and I promise there is nothing Freudian going on here), the physical interaction between tablets and the Wacom Bamboo pen feels a little superficial. Like somehow the thoughts I jot down are not as profound as those facilitated by my deep and meaningfuls with my other lover. Hmmm.

Hence my indecision. In contrast, my Moleskine diary is a soul mate. We know each other intimately. She has been there during the good times and the bad times. She has listened to my deepest emotions and strangest thoughts, without judgement. Her pages have depth and the very process of connecting with her blank pages is intimate. I feel like I am a better, more creative, more expressive person when I am with her. We have deep and meaningful, existential conversations. My hand feels light as the Lamy pen dances across her pages. The thoughts seem to connect in a timeless continuum, and I feel connected to the dignitaries who have danced across similar pages before. The pages absorb the Lamy ink in a way that awakens the artisan in me. I let the creative side of me come out. Time stops occasionally. Sometimes, I will break up our in-depth conversations with a quick to-do list. I take her to exotic locations around the world. (She is much better in the sun and on the beach than Penultimate, who tends to be a bit high-maintenance in the glare.) A coffee stain, some red wine or a couple of sand corns doesn't much concern Moleskine. She has timeless wisdom. She is a great analogue mirror, providing emotional and intellectual depth. But sometimes she is a bit conservative, a bit introverted. I can't show her off in a big way. Sometimes I need her to be more easy-going.

See — I am torn.

Yet, I keep dating both. Why? One, Penultimate, provides value to my digital mind. The other, Moleskine, connects with my analogue heart. Penultimate can't fulfil both needs — it's trying too hard to connect with my analogue heart. Think about it. Tablet. Touch screen technology to mirror analogue handwritten notes. Handwriting on a computer. Technological high-touch, yes — but is it enough?

Penultimate encourages me to engage with her often, and is hoping that I will forget Moleskine and commit to pure, monogamous engagement with Penultimate. But Moleskine is on the counterattack, and is trying to provide value to my digital mind. Moleskine (at the time of writing) has just released a co-featured digilogue version of the notebook, in collaboration with the cloud-based Evernote. The Evernote application enables me to take notes and to digitise them across a variety of devices. Moleskine is fighting for modern relevance. Here is a note she included recently.

Today, Moleskine is synonymous with culture, imagination, memory, travel and personal identity—in both the real world and in the virtual world. It is a brand that identifies a family of notebooks, journals, diaries, and innovative city guides, adapted to various functions. With the diverse array of page formats, Moleskine notebooks are partners for the creative and imaginative professions of our time. They represent, around the world, a symbol of contemporary nomadism, closely connected with the digital worlds through a network of websites, blogs, online groups, and virtual archives. With Moleskine, the age-old gesture of taking notes and doing sketches—typically analogue activities—have found an unexpected forum on the web and in its communities.[1]

Both Penultimate and Moleskine are going digilogue in their ambition to provide value to digital minds, and to connect with analogue hearts — and for brands such as these to survive and thrive, they need to. This is a lesson all brands and organisations must learn. Now.

The digilogue mix

So what is the right mix between the digital and the analogue? *Digilogue*, of course. Figure 4.1 outlines the features of each of the three possible approaches.

Figure 4.1: analogue v digital v digilogue

Analogue	Digital	Digilogue	Examples
Hearts	Minds	Hearts and minds	TED
Face to face	Interface to interface	Multichannel	Zappos
Facial recognition	Facebook recognition	Social to commercial	Toni & Guy
High-touch	High-tech	Brand amplify	Apple
Old school	New school	On/off loop	GoGet
Personal touch	Digital connect	Respect and innovation	Qantas
Nine to five	24/7/365	Personal connect	LEGO
Slow	Fast	Customer DIY	iPad launch
Word-of-mouth	Word-of-mouse	Brand alignment	Woolworths
Offline	Online	Omnichannel	Virgin Airlines

In figure 4.1, you can see how digilogue and the examples thereof compare with pure analogue and digital channels. (Some of these examples are covered later in this chapter. Woolworths is covered in chapter 5, Qantas is covered in chapter 6, iPad and LEGO in chapter 9.) Digilogue is an adaptive mindset, and it is required of both digital new-school businesses, as well as old-school analogue businesses. It is *the* way of winning hearts and minds in today's brutal business environment. To be the industry leader, the thought leader and the market leader, you need to think digilogue and embrace this paradigm shift in doing business. The research backs this up. In a 2011 Forrester survey of retailers, one of the most digitally disrupted industries, the question 'What impact has the transition from a single-channel focus to a multichannel strategy had on your company?' generated the following results:

- 78 per cent of respondents saw increased online sales
- 59 per cent saw increased profitability
- 52 per cent saw increased customer satisfaction
- 36 per cent saw increased offline sales
- 3 per cent saw no impact whatsoever.[2]

The examples of companies in this chapter (and in this book) are just an initial group of brands that are embracing the idea of providing value to digital minds and connecting with analogue hearts. Inspiration abounds. And when you create a virtuous loop between the digital and analogue, and weave them seamlessly together, digital disruption will not rock your market position.

Our monogamous love affair with analogue or digital has to end. This is not an either/or proposition. Businesses today have to embrace the and/also mindset, and seamlessly integrate the digital with the analogue. You might think of this as polyamory and perhaps it is. But I say it again, to win the hearts and minds of customers today, you must provide value to digital minds and connect with analogue hearts. Whether your business is old school or new school. Take note of this truism — in both your Moleskine diary and your Penultimate app. You can now have your cake and eat it too.

The Digilogue Strategy Map

So how can you ensure your present and future strategies provide digital value and connect with analogue hearts? Introducing the Digilogue Strategy Map, shown in figure 4.2.

Figure 4.2: the Digilogue Strategy Map

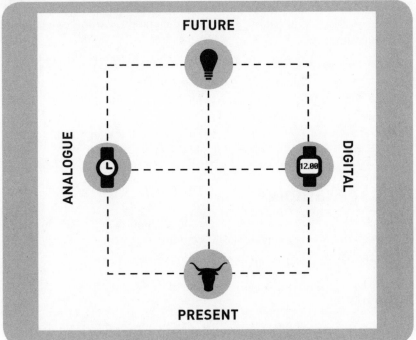

On the strategy map, you'll notice a simple X and Y axis, or a 2×2 matrix. On the left side on the horizontal you can see the word *Analogue*, symbolised by an analogue watch; on the right side on the horizontal you can see the word *Digital*, symbolised by a digital watch. On the vertical axis, at the bottom you can see the word *Present*, symbolised by a Texas Longhorn (remember the origins of branding from chapter 3). At the top, you can see the word *Future*, symbolised by an idea light bulb.

These two axes and their extremes form four unique quadrants, and a well-developed Digilogue Strategy Map needs to have ideas and activity in each quadrant. The great thing about this strategy map is that it can be applied across industries, and as my examples (chapters 6, 7, 9, and 10) will show, Thinque clients from various contexts are using it to better map and execute on their client touch points. (See the end of this chapter to have a go at drafting your own Digilogue Strategy Map. You also have another chance to do this at the end of chapter 10, when you've read through the whole book.)

To help you on your strategy-planning way, take a look in the following sections at what digital brands are doing to provide value to digital minds and to connect with analogue hearts, and respectively what analogue brands are doing to provide value to digital minds and to connect with analogue hearts.

Recall Google's analogue, printed $75 voucher from chapter 2. This voucher provided value ($75 digital advertising) to digital minds. By sending out an analogue voucher, Google gave the digital brand a human, analogue face. This connected with analogue hearts. The question for all brands is how you can optimise your digilogue thinking — how you can provide value to digital minds and analogue hearts.

Digital examples

In this section I give examples of digital companies adopting analogue qualities — and so providing value to digital minds while connecting with analogue hearts.

Instagram

In April 2012, Facebook bought the digital photo-sharing application Instagram for $1 billion. Let me expand on those details a little. This valuation meant that a company with seven employees in Silicon Valley, and which had never turned a profit, had received a higher market valuation than *The New York Times*. Murmurs of dotcom bubble 2.0 immediately began. Why did Facebook buy Instagram? A few reasons. At the time, Instagram had 35 million members in its social network, many of whom are more active on Instagram than on Facebook. Also, Instagram's graphic user interface is one that has high appeal on mobile devices — unlike Facebook, which was spawned in the laptop era, Instagram began its young life on the mobile device, and thinks differently about a device that Facebook hasn't quite figured out yet.

Instagram is also a little more artiste than Facebook. It has re-enabled the way we used to take pictures. More economically. When we used to use a Kodak or Polaroid, and we used film, we tended to think more about composition, light and whether we actually wanted to capture a scene. Twice. Digital photography changed that, as we happily snapped away without having to think about expensive film and development. This meant more spontaneous photography and, with the advent of digital photography via mobile devices, we started snapping everything. When Facebook then enabled you to post all of these digital photos onto your Wall, this added up to lots of digital quantity and little analogue quality. Instagram enables quality. If Facebook is an unfiltered, reasonably direct reflection of someone's social status and life, Instagram's filters allow the average punter to capture images in a more creative way. Typically, this means that there are fewer humans with beer bottles in the pictures, and the shots tend to look more professional than your average Facebook shot.

If digital sharpness is instant and nice, Instagram requires some thought and emotion. It has a range of filters that you apply to your photo, to give it a certain mood. Interestingly, these are all old school — vintage, black and white, sepia, antique, blurred edge. These filters give the photos an analogue feel to them. Check out my Instagram feed on @funkythinker and you'll find examples. Instagram enabled me to think about how I wanted to feature the Beach Shack in Cape Naturaliste National Park in Western Australia, and instantly gives me the false (digital) sense that I am actually analogue developing the photos in a mobile (faux) lab — without running the risk of losing a batch of photos, or having to invest the time in the actual analogue development of them. For those of us who feel like we have lost some creativity and quality with the advent of digital photography, Instagram offers a vintage, analogue escape. It connects with our analogue hearts. But it also provides value to our digital minds.

It is convenient. It's fast. And the mobile interface enables you to shoot digital photos straight through the application, so that if you choose, the creation can be visible immediately to your Instagram social network. Digital minds, analogue hearts. Tick.

Zappos

Zappos — another $1 billion investment, this time acquired by Amazon. The site sells shoes. Online. Weird, yes. Does it work? Yes. This is another digital example of a company that provides value to digital minds, and connects with our analogue hearts. Jeff Bezos admitted that he acquired the company largely because of its customer service culture. Wait. You mean this is a digital company that actually wants to speak and interact with customers? You bet. Zappos is part of the digital retail disruption I talked about in chapter 1. It also knows it has to have analogue soul.

There are some things we never thought we'd buy online, and shoes are one of those things. I mean every brand has a uniquely different fit — every brand's size range varies. If I am going to buy a pair, I need to go analogue and test it for fit, in-store. Big challenge for an online retailer. So let me tell you a little about its business model. Zappos asked their customers a simple question: why buy a single pair of shoes? Buy 10 pairs — obligation free. Browse our site and find, say, three different models of stilettos, and then select each in the two or three different sizes nearest to your own estimate. Just like you might in real life. But, get them all delivered overnight or on the same day. And simply send back the shoes that you don't want to keep — at Zappos's expense.

This provides value to your digital minds, and it reduces the risk of buying shoes digitally. But, hey, it's still shoes and sometimes we need taste advice, right? Well, Zappos have nailed this piece, too. Set up in Las Vegas, an area renowned for its availability of customer service talent, its call centre is second to none. The staff have no scripts, and are asked to rely on analogue emotional intuition in the way they communicate with clients. They have a month's induction program, after which new recruits are offered $2000 to walk away. Why? Because they want to retain the unique culture they have created. Beyond the customer service, Zappos sees every call as a sales opportunity. At a time when most digital companies hide their phone numbers on a back page on their digital website,

Zappos has it front and centre. It wants us to engage analogue. In one of my favourite Zappos YouTube Ads (where Zappets — Zappos Muppets — deal with real in-coming customer calls that customer service staff have experienced), a caller tells a customer service representative that the dress she ordered the day before has already arrived but that she is 'not really emotionally ready for it yet' and has hidden it in the garage under a tarp. When the caller asks if she can return the dress, the employee says yes, and then adds gently, 'You will have to touch the box, though'. True story. And interestingly, this analogue touch has meant that the most profitable clients Zappos has are the one who return the most items.[3] Ironic, yes, but being a digital company with an analogue heart can pay off.

Apple

Apple. Not the fruit, the company. In chapter 2 I covered Apple imitating Ritz-Carlton's analogue customer service, and how this has differentiated the company vis-a-vis its competitors. This is its bricks and mortar story. What about the clicks in 'clicks and bricks', though? Does Apple provide value to digital minds? Yes, through education and digital convenience. Go online and find any number of educational videos. Thinking about switching from PC to Apple? Here is how to switch your mindset. Thinking about syncing your iPad, MacBook Air and iPhone? Yes, please. Here is how. Wanting to ensure that all your mobile data is backed up in the cloud? That would be good. Want your ecosystem of media integrated across your digital platforms? Okay, let us show you how. Want to troubleshoot Mac challenges? Check out one of the forums curated by Apple fans from around the world for peer-to-peer advice. (Okay, the last one is kind of lazy, but it's a very efficient form of customer service.) Want a great digital device that just works? Here is an Apple. The company provides digital education and convenience, and matches its analogue brand touch points across the digital sphere. Impressive.

Three digital examples. Done.

Analogue examples

In this section, you can take a look at some analogue businesses that are equally capable of providing value to digital minds, and of connecting with analogue hearts.

TED

I love TED. Not just the 2012 movie with Mark Wahlberg and his weird teddy bear. I love the TED Conference. TED stands for Technology Entertainment Design. It's an analogue conference that's historically been held in two physical locations each year: Long Beach and Edinburgh. I have had the privilege of attending — twice. (Yes, I re-offended.) People come together to share ideas, to physically interact, to strategically network and to be moved by inspirational ideas. At $7500 per ticket, invite only — or $15000 if you are selected to be a VIP donor, by application. But why? All the talks can be seen digitally, online — via your tablet, via a smart phone app or via your desktop. They are each 18 minutes. Why travel to Edinburgh or Long Beach to see someone speak for 18 minutes? It's for the analogue experience. It's for the status. The emotional connection. The experience. The intellectual stimulation. The analogue network. The new friends. What's important here is the demand creation that TED has been able to produce. Digitally.

Let me go into a bit more detail on how an exclusive, analogue conference managed to digitally entice ever-increasing audience numbers at an ever-increasing ticket price. High volume, high fee. It's what we all aspire to, right?

Digital drove analogue demand. In 2005, TED started letting its TED talks escape digitally. For free. All of a sudden, we were given an insight into this exclusive club, and access to some of the world's greatest content was digitally democratised. The videos from TED — 'Ideas Worth Spreading' — now became easily tagged, shared, commented upon, embedded on blogs and spread via social networks. TED had gone from a small collection of 1 per cent intelligentsia, social entrepreneurs, tech heads and venture capitalists to the mainstream.

All of a sudden, we could tune into Al Gore's presentation on climate change, we could be inspired by Richard Branson's talk on space tourism, be mesmerised by Dr Hans Rosling's quirky take on statistics, and be given a real-time insight into the mechanics of a stroke by Jill Bolte Taylor — a neuroscientist — in a talk aptly named 'A stroke of insight'.[4] All of a sudden you and I had access. The veil had been lifted. Digitally. And when we got a digital taste, our analogue hearts wanted more. Digital hors d'oeuvre? Yes. Analogue cravings for the whole degustation menu? Absolutely. This digital distribution of ideas worth spreading gave TED's brand a massive boost. But the barrier to analogue entry was still high. There was the cost. There was the invite-only aspect, the required endorsement by previous members of the TED club. There was the travel. So, on the back of the successful digital distribution, TED decided to start launching TEDx independent events around the world. These have very little to do with the central TED organisation, other than the fact that TEDxNYC, TEDxSydney, TEDxMacquarieUni or TEDxStockholm (for example) have received licences granted by HQ. Fans, members and even non-participants of the main stage TED events could all of a sudden organise a TED-in-a-Box event closer to home, locally. As long as certain quality measures were in place. Think global, act local. The brand had entered our hearts and minds in a big way.

Smart moves. Great feeder database for the main stage events. Both digital and analogue. Here is how the marketing funnel or cycle works in rough measures (and I know because, as I mentioned, I am a re-offender). Watch a few videos online. Download the application. Watch some while in the gym on your iPhone. Better than MTV. Comment on a video while on the subway on the way home. Share a video on Facebook. Sign up to the digital newsletter. Receive regular updates and news on the latest videos. Get an update on a local TEDx event you can apply to. Be somewhat interested. Get lucky and have the right local connections to endorse your application. Attend the TEDx event in your city. Get an analogue taste. Aspire to go to the main stage TED event in Long Beach or Edinburgh. Write it down as your goal for next year. Fill in your digital application, statement of reasons, TED endorsers and quirky attributes. Be rejected a few times to whet your analogue appetite even further. (*Maybe next year*, you think.) Finally—big moment—get accepted. Part with $7500—and then add to that an airfare (somewhere between $500 and $10 000, depending on your location and class preference) and accommodation at TED-endorsed hotels ($1500 for the week). Attend. Be amazed. Get inspired. Meet some incredible people. Lose your TED virginity. Kick the bucket list. Get digitally added to the VIP list of the past guest segment and so receive newsletters offering prioritised seating in the future. Attend again. Sorted.

But don't just take my word for it—take a look at these digits. TED has gone from roughly 600 attendees in 1999 to 1500 in 2009. But now check out the price point differential: tickets have gone from $600 to $7500. That's a 12.5 times increase—in 10 years. And the conference has been in existence since the early 1980s. What happened were digital hors d'oeuvres, which whetted our analogue appetites for more. For the analogue inner circle. A not-for-profit managing to charge seriously corporate dollars. Why? Because it provides value to our digital minds, and connects with our analogue hearts. That's an idea worth spreading.

GoGet

Another idea worth spreading is that of car sharing. Let me give you some quick context. Car sharing as a concept has been around since the car was invented, but we can trace sharing back to co-ops and its agricultural roots. Sharing means resources can be pooled into a buying group, and that the thing that is bought is then passed between the people who have pooled their resources. For example, if there was one tractor in the village buying group, the farmers could all use that tractor in a phased fashion. This would maximise utilisation, and mean that less capital would be tied up in capital expenditures. Access was more important than ownership. The old adage that we don't want to own a drill, we want a hole, holds true here.[5]

Now apply this to the idea of cars. If you think of automotive companies, old steel behemoths in Detroit and Bavaria probably come to mind. Old school, right? Car brands have often used tech language in their promotion, but have also appealed to our analogue hearts in their references to history (Aston Martin — James Bond), provenance (BMW — Bayerische Motor Werk), engineering precision (Audi — Vorsprung Durch Technik), and design aesthetic (Maserati — Italian).

Cars became a personal thing, and they became a personal brand extension of you. Ownership, or the perception of ownership through car financing, became ingrained in the analogue story. Then the global financial meltdown happened, petrol prices went through the roof, and Al Gore launched his inconvenient truth at TED. Cars became unsexy. And I don't mean in the Volvo way. Car sharing re-emerged on the scene in response to disruption.

Let's take a look at GoGet. The product they use isn't new and it's highly analogue. But the way the company adds value to digital minds is impressive. Within three minutes of my apartment in Sydney are 10 GoGet cars available for me to use. I share these with my neighbours. I have no idea which ones. I don't care. I can pick any of these cars whenever I need one. They are cleaner than any car I have ever owned (think espresso cups, napkins, scratched CDs without cases, adapters and, yes, the occasional banana peel). I don't need to pay for petrol. I don't need to get the car registered or stamped as roadworthy. I don't pay for parking in my local area. And booking it is a hoot via GoGet's mobile app or via its website. I pay $29 a month for the membership and $6 per hour for hire. For every GoGet car, 7.5 owned cars are removed from streets. No — not physically by GoGet, but by owners opting for this better economic option, and divorcing themselves from car ownership. For me, GoGet is perfect.

In the United States, the same holds true for Zipcar. As I said, the idea of sharing cars is nothing new, yet before digital technology, the arrangement of it could be difficult. You would have had to create a roster, put in a bunch of phone calls, deal with messy passengers, and continue to work out the kinks in the system. It would have been like a timeshare scheme on a tequila hangover. Messy. Now, trust is built digitally. If you return the car late, you get charged $50. If you return the car to the wrong location, you get charged $50. If you don't fill up the car (at GoGet's expense, with the GoGet-provided credit card) when it's running low on petrol, you get charged $50. For me, I need the car to get me from Point A to Point B, and given that I am only in Australia for a few months each year, it doesn't make sense for me to pay for latency. Even for people who mostly stay in the same country, the average car stands idle for 23 hours a day,[6] so why would I (or most people) want to pay for a high-tech horse that has incredibly expensive taste and costs me money while I sleep? I wouldn't. The idea of sharing resources is very analogue, and it appeals to our agrarian, cooperative, fiscally conservative, environmental souls. But it used to be hard. Now it's digitally easy. And it enables me to make smart financial decisions. Its digital booking system, damage reporting, invoice system and geo-location apps provide its members and my mind with digital value. Hearts and minds. Tick, tick.

Toni & Guy

The big question is whether the digilogue principles apply to analogue retail. Bricks and mortar. Who, if anyone, is adapting? At Thinque, we've been consulting with many of the world's major professional hair product brands including, Kerastase, Schwarzkopf and Wella. These brands rely largely on professional retail channels — upmarket hairdressers — in their go-to-market strategy. In other words, these brands are reliant on the continued existence of bricks and mortar salons, and hairdressers at the salons marketing and selling the individual brands generally and their products specifically.

Over the years I have become good friends with some of the leading hairdressers in the Australian market, and many of these are very good business operators. Others are highly aesthetic and creative, and sometimes their business acumen leaves something to be desired on the numeric side of things. The hairdressing business is a margin business, and the old industry wisdom says that you break even on the service (the cut and colour), and you make a (small) margin on the product that you want the client to leave with. The service is the bread and butter that makes the salon (just) go around, and the product gives it a margin boost. Add to this that many hairdressers and salon owners suck at marketing and driving demand, and you have a ballerina balancing act on your hands. Imagine the large-scale effects if Kath chooses to book her cut and colour every eight weeks instead of every four weeks, and instead of getting her hair professionally coloured, gets her pal Kim to do it — with retail product from Duane Reade or Priceline, or online. Obviously, this affects the salon. (It also definitely affects the quality of hair of Australia.) And it affects the professional brands the salon sells. Amplify the Kath and Kim example into mass customer behaviour, and the salon business, totally reliant on analogue visits, is in big trouble.

The key to retail renaissance is clicks and bricks. Think about how FSC Barber in the West Village, NYC (see chapter 2), enticed me to prioritise getting there as a must-do during my 43-hour visit to the city. Strategies included digital collaboration with The Sartorialist, tweeting of a special offer, a sexy digital website, using an Instagram feed and maintaining a great blog. Digital value to digital minds. This digital window-shopping drove me to do analogue window-shopping and, ultimately, to part with a sum of money I would gladly part with over and over again. Online drove offline behaviour.

So it is that Toni & Guy, another leading professional hair brand, staged their Blogged & Bound marketing campaign, a fashion shoot and interview series featuring Australia's top 20 fashion bloggers, styled by Toni & Guy hair professionals. The fashion shoot and interview series was distributed via the Toni & Guy website and, ultimately, via the 20 Australian fashion bloggers, reaching their digital databases, with a soft endorsement by the blogger. Viewers could get inspired by their favourite blogger, check out the blogger's cool haircuts and access styling tips digitally. Cool. Additionally, Toni & Guy printed a version of the fashion shoot in a beautifully bound photo album book featured in salons. Remember those? You can flick through the books in the salon and check out the campaign, and show your analogue hairdresser what colour you'd like (or perhaps like them to restore after having visited Kim). Value to digital minds (blogs, digital photos, online buzz, digital ambassadors), plus connection to analogue hearts (printed book, brand ambassadors, events, newsletter sign-up). Digital drove analogue, and analogue drove digital in a beautiful virtuous retail renaissance loop.

On an even more local level, Toni & Guy do photo shoots in well-known local hot spots. For example, to promote their Cotton Tree salon on the Sunshine Coast in Queensland, Australia, Toni & Guy did a photo shoot featuring models in the local restaurant The Boat Shed. This digital value to the local community, and potential clients, showcases the analogue cementation of the brand in the locale, while giving prospects a taste, or digital hors d'oeuvre, of the salon. This sense of analogue, heartfelt connection to brand augmentation in real life, can effectively be digitised, and drives local buzz — both on and offline. In fact, one of my publisher's collaborators on this book describes her digital due diligence of the Toni & Guy brand presence in her town of origin as an 'analogue moment'. You know that a brand is providing value to digital minds and connecting with analogue hearts when you hear emotional exclamations like this.

Round 4. Draw.

The dance continues. It seems both parties are observing each other, learning from each other, being inspired by one another. Analogue Anachronism is adding flexibility and speed to his moves, while Digital Disruption is learning to connect more deeply in his punches. If we zoom out for a moment, we canvass a round that has showcased the combinatorial strengths of both the digital and the analogue, and how each is trying to adopt the core strengths of the other. We illustrated the comparative strengths and features of my two note-taking devices, the

Penultimate app and the Moleskine diary, and established that the future is neither one nor the other, but a polyamorous affection for two note-taking lovers. We then investigated how digital brands like Zappos provide value to digital minds and connect with analogue hearts, while analogue brands like TED capture digital imaginations and analogue heartstrings. Both result in connections with the customer's purse strings.

Ask yourself. . .

Get yourself thinking with the following questions:

- Which parts of your customer interactions can be digitised, and which ones can never be?

- What are you currently doing to provide value to digital minds, and to connect with analogue hearts?

- Which bits of timeless wisdom and analogue interactions are unique to your business, and ought to be continued?

- What digital initiatives are already reaping results, and could be amplified?

- What new analogue touch points and digital connections can you make to better position your business in the hearts and minds of your current clients, and tomorrow's prospects?

Endnotes

1   Unknown, 'Moleskine — the history', Moleskine, mymoleskine.moleskine
    .com/journalist/press/the_history.pdf, last accessed 9 April 2013.

2   Unknown, 'The state of retailing online 2011: Marketing, social, and
    mobile, Forrester Research, www.forrester.com/The+State+Of+
    Retailing+Online+2011+Marketing+Social+And+Mobile/fulltext/-/
    E-RES58625?docid=58625, last accessed 9 April 2013.

3   A. Dugdale, 'Zappos' best customers are also the ones who return
    the most orders', Fast Company, www.fastcompany.com/1614648/
    zappos-best-customers-are-also-ones-who-return-most-orders, last
    accessed 9 April 2013.

4   J. Bolte Taylor, 'Jill Bolte Taylor's stroke of insight', TED,
    www.ted.com/talks/jill_bolte_taylor_s_powerful_stroke_of_insight.
    html, last accessed 9 April 2013.

5   R. Botsman, 'The case for collaborative consumption', TED,
    www.ted.com/talks/rachel_botsman_the_case_for_collaborative_
    consumption.html, last accessed 9 April 2013.

6   R. Botsman, 'The case for collaborative consumption', TED,
    www.ted.com/talks/rachel_botsman_the_case_for_collaborative_
    consumption.html, last accessed 9 April 2013.

Taking stock

The Digilogue Strategy Map
and you (part one)

Here's your first chance to apply digilogue principles to your situation, first by outlining your digilogue strategies and then by applying them to the Digilogue Strategy Map. As you go through this book, I provide more insight into the strategy map and how I've applied it to a number of businesses, including my mum's business, Georg Sörman. So you'll also have a chance to revisit using the strategy map for your own business in chapter 10.

Digilogue strategies

How do you as a brand best provide value to digital minds and connect with analogue hearts in your communications? Take a moment and write down your top three priorities, both for providing value to digital minds and for connecting with analogue hearts so far.

Here are some quick ideas, first for providing value for digital minds:

- Provide thought-leading hindsight, insight and foresight through visual communications online — for example, via well-designed visual comparisons or graphic overviews based on digital data. (See Mint.com, Roambi's app or American Express's spend calculator for some specific examples.)

- Provide real-time data, analysis and advice to your clients — see, for example, Nike+ Fuel and its personal coaching arm band or Fly Delta's iPhone application.

- Provide free thought-leading insights via your blog and digital newsletter on trends that may affect your clients — see, for example, retailer Haberdashmen.com's website for fashion advice and event updates on things relevant to its clientele.

And some ideas for connecting with analogue hearts:

- Host events for your key clients with unique, interesting content and entertainment, and curate the guest list to provide networking opportunities — for example, have a look at *Monocle* magazine's geo-location events tailored to local cities or attend a TEDx event in your city.

- Remind yourself that 20 per cent of your clients generate 80 per cent of your revenue, and provide them with visible status — check out, for example, Qantas Airlines' high-end personalised Platinum packs with impressively heavy luggage tags and a welcome letter signed by the CEO.

- Send a handwritten note, which goes a long way to generating loyalty from clients — see, for example, the W Hotel in Hong Kong's Welcome Ambassador program and the impact handwritten notes can have on customer service.

Check out our blog on www.thinque.com.au/blog for more examples.

The Digilogue Strategy Map

Now use your two sets of top three priorities to complete your personal Digilogue Strategy Map, included below (you can download and print a strategy map from www.thinque.com.au/digiloguemap).

Remember — a well-developed Digilogue Strategy Map needs to have ideas and activity in each quadrant. Include the digital and analogue activities you're doing now in the Present half, and use your digilogue strategy priorities to complete the top two quadrants. Make sure you write at least three things in each quadrant.

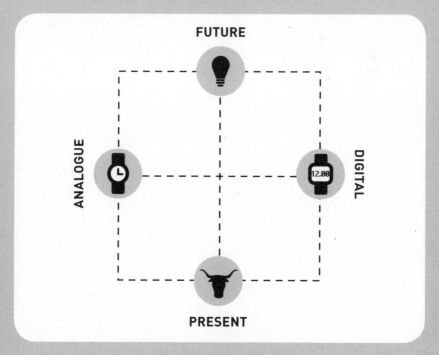

PART II
The battle heats up

When we thought things were hunky-dory, the contestants spark up again. It seems the co-existence and the capoeira-inspired moves were a charade, a diversion. The battle is back on.

Chapter 5 sees Digital Disruption engaging in some foul but effective play. Here we investigate how Big Data is leading us toward Big Brother of *Nineteen Eighty-Four* fame, and why we may soon find ourselves walking through shopping malls that scan our retinas to deliver tailored messages. While the chapter's futuristic exploration of technology and its communication processes may make you a bit unsettled, it also offers hope that new-school technologies can enable old-school communications, recognition and care for the customer's personal context.

In chapter 6, we will see how the seesaw competition unfolds. We investigate how customers and consumers seek to reconnect with themselves by periodically disconnecting from the digital world, and seek to get away from the digital doldrums that technology is also responsible for. To our surprise we see how analogue escapism and the customer craving for work–life balance and reconnection with their analogue souls create commercial opportunities for centred business owners.

Chapter 7 case studies how ideas, services, products and solutions can be digitally amplified across society. The chapter strategically maps how Steve Jobs launched the digital idea of the iPad, and how Apple was able to diffuse this disruptive idea across various groups of adopters in society, to the point where an idea virus became a mainstream accepted device now seen in boardrooms around the world. The chapter gives concrete advice on how the mobilisation and ubiquity of technological devices give both analogue and digital businesses incredible opportunity to provide value to digital minds and to connect with analogue hearts, and why that ubiquity puts pressure on all future-focused business leaders to start thinking like flexible media companies.

This leads us to chapter 8, which in an ironic twist illustrates that it is analogue physical location that drives digital innovation. We case study the innovation and creativity flowing from Silicon Valley, and the chapter reaches the conclusion that at the end of the day, all business is local and that, despite the digital advances we have made as human beings, we still crave analogue connection with our co-workers. At the end of chapter 8, the judges will deliver their verdict on who has thrown the most vicious and impactful punches, and we will see the referee raise one paw for a sweaty contestant (or perhaps not . . .).

5: Digital grapevine

Round 5. After two unorthodox rounds of co-existence the battle is back on. In this round, the big guns come out. Digital Disruption introduces a series of hits called Big Data, and we look at why we're nearing life in *Nineteen Eighty-Four*. You'll see why artificial intelligence is getting increasingly emotionally intelligent by virtue of big corporate analysis of your digital data trail, why sabermetrics leads to better decision-making and why scientific marketing spells the death of demographics. Despite seeing attempts at foul play, including incursions into privacy, this round digital will be poking analogue in the face. With glee.

Big Brother, big data

In the 2002 movie *Minority Report*, the pre-crime detective John Anderton (Tom Cruise) is rushing through a mall in the year 2054. Inadvertently, he has been caught up in a government conspiracy and has been turned on by his pre-crime unit as a potential future criminal. In this neo-noir sci-fi, people can be arrested well before even carrying out a crime, based on the predictions of three precog(nitive) oracles who can see into the future. In this particular scene, John Anderton runs through a mall, and the high-tech advertising displays on the sides of the mall scan his retina and deliver advertisements that are uniquely tailored to John Anderton. Lexus, Guinness, American Express and Fiji Tourism all make an appearance, based on John Anderton's consumption habits and the digital data trail he has left by using credit cards, by sharing his financial information, by browsing the internet, and by shopping online and offline. He has few places to escape in a world dominated by screens, traceability, facial recognition and digital data. Digital doldrums.

Sound familiar? We may not be quite there — yet. You may not yet walk down a mall and see the actual walls of the shopping centre tailoring advertisements to you based upon your digital data trails. But we are seeing the early signs of a shift toward *Minority Report* type marketing. Advertisers can now push information to you via your mobile device based on your location, so that when you enter a mall, the retailers in that mall can send you targeted ads, enticing you to come and spend money with them 'seeing you're in the area anyway'. Freaky. In London, the grand old dame of London branding — the London double-decker bus — has been fitted for some years with GPS-enabled LED screens that shift the advertisements on the sides of the bus based on which demographic neighbourhood it's passing through. In other words, as the bus runs through Kensington to South Kensington to Notting Hill, programmers know that Julia Roberts and her friends are a little more high maintenance than their neighbours down the road, and will change up the ads to suit this unique demographic. And today, when you communicate via Facebook or Google Chat with a loved one about your next family holiday, don't be surprised if Facebook or Google throw you tailored ads, like 'Thinking about a holiday? Take a look at our packages', 'Going to Hawaii? Specials on Waikiki beach', and 'Fly Qantas to Hawaii'. This is exactly what happened to one of my clients as he and his wife were brainstorming and using certain keywords like 'Hawaii' and 'holidays' in their holiday planning chat online. We may not be in 2054 yet, but in many ways we have landed in 1984.

As in *Nineteen Eighty-Four*. As in Big Brother and screens that watch you — and not as in reality TV shows. As in Orwell, doublethink and Winston Smith. As in thought crime, newspeak and constant surveillance. In *Nineteen Eight-Four*'s Oceania, Big Brother is the demagogue and the centre of the personality cult that the propaganda machine of the nation is building. Watching television is a popular pastime of the citizenry of the nation, yet the citizens are strangely aware that the screens are watching them back. Digitisation and screen addiction is moving us one step closer to this reality, and while the surveillance may or may not be performed by a government demagogue today, the data soundtrack of our lives is playing right now in the corporate headquarters of a brand trying to reach you.

Big Brother and your credit cards

The ecosystem of loyalty is built on this consumption data. When I refer to *loyalty* here I mean loyalty cards, and not the brainwashed adherence to a nation's policy and Ministry of Love. The whole loyalty card industry is centred on data gathering. Take my Woolworths Everyday Rewards card and its network of data-gathering devices. Here are the specs. If you do your grocery shopping at Woolworths and swipe your Everyday Rewards card each time you make a purchase, you can accumulate frequent flyer bonus points. For a frequent flyer fanatic like myself, this is great news. Sign me up. Frequent flyer miles is big business. According to analysis by *The Economist* magazine, by 2005 the global stock was worth more than US$700 billion (£370 billion), more than all the US dollar bills in circulation, and streets ahead of Britain's £42 billion of notes and coins. In 2012, Macquarie Equities analyst Russell Shaw valued the Qantas Frequent Flyer program at more than $2 billion — more than half of the airline's market value. According to this analysis, the world has a new global currency. Business people flock to frequent flyer rewards schemes like moths to a light. In the Woolworths example, you only start accumulating these points after a $30 spend. If you only spend up to $30 you have absolutely zero reason to swipe your card. I mean, the company tells you that swiping your card every time means they can capture the exact items you have bought, on which particular day of the week, on which date and in what quantities, and this will help them make their offers and discounts even more relevant to your needs. Wow. No thanks. Yet many people do swipe every time. You have to at least bribe me to make me part with that data.

Let me dive in deeper into this data-gathering activity. In August 2012, I was invited to spend the weekend with a friend and the famous Australian pâtissier Adriano Zumbo at my friend's place in the Blue Mountains outside of Sydney. On the Saturday afternoon, we ducked off to the local Woolworths store in Leura, and I picked up a few items, including sourdough bread and roast chicken. The total bill was close to $80 so I decided to swipe my Everyday Rewards loyalty card, and collect some frequent flyer points, which integrate with my Qantas frequent flyer card and which I can redeem for flights and upgrades. Swapping 50 frequent flyer points for a bit of a privacy invasion? Fair trade-off, I reckon. Unsuspecting, we leave the store, and enjoy a nice old-school, baking and foodie weekend at my friend's place. A week later, at that exact point in time in the day, I receive a special offer from 'Woolies' to buy sourdough and roast chicken (based on my previous weekend's consumption pattern). I respectfully digitally declined, by removing the email to the trash can.

This digital trail is redefining business and how we communicate. It is also a massive future play by retailers and brands like Woolworths, Tesco and Walmart. These retailers have a direct insight into our analogue behaviours, because of the digital trails we leave. This is why they integrate their private-labelled credit cards issued by Visa, MasterCard and American Express, sometimes backed by banks, and often co-sponsored by airlines' frequent flyer schemes. Yes, I am guilty, I am the owner of a Woolworths Everyday Rewards Qantas MasterCard card. Tricky name, but indicative of where my data flows. Woolworths has my data, Qantas has my data and MasterCard has my data.

I'm reasonably judicious in what I share but, in this case, I willingly gave up some of my data. What had to happen for me to give it up? The promise of 8000 Qantas frequent flyer points for signing up, 1.5 additional frequent flyer points for each dollar spent, and an additional 1 frequent flyer point for each dollar over $30 spent with Woolworths. Bribery, yes — and I was open to it. So were 3.8 million other Australians who signed on for the Woolworths Everyday Rewards card between 2008 and 2009.[1] (I should add here that the CIO of Woolworths is a good friend of mine, and I'm only referring to this company, of which I am a fan, to illustrate an example.)

Retailers can do a few things with this data. They can choose to tailor specials to particular customer segments — who, say, happen to eat roast chicken and sourdough bread on Saturdays. They can segment and cut and slice their database for their own revenue calculations, and they can on-sell this data to vendors and partners. They do all three. For example, Tesco will on-sell data about consumption patterns to Coca-Cola, Foster's and Pampers, so that these fast-moving consumer good giants can work out price promotions and point of purchase specials. Using the three strategies in tandem, Tesco might figure out that customers who start buying Pampers also start buying more beers. Why? Because dads of toddlers don't have as much time to go to the pub, and instead need to drink beer at home (hopefully not while changing the nappies). Thus, a co-promotion from Tesco offering Foster's and Pampers products to this particular segment might make sense. Sound futuristic? It's not.

These digital promotions tend to be disturbingly relevant. While at first we might marvel at the thought that *Hey, Tesco really cares for me and knows my preferences*, that feeling quickly dissipates when we realise that the only way they could know is because they have been digitally eavesdropping on our analogue behaviours and shopping-aisle conversations. And it can be awkward (and disturbing). This is the digital equivalent to a stranger walking up to a group of friends in a pub and asking if they're happy with the dishwashing liquid they're using. But brands still don't quite know how to utilise the data and tailor communications in a way that seems analogue authentic. Even though we may think that this enables us to be 'local' on a 'global' scale, we are not quite there — yet. My roast chicken and sourdough offer is still not the same as what I experience when I walk around my analogue neighbourhoods in Surry Hills, Sydney, or Williamsburg, NYC.

Digital awkwardness versus analogue connection

One of my favourite pastimes when I get to spend time in Sydney is to spend a day in Surry Hills. This is how a typical day might unfold. I head up for a morning meeting at Kawa Cafe on Crown Street, where the French waitress knows that I like a freshly squeezed apple, carrot, ginger, beetroot, celery and mint juice, followed by their homemade granola muesli with organic poached fruits and yoghurt. Only once I am finished do I like my coffee — that's a *piccolo latte*, please. (There is no masculine way of ordering it.)

After seeing a couple of clients, who usually comment on the high-cosy factor at Kawa, and doing some blog posts at one of the wooden tables, I might go for a 75-metre stroll up the road. I stop by the Barberia where Nathan, who knows exactly how I like my hair cut — 'not marines, but neat' — will seat me promptly, because he also knows I don't want the haircut to take any longer than 45 minutes. After Nathan has flicked off a few stray hairs that were tickling my neck, I continue towards Cleveland Street to pick up some flowers from Jay Jay at Buds and Bowers florist store for Thinque's COO, Emma (I should do this more often). Jay Jay knows what flowers Emma likes, and I've asked him to save my Amex details in his scrapbook in case I am overseas, so he can charge me at month's end for the flowers, together with any other purchases. I ask Jay Jay whether he is singing in church this weekend or if he has plans to go to his native Hong Kong any time soon. Good chats, analogue connection. I stroll another 30 metres past the popular gelateria, Messina, and its 10-metre line, and go into the design store Chee, Soon & Fitzgerald and chew the fat with Liam about what cool new Scandinavian designs they have. He talks of Marimekko and Resteröds. He asks me how my German Dibbern porcelain cups are going, and whether I am ready to complement the collection (he knows exactly what pieces I have accrued from his scrap book). Analogue connection, involving personal memory and face-to-face recognition. Unsophisticated data collection, sure. Genuine connection, yes.

Digital grapevine

It is this sort of analogue connection that data miners are looking to create. But while it is possible to manage and execute this on a small, local scale, it's definitely harder on a large, corporate scale when you have millions of customers. The other thing that is important here is that digital disruption is digitising the 'grapevine', or what Aussies call the 'bush telegraph'. Nowadays, people don't hear on the grapevine what the latest gossip is, they read it on Facebook. On numerous occasions, I've run into distant acquaintances around the world and had them ask intimate questions about my latest holidays, only to realise they must have spent reasonable amounts of time on my Facebook feed and holiday albums, and so be able to retrieve enough data to have an intimate conversation starter ready. Disturbingly relevant and entirely my own fault (must remember to defriend...).

So, if it seems weird that distant acquaintances can have this type of insight into your life, because you volunteered the information, it would likely seem even more absurd if Woolworths were to send you a 'Welcome back from Thailand' eCard, with specials promoting products that you should use to fill up your fridge with — because the milk you bought three weeks ago would be off by now. *Anders, we heard it on the digital grapevine.* It's very kind of you, Woolworths, but I'm a little freaked out right about now. Digital creep. While this hasn't happened to me personally just yet, owners of the LG Smart Manager fridge have seen their appliance morph into a food management system integrated with the digital ordering system of their local grocer. The fridge's 'brain' monitors its contents and automatically adds food to a user's digital account when milk and other ingredients are running low or going off. Every individual can set up an account in the fridge's brain, which also respects dietary requirements, age, gender and BMI. It can compile shopping lists for its owners based on their particular tastes and recommend recipes based on the combination of foods it is chilling.

What is most important about this type of digital tailoring is that it is happening — right now. The technology is in its infancy, and the way it is being delivered at the time of writing does seem more artificial intelligence than emotional intelligence. Aside from the LG Smart Manager fridge example, delivery of bespoke offers is a bit invasive, a bit over the top, a bit socially awkward. What makes it awkward now is the seeming lack of humanity behind the offer. Imagine the same offer being made to me at Kawa Cafe in Surry Hills (and with a French accent). If I'd been asked there if I needed to replace the milk in my fridge since I had just arrived back in the country from the holiday that I had told them about, it wouldn't have been strange. And I would feel so honoured she remembered me. The same information, with one translated into emotional human caring, and the other translated into a technologically optimised message. Vastly different outcomes in our own perception. With the help of Moore's Law (that is, the doubling of computer power every 18 to 24 months — see chapter 1), the odds are on Woolworths and LG starting to get this right with time (unless privacy regulators start stepping in in a serious way). However, at the moment, French accent = big heart. Woolworths eCard = Big Brother. Digital disconnect.

Death of demographics, and marketing as a science

Big data could mean the death of demographics, meaning the second important aspect here is that this digital data trail fundamentally shifts marketing from an art to a science. We have traditionally ascribed huge kudos to marketers, brand gurus and advertisers for their creative genius and their knowledge of demographics. We applaud them for their witty campaigns, we interview them at Cannes and we love to try to figure out how they could so cleverly get inside the minds of consumers. Marketing used to be more art than science. Develop some clever headlines, add some well-chosen images, compelling calls-to-action and some cool copy, and away you went. As long as you referred loosely to demographics, and some market testing of generation X females, the likelihood is someone would sign off on the campaign.

Demographics tracks statistical data relating to the population and particular groups within it — for example, the demographics of book buyers, religious Christians, baby boomers or high net worth individuals. It's a pretty loose term, but it's what we used to have to work with at a time before big data and before the digital data trail we now leave everywhere around us. Now things are different. While focus groups, market research, demographic tailoring and split testing have been part of marketing return on investment (ROI) studies for a while, digitisation has enabled a rigorous focus on digits and measurability. This kind of analysis used to be clunky. It used to be a mix of quantitative and qualitative inputs, with some analysis and a few sprinkles of gut feel on top. Marketers tried desperately to make it scientific, but it was a hard slog. I remember working for a marketing agency on a direct mail campaign. Direct mail was perhaps one of the most numerically focused marketing schools of thought back in the day of snail mail. Unlike its fancy communications cousin branding, it was less emotional and more focused on immediate sales results. It was also more measurable — at least in theory.

Here is what we did. We had a database of around 5000 prospects and clients. We wanted to test certain variables and check which combinations of headlines, offers, calls-to-action and images would generate the most dollars, so we printed five different versions of what was ostensibly the same brochure. The five slightly tweaked brochures were broken down into 1000 print copies and were sent to the database. Each was tagged with a unique code, so that when the recipients who wanted to take up the offer responded by snail mail or fax, we would know exactly which of the five brochures was most successful. We tallied the results religiously, with a view to using this information in our electronic direct mail (EDM) campaigns that followed the snail mail campaign. This process was very expensive because it was done on analogue, offset print machines and we were still some way away from getting proper economies of scale based on the quantity of print we required. The campaign ran close to $20000 in print costs alone. And it was clunky. Some of the data that we received back was unclean and entries were sometimes duplicated, but we could glean some trends from the responses. Overall, we could roughly measure the campaign and, based on sales results from the snail mail campaign and ensuing EDM, we usually broke even on the campaign. This was considered decent as an analogue test case. We felt that we had come as close to scientific marketing as we could get.

Digitisation is shifting the game. Today, this same campaign would look very different. First of all, it would begin with an EDM campaign, meaning the testing takes place electronically. Headlines, calls-to-action, copy and imagery would all be tested inside an EDM program, and then the responses, and perhaps requests for analogue, printed materials, analysed in a controlled environment, with direct and immediately measurable actions by the target group recorded. Marketers would be able to see which individual members of the database had opened the correspondence, which ones had clicked on what individual link, which ones had responded to which unique call-to-action and, ultimately, which ones had converted into paying clients. They might then follow up this campaign with a much more cost-effective digital print campaign via snail mail.

Print costs have also been digitally disrupted. When I say *digital print*, what I'm referring to are digital printing presses, which allow fast-turn-around, highly individualised and short-run campaigns. Economies of scale kick in early in the piece, and you might find that printing a single book even just once could cost $50 or $100 (or less). Anyone can now be a published author at low cost! Printing five different brochures in quantities of, say, 134, 178, 213, 217 and 182 is possible and would be extremely cost effective. This differs from the old-school analogue presses, where often over 2000 copies of the same material need to be printed before the per unit cost makes economic sense. For example, you wouldn't print one single copy of a book on an offset printer because the individual unit cost might be $5000 (or more), because individual analogue press sheets have to be prepared to ensure correct delivery of ink. Equally, demanding five unique brochures at quantities of 134, 178, 213, 217 and 182 would be phenomenally expensive per unit. Analogue, offset printing is bulk printing, long print runs and mass marketing in the true sense. The scattergun approach. Today, digital enables low-cost, targeted, effective communications. Marketing thus is becoming more science than art.

Imagine what this means for the holy troika of Woolworths, Qantas and MasterCard. Based on my frequent flyer activity (295 387 kilometres, 42 cities, 15 countries in 2012), my spend in those countries (business and leisure) and my shopping habits across the Woolworths ecosystem (groceries, petrol, liquor and hotels), I can expect some very targeted promotions. Digital direct mail and low-cost digital printing enables scientific marketing based on my psychographic profile. In other words, *psychographics*, the study and classification of people according to their attitudes, aspirations and other psychological criteria, is starting to replace demographics as the primary segment in marketing. My psychographics are made evident by the digital data trail that I leave. Thus, my next MasterCard bill might use the data captured on my Woolworths Everyday Rewards Qantas MasterCard to send me a uniquely tailored, digitally printed billing statement with my personal information and bespoke promotions based on my empirically validated tastes. Beyond a statement of spend, liabilities, points accumulated and frequent flyer points earned, the company might include relevant, variable imagery (unique to each psychographic of one person), hotel promotions based on my upcoming, validated travel destinations, and food discounts based on my grocery shopping habits when I am back in base camp in Surry Hills. Equally, Qantas's next EDM may be attuned to the fact that I am likely to be in Sydney for a couple of weeks, and it might do a cross-promotion with Woolworths for a nice bottle of wine. Impressively or disturbingly relevant? It's in the eye of the beholder. Either way, it inches us closer and closer to the screens of *Nineteen Eighty-Four*.

We used to have to rely on demographics. Now we don't. We have psychographics instead. Now we have all become segments of one and, increasingly, brands are using this to have a (sometimes) awkward, yet (sometimes) profound one-on-one conversation with you. This freaks out privacy groups, and is one of the main reasons both Facebook and Google have met staunch political resistance in the EU to tweaks of their algorithms and technological advancements like automatic facial recognition of your Facebook photos. The more information they have, the more relevant the promotions you receive. Spam is starting to sound like scattergun bliss. The more we know our clients, the more personal we can get. Until we start getting chased through malls with cameras that scan our retinas and deliver targeted ads in real time, right? You can see that we're heading in the direction of 2054 in more than a temporal way.

So, how do you know if you're leaving a digital data trail? One scary way of thinking about this is to consider the following. Every time you don't pay for a service or product online, you are the product. For example, you don't pay for your Facebook membership. That's because Facebook sells your data to brands who are marketing to you in a tailored fashion in your Facebook feed, and along the sides on your Wall. Facebook will sell pay-per-click advertisements to brands that, say, want to target 45 to 55 year olds who like to travel, who are Qantas frequent flyers and who are keen to travel to Hawaii. Got one? Boom. They can deliver an ad in real time to the couple chatting about Waikiki. You don't pay for Google? Yeah, right. Every time you search, click through on a sponsored or organic listing, send an email via Gmail or visit a YouTube video, its algorithms get tweaked, and your information makes you increasingly likely to receive highly relevant promotions to your real-time activity. You don't part with money, but someone else is. They are paying for your personal data. And the foreign exchange transaction you are participating in is that you pay with your personal data and the digitisation of your analogue consumer behaviour. Every step you take, every breath you take, someone is watching you. Big Brother. (Or Big W.)

Big data

Big data is on the side of Big Brother. And we are producing astounding amounts of it. Consider the following pieces of data about data:

- 90 per cent of all of the data created in the history of man has been created in the last two years alone.[2]

- Five exabytes of information were created by the entire world between the dawn of civilization and 2003. Now that same amount is created every two days.[3]

- Walmart handles more than one million customer transactions every hour, which is imported into databases estimated to contain more than 2.5 petabytes of data — the equivalent of 167 times the information contained in all the books in the US Library of Congress.

- We are projected to move from 2.75 zettabytes of data in 2012 to 8 zettabytes of data in 2015.

- The 2.75 zettabytes of data in existence in the digital world at the time of writing include:
  - 107 trillion annual emails
  - 566 billion objects stored on Amazon's S3 cloud service
  - 340 million tweets posted to Twitter — every day
  - 82 petabytes stored on the largest Yahoo! Hadoop cluster
  - 60 hours of video uploaded to YouTube every minute — that's one hour every second
  - 15 terabytes generated by 845 million Facebook users — each day
  - 300 million new photos on Facebook — each day
  - 3.2 billion new likes on Facebook — each day.[4]

- To store the 2.75 zettabytes of data would require 57.5 billion Apple iPads (each with 32GB memory), which would build a mountain of iPads 25 times higher than Mt Fuji.[5]

Lots of 1s and 0s. *But who's benefiting, Anders?* you may ask. Well, one of the most inspirational examples of turning data into information and into insight are the Oakland Athletics. *Huh?* Think of sports and all the seemingly useless statistics that you as the viewer receive on the TV screen. As it turns out, teams don't find these statistics useless. In *Moneyball: The art of winning an unfair game* by Michael Lewis, we get an insight into how data can be used for smarter decision-making. The book covers the use of sabermetrics by the Oakland A's in their selection of players. *Sabermetrics* is the analysis of baseball performance through objective evidence, especially statistics that measure in-game activity. Being forced to outsmart their well-to-do competitors like the New York Yankees, with payroll budgets three times as large as the Oakland A's, the A's selectors turned to data and analytics to ensure it could field a competitive side.

The book was turned into a 2011 film featuring Brad Pitt and Jonah Hill, and the film's focus is the team's analytical, evidence-based, sabermetric approach to assembling a competitive baseball team, despite their disadvantaged revenue situation. The central premise of the book is that the gut feel and amassed wisdom relating to baseball over the past century was subjective and often flawed, which meant player selection was not optimised and was often uneconomical. Rigorous statistical analysis demonstrated that on-base percentage and slugging percentage were better indicators of offensive success, and the A's became convinced that players with these qualities were cheaper to obtain on the open market than those with the more historically valued qualities such as speed and contact. These observations often flew in the face of conventional baseball wisdom and the beliefs of many baseball scouts and executives. However, with the repeat successes of the Oakland A's, this data-based approach to the game of baseball has been adopted by the major clubs, with the New York Yankees, Boston Red Sox and Cleveland Indians all playing Moneyball by hiring full-time sabremetric analysts. In business today, if you rely on pure gut feel you might go hungry. If you turn data into insight, you can compete with the big boys. Digital smarts.

Big data provides us with a veritable data explosion. While big data promises to deliver better business decision-making, smarter science, and increasingly scientific marketing, it's also moving us into the realities of *Nineteen Eighty-Four* and 2054. In 2009, the Indian government established the Unique Identification Authority of India to fingerprint, photograph and take an iris scan of all 1.2 billion people in the country, and assign each person a 12-digit ID number, funnelling the data into the world's largest biometric database. For better delivery of government services supposedly. Let's hope. With all of this digital data around, and so much virtual memory space to store it, it's becoming harder and harder to forget. (And let's face it, some things we'd prefer to forget — like fashion crimes committed in the 1980s, old heartbreaks, faux pas and instances of foot in mouth disease.) Humans deserve to forget some things, but with the digital data trail, your friends, family and the brands that increasingly are getting to know you will always be able to remind you of your habits, your ex-partners and your transgressions. It may not be long until our individual unconscious has been thoroughly digitally colonised. This is another consequence of the digital move towards psychographics.

Big business versus small business

Big data may be big business, but how the heck do you make money from it if you're a small business? I mean you can't eat big data. So how do you turn it into dollars? How do you turn the disturbing elements of big data into an advantage? Big Data is applicable to big business and small business alike. Targeted campaigns tend to get greater response rates, and personalisation of offers result in greater marketing ROI for the business that taps into the Big Data, like customer preferences, that they have amassed over the years. Let me give you a couple of examples.

Despite my occasional criticism of Georg Sörman, the business has been an (un)sophisticated user of big data in a small way for decades. One of the beauties of a family business is that you can delegate all the most menial labour to your family members, so my brother Gustaf and I didn't just get to enjoy polishing an outdoor copper frame in –10 degrees Celsius, but occasionally we were also sent to the Swedish Tax Authority to collect data for the store — data that helped build their now substantial database. The data we collected was the names and contact details of any men above a certain height. How does the Swedish Big Brother state have access to this information, and why do they make it publically available? Well, the latter question I cannot answer, other than naively offer transparency. I do know the answer to the former. Any man who completed his mandatory physical test for military service was measured, and his data collected. This means that the Swedish state knew exactly which men were above 190 centimetres in height, and these men were interesting for my mum's store because the business offered a clothes tailoring service and also catered for large sizes. Consequently, Gustaf and I used to do detective work, and build a database of contacts and prospects that the store could send direct mail campaigns to. If you think this is kind of freaky, you could be right. I am just saying that this is use of big data, in a small way. Because these campaigns were psychographically targeted, response rates were good, because the offers and promotions were always relevant to those Swedish men who were above 190 centimetres.

The second example, this time definitely in the interest of transparency, relates to what we do at Thinque. And it's also something we advise our clients to do. In other words, we walk our big data talk. If you check out www.thinquetank.com or www.anderssorman-nilsson.com you may find that there are all these cool fields where you can fill in information, like your credit card details. Yikes. Nonetheless, there are fields where we offer downloads of thought-leadership reports, presentations, strategy models, video series and access to VIP information. We offer these in exchange for your information and being allowed to market to you. It's called permission-based marketing. We give you something, and you pay with your information. This enables us to know who is browsing our website, from which company's IP address, and with what digital intent (as described by the sign-up forms), which indicates specific psychographic interest. This data we store securely, and it becomes a treasure trove for our marketers and business development managers, because they can decipher who is a cold and who is a hot lead.

The back-end engine that drives this exchange of content is called Hubspot, and it's a digital marketing-in-a-box system that integrates with our customer relationship management system, Salesforce. And it provides us with some great information for our business. Beyond the macro analysis of our website's efficiency in converting leads into prospects into clients, it also gives us interesting information on your browsing habits in our website. Hubspot might tell us, for example, that you found us via a Google search of the keyword 'futurist', that you checked out a speaking video, downloaded a Scenario Planning report, read a blog and then you bought this book. Based on the amount of time you spent on the site, and how 'sticky' you found it, Hubspot then provides a rating for how hot you are as a lead. Yes, you. If at any stage you provide us with your phone number and email address, we may contact you based on Hubspot's recommendation and rating for a more personal, human interaction, where hopefully you'll choose to engage us for strategic consulting, digilogue marketing strategies, or a keynote presentation in an exotic location for your next leadership conference. Sold? Do we do this? Yes, like smart digital players and as a small business, this is a great tool for getting closer to our clients and being impressively relevant in the information we provide. Digital data makes us sticky, and brings us closer to those we want to speak with. Big Brother-esque, yes. Highly relevant and tailored information, yes. As a research project, spend some time on our website, and reach out to us on info@thinque.com.au and ask what rating you received based on your browsing habits on our website.

The precursor to the digital grapevine — Wagga Wagga

Because of digital incursion, we all veritably live in Wagga Wagga. What? A colleague of mine, Chad, grew up in this interestingly named rural town in western New South Wales, Australia. Wagga Wagga is largely accepted to have meant 'the place of many crows' in the original local Aboriginal language (although there is some dispute about this), and is today a city of 46 913 people. My colleague is homosexual — something that wasn't necessarily appreciated in this all-macho, sport-loving town in the Australian bush. Growing up in Wagga was tough for Chad. His every move was watched, with the ladies in town gossiping at the hairdressers about his 'indiscretions'. Word about the mischievous Chad spread like wildfire along the bush telegraph, as inquisitive neighbours and Victoria Bitter–sipping sheep farmers accelerated its speed. Paparazzi were not required to authenticate rumours, and each stage of his teenage years was carefully monitored by the Parents & Friends Association (without his personal approval). As soon as he could, Chad moved to the big city, away from the bush telegraph and the gossip, and the inquisitions into his private life. The difference today is that Chad can make a choice about what pieces of his life he chooses to share — both in the analogue world and in the digital world.

While you may both fear and welcome the digital grapevine, there is an element of personal choice inherent in this aspect of digital disruption. While you need to remain cognisant that people don't flee the system and that privacy isn't invaded in this digital realm, you also need to recognise the opportunity that the digital data trail entails for you to know more about your clients and to be more emotionally intelligent in your communications, even if it is artificially enhanced. One day, Woolworths, Qantas and MasterCard may be as enticing as my French waitress who knows how I take my piccolo latte. Let us hope that it's in a way that is respectful of our privacy.

Round 5. Digital. With warnings for foul play. In the analysis, it becomes clear that analogue memory and human computing capacity could never handle this amount of data, which limits business scale and opportunity. Yes, the French waitress might remember the VIP clients of Kawa Cafe, and they might number 20, 30 or 50. However, beyond these numbers humans start struggling. No matter how good we are at crunching emotional numbers and data in our head, it's challenging for a human to juggle the shifting tastes, the diets, the preferences and the eccentricities of large numbers of clients. Customer relationship systems like Salesforce do this via the kind of optimised algorithms that we could never replicate in our brains.

On balance, this round belongs to Digital Disruption, which despite our concerns around foul play was able to move marketing from an art to a science, deliver a deathly blow to demographic analysis, and be impressively and sometimes disturbingly relevant. Digital communications and digital print enables an almost sabermetric approach to better marketing decision-making. Supported by Moore's law of exponential returns, this is likely to see artificial marketing intelligence become ever more emotionally intelligent, as we accept this digital creep into our lives.

So digital takes this one out. But the crowd is increasingly respecting Analogue Anachronism for its ability to connect in an analogue authentic way, and for remembering my preference to drink a *piccolo latte*.

Ask yourself...

Start thinking data tactics with the following:

- How are you currently using data and customer information in your business?
- Are you maximising your insights about your clients by utilising a cutting-edge customer relationship management tool with real-time data?
- What data or correlations could you dig out to turn data into information into knowledge about your core clientele?
- What new insights might you be able to attain if you visualised the data you own?
- How could you boost your client loyalty by continuously tweaking your communications based on the information you gather?

Apply it...

Here's how to use data to your advantage:

- If you don't already have a cutting-edge, web-2-lead customer relationship management system like Salesforce, get one. Put data crunchers and data analysts to work updating your client records, and deriving insights about psychographics, demographics, purchasing patterns, sleeve lengths or their equivalent, and digital and analogue contact details for your clients and prospects.

- Stage a digilogue campaign using both digital communications and analogue features based on digital print. Measure and test what headlines, copy, photos and calls-to-action work best across your database. Tweak and optimise for the next campaign.

- Create a loyalty program. Most likely 20 per cent of your clients are responsible for 80 per cent of your business. Treat them like the platinum members they should be. Gather more data about them and use this information to pursue lateral opportunities that mirror their data, and over time create more platinum clients or members. Provide exclusive offers to your membership.

Endnotes

1 Unknown, 'Woolworths Full Year Report', Wikipedia, en.wikipedia.org/wiki/Woolworths_(supermarket), last accessed 9 April 2013.

2 U. Friedman, 'Big Data: A short history', *Foreign Policy*, www.foreignpolicy.com/articles/2012/10/08/big_data, last accessed 9 April 2013.

3 E. Schmidt (CEO of Google), in U Friedman, 'Big Data: A short history', *Foreign Policy*, www.foreignpolicy.com/articles/2012/10/08/big_data, last accessed 9 April 2013.

4   Unknown, 'IDC Worldwide Big Data and technology and service 2012–2015 forecast, March 2012', IDC, www.idc.com/getdoc. jsp?containerId=prUS23355112#.UWKorBz-HUk, last accessed 9 April 2013.

5   Unknown, 'World data more than doubling every two years — driving Big Data opportunity, creating new IT roles', EMC-sponsored IDC Digital Universe study, australia.emc.com/about/news/ press/2011/20110628-01.htm, last accessed 9 April 2013.

6: Analogue escapism

We saw Digital Disruption take out Round 5, but with hints of foul play. Can Analogue Anachronism resist the riptide of Big Data and digital combinations and counter-punch back into a balanced scorecard for the bout? In this round, we will travel in an RV to the Nevada desert to explore a temporary city built on the notion of reconnection through disconnection, we will explore why the analogue outdoor sports industry is booming in response to increased digitisation of our lives, and why the spa industry has grown healthily during the global financial meltdown. Finally, we will explore why analogue hobbies like recycling have become cool for Swedish gentlemen, and why knitting is retro chic among today's digital natives. It seems that in this chapter the old tortoise gets to the finish line before the frantic hare.

Reconnecting by disconnecting: Burning Man

The Nevada desert dust is kicking up a desert storm (see figure 6.1, overleaf). It's dusk. I have been driving for 10 hours and I'm 511 miles north of Las Vegas. I'm in the middle of nowhere — off-road, no GPS signal. I am greeted by signs encouraging 'radical self-expression', and advising 'leave no trace' and 'don't be an observer, be a participant'. I'm in a Ford recreational vehicle and, like 56149 other people, I have made the pilgrimage to Burning Man at Black Rock City, Nevada. The mutant art cars, RVs and SUVs in front of me are driving at five miles an hour, yet their tyres are still disturbing the finely cut sands, and the winds pick up the fine grains and cover everything in sight in a grey mist. I am greeted by Burner officials in seriously weird outfits. Cyber bunnies. Dreadlocks. Magic wands. This is *Mad Max* meets Woodstock meets TED. In August of 2012. Fur, neon, leather, bandanas, Oakley goggles — and gas masks. For the dust. The theme — Fertility 2.0. Strange? Yes. Wonderful? Kind of. Connected? Absolutely.

Figure 6.1: Burning Man from the air

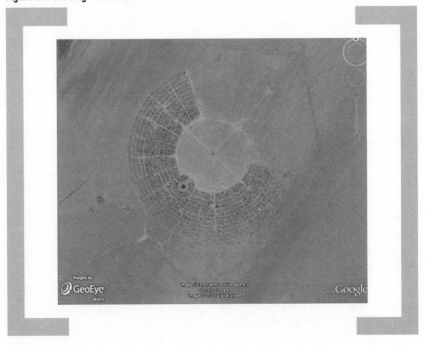

Connected not in the technological sense, but more in the human sense. You know the type of connection that your strange Californian aunt speaks of, the one your cousin tells tales of from an Indian ashram, or the one your astral friend from Byron Bay, Australia, fondly romanticises about. Batique, flower power, polyamory, expression, incense, yogis — that kind of thing. For those of you not familiar with Burning Man, let me bring you up to speed. Each year an increasing number of Burners descend from around the globe on Black Rock City, Nevada, a town that only exists for one week a year. The location is a vast dry lakebed (a playa) wedged between two imposing mountain ranges. The Burners' pilgrimage here is an expression of community, artwork, absurdity, decommodification and revelry. The city's street and block infrastructure is a half-moon shape, with concentric circles forming avenues with names like Alyssum, Begonia, Dandelion, Hyacinth extending in alphabetical order from the main drag — the Esplanade. The streets are named after the 12 hours of the day, and oriented and aligned similar to a sun dial, and the city is accordingly arranged to align with solar angles. Very hippie, I know. Upon arrival, I am told that my camp, Palace of Balunsia, is located in the right fallopian tube at 4.15 pm. The city of Black Rock, once conceived of on a beach by 20 burners setting an effigy alight, has now become an annual mecca for people wanting a week's escape from digital connectivity, the rat race, commercialism, social norms and traditional structures, and seeking to reconnect with themselves and others in an alternate dimension. Human connectivity — no broadband required.

I think I am prepared for this. I have decided to do Burning Man in style. My Apollo rental RV is decked out with air conditioning, shower, toilet, fridge, double bed, DVD player, and a TV — just in case. On my last city visit in Reno, three hours from the gates of Black Rock City, I stocked up for my five days in the Playa on gourmet food, microbrewery beers and supplies at Macro Wholefoods. Some Burners look at RVs in disgust from the safety of their hard-core tents but, for me, I'll claim the RV as part of my radical self-expression — and certainly an expression of my desire for some level of comfort.

I head in the direction of 'the right fallopian tube' and arrive as one of only two RVs in camp Palace of Balunsia, 'an ever-shifting Bedouin nation far far away, who hold strong the notion of perfect unbalance'. Balunsia is a place for 'kings and sultans, gnomes and fairies, musicians and philosophers, historical figures and sparkle ponies, fireflies, butterflies and all the creatures of the sea' (see figure 6.2, overleaf, for an example of the strangeness). I wonder whether I as a futurist and innovation strategist will fit the bill. (Last time I checked, I was not a sparkle pony.) Therese, my first-ever girlfriend from Sweden who is also a 'Balunsian', and her tent friends, Scott and Emma, had encouraged me to pick out some weird gear in Reno to wear — and to err on the side of the absurd. I come ready with a big furry top hat, long leather jacket, neon buttons, green leggings, cyber sailor gear, silver wrap-around sun goggles, and glow sticks. The beach cruiser bike in turquoise from the Vegas Walmart is in the back. By my standards this would be the weirdest I have ever looked. No cell reception. No wi-fi. No way back.

Figure 6.2: Burning Man transport option

This is my creative break in 2012. A way of escaping the fast pace of everyday life. A way of seeing something totally different from board rooms, hotel suites, tailored suits, strategy maps and airport lounges. Somewhere totally outside of my regular comfort zones. I have no idea, really, of what to expect, but I have been told the basic principles in the cashless economy of Burning Man: radical inclusion, gifting, decommodification, radical self-reliance, radical self-expression, communal efforts, civic responsibility, leaving no trace, participation and immediacy. And that nudity is common, but not required. In other words, Burning Man is creative escapism. And that's totally okay. I am here to reconnect by disconnecting. I arrive at the camp site and walk to the main tent, past the grey water evaporation site and the camp's kitchen. I see Therese and Emma in their neon wasp outfits, and strange head gear. I have arrived, and I am ready to radically self-express. Let Burning Man begin.

Burning Man is a reminder for me that the more some things change for real, the more other things stay exactly the same. The more teched-up, tuned in, and mobile enabled we become, the more we crave to escape this digital speed. The more our smart phone connects us technologically to the rest of the world, and particularly to work, the more we seek to escape its pinging, reminders and vibrations. We crave the stimulus of knowing how many likes our recent holiday post got us on Facebook, whether our latest Tweet garnered viral reception, and whether our Klout score improved overnight. Digital distress. Our tech devices tend to wind us up, while we want to wind down.

Digilogue balance

Ironically, we crave the human connection of a digital Christmas SMS, yet sometimes forget about the analogue people we're actually spending Christmas with. Technology has fundamentally shifted our behaviours, and enabled us to text or tweet someone while sitting next to that someone, and we're becoming push notification stimulus addicts. But simultaneously, we derive increasing value from analogue escapist experiences like Burning Man, yoga, meditation, holidays, religious groups, slow cooking, farmers' markets, and, yes — cross-generational family time. The more technology changes our communication patterns and our attention spans, the more other things stay exactly the same — things like our need for touch, for timeless patience, for community and for meaning. As human beings, we're constantly balancing the fast and convenient aspects of the digital world, with the deep and meaningful of the analogue world.

Here are some ways to canvass our digital connectedness:

- In the United Kingdom, Britons spend 128 minutes a day on their smart phones.[1]

- Mobile internet traffic in the United Kingdom increased 100 per cent between Q3 2011 and Q3 2012.[2]

- Mobile internet traffic in the United Kingdom is projected to increase 40 times by 2016.[3]

- Swedes looked at their smart phones on average 150 times a day in 2011.[4]

But balance this by canvassing analogue escapism:

- Between 2000 and 2011, the total US private job market shrunk by 1.5 per cent, while the American salon and spa job market grew by 18.3 per cent.[5]

- Over a decade, the number of US spa locations has grown from 4140 in 1999 to 19900 in 2010, with industry revenue growing from $5 billion in 1999 to $12.8 billion in 2010.[6]

- In 2008, the 6.8 per cent of Americans who practised yoga spent $5.7 billion on yoga classes and products, including equipment, clothing, vacations and media. This figure represents an increase of 87 per cent compared to the previous study in 2004.[7]

- US mega churches (those with congregations that draw 2000 or more adults and children in a typical weekend) continued to increase in the number of people they drew between 2000 and 2008. Their average rate of growth for five years was around a 50 per cent increase in attendance.[8]

- Between 2002 and 2007, the number of American farms increased by 76 000, compared to a decline of 87 000 in the five years before that. But half of all farms in the United States have sales of less than $5000, which means many farms exist for non-financial, hobby-based, reasons.[9]

- National farmers' markets listings in the United States increased from 2863 in 2000 to 7864 in 2012.[10]

- Slow Food is a grassroots, international non-for-profit organisation that has over 100 000 members in 1500 *convivia* — local chapters — worldwide, as well as a network of 2000 food communities who practise small-scale and sustainable production of quality slow foods in a reaction against fast food.

- Americans spent $11 billion in 2008 on self-improvement books, CDs, seminars, coaching and stress-management programs — 13.6 per cent more than they did back in 2005. Latest forecast is for 6.2 per cent annual growth through 2012.[11]

- Americans spend nearly as much on snow sports ($53 billion) as they do on internet access ($54 billion).[12]

- Americans spend more on bicycling gear and trips ($81 billion) than they do on airplane tickets and fees ($51 billion).[13]

- The outdoor recreation economy grew approximately 5 per cent annually between 2005 and 2011 — during an economic recession when many sectors contracted.[14]

These statistics indicate that we are more Zen than we give ourselves credit for. I know, I'm getting a bit Yin and Yang on you here. For everything there is a counterbalance. Good and evil, Luke and Anakin Skywalker — all of that. Digital balanced by the analogue, hearts and minds, the rational versus the emotional, urban versus rural, global versus local, iPod touch versus haptic touch. If the digital lets us escape to anywhere, anytime at a time when even instant gratification is too slow, the analogue lets us escape to the nostalgic, the old school, the natural and the slow. It helps us rewind, recharge and reconnect with ourselves, while disconnecting from the digital world. Critical.

Reconnect by disconnecting

So it was that in June 2012, I jumped on my bike and rode up to the Surry Hills farmers' market. It wasn't so much that I needed anything in particular, but I wanted to escape the urban feel of my apartment, and bypass the food duopoly where I conveniently access supply-chain optimised and mass-produced fast-moving consumer goods most of the time. It was a sunny, fresh Sydney mid-morning and, immediately upon entering this bazaar of stalls, I was met with the aromas, smells and visual attraction of farmers displaying their wares and showing off their artisan skills. For three hours every Saturday in this spot on Taylor Square, rural meets urban. When you pass through the busy lanes and feel the herbs grown in the Blue Mountains slip between your fingers, when you speak to John who roasts and grinds the coffee beans, and then makes you a handcrafted espresso with a personal history, or when you're given a family insight into the origins of a rye bread recipe, you realise that business can still be local. And visiting a farmers' market again makes me think that the more some things change for real, the more other things stay exactly the same.

The feeling during this shopping experience was different. This time it wasn't about efficiency and speed. When my family spends time together at Christmas, visits to the supermarket tend to be about digital lists on iPhones, and time-critical missions of hunting and gathering. Although Mum usually slowly meanders up and down the aisles, the three boys — Lars-Olof (Dad), Gustaf (my bro) and I — tend to rush off to gather the ingredients. We all want to get the job done, and get out of the machinery as quickly as possible. It's about speed. And it's the opposite of what food should be about — enjoyment, connection, sharing, caring, slower vibes, personal conversations, savouring, lingering, siestas. The farmers' market slowed things down for me. I had a chance to connect. A chance to buy things that weren't on a digital list, and to go with my gut instinct — literally. To be inspired by an artisan honey, to stop for a sinful scone with jam and homemade marmalade, to sample a quince paste with blue cheese. It allowed me to go into analogue tortoise mode instead of digital hare mode.

Today, people around the world connect by disconnecting. Some turn to meditation, some to downward dog poses, some to extreme sports, some to churches, synagogues, temples and mosques. Others make pasta from scratch, or make and hang their own chorizo. Some join forces and participate in Tough Mudder events. Others keep bees. Some do bonsai, while others have urban gardens. This is not just about hobbies, it's also about creating mind space away from the 24/7/365 pressures of digital mania.

During a writing retreat in 2012, I was reminded of the value of analogue escape. As a futurist, I am constantly connected to my devices, and I admit I am often one of the first people on a flight to turn my phone back on to check messages, emails and status updates. As soon as we hit the tarmac, and most often before the pilot gives me the green light. (A digital maverick I know — I am sorry.) The week prior to this writing retreat I was in Las Vegas for a speaking engagement — a city of technology, digital bling and 24/7 entertainment — and I was as far away both geographically and mentally from the slower pace of Injidup in Cape Naturaliste National Park, Western Australia, as I could conceivably be. Still wired I emailed the landlord of the Injidup Surf Shack that I was about to rent. My question was whether wi-fi was available at the house. Kate, the artist slash landlord, responded kindly but firmly. There was no wi-fi, there was no TV, there was no DVD player, there was no mobile coverage — that was the point. Ahaaaa. Solid reminder that sometimes creativity gets boosted by being disconnected from our fast-paced world, and that boredom sometimes equals relaxation and winding down. Now you've reached this far into *Digilogue*, hopefully you've found that some good, solid writing was done at Injidup while I was overlooking the Injidup Spa Retreat and the Indian Ocean sunsets. We unwind by being unwired. We all need to escape the digital and go analogue from time to time.

Analogue hobbies

Meanwhile, on the other side of the world, another analogue grassroots movement is picking up speed (and not in a stressed-out way). In Sweden, recycling is super hip, and my parents are totally into it. In fact, I'd be prepared to say that my father, Lars-Olof, loves it. He is part of a movement that is producing significant results for Sweden's environmentally friendly national brand. Sweden's recycling system, dependent on grassroots activism and family-based support, is now so efficient that it's importing trash from other European countries to power its renewable energy plants, which burn trash. (Burned waste powers 20 per cent of Sweden's district heating and provides electricity for roughly 250 000 Swedish homes.) But Sweden doesn't have enough trash. Because of family-based recycling, only 4 per cent of Sweden's trash ends up in landfill, compared to 34 per cent in the United States. So now Sweden has to import 800 000 tonnes of waste from countries like Italy, Bulgaria and Norway annually.[15] And this is because of people like my dad, who spend their time sorting through cartons, bottles, Tetra Paks, newspapers and food waste. While urban living offers its citizens roadside collection, rural citizens like my parents take their recyclables to a recycling station. In many nations this would spell the death of recycling efforts, but the compliant Swedes, and particularly men whose gender role it tends to be, love the weekly outing to the recycling station.

The success of this program is largely because of people who are reacting to the digital consumer society we now live in and want to help make an impact on the environment. Who spend time diligently placing things into their proper place and compartments, so that once a week they can make an excursion to the local SITA recycling station. I have had the privilege of having to help out in this modern pastime. (While recycling in Sweden is a voluntary activity, once your family decides to recycle, intra-family the activity is no longer voluntary.)

On any given Saturday morning, Dad fills up the olive-green Volvo (of course) with the week's recycling. Washed out Arla milk bottles, neatly stacked SvD and Dagens Nyheter newspapers, wine bottles (sorted into colours and transparents), batteries and plastic packaging sit in individual re-usable hard plastic containers, and are secured in the back seat and in the station-wagon luggage compartments in our hound Cicero's secure dog cage. We are off on an expedition. We arrive at the local, and very high-tech, recycling station, and Dad swipes his membership card at the boom gate. Other Swedish men have had the same idea this Saturday morning, and there is a neat, orderly line of vehicles that make their way around the circle, as people diligently (and correctly) place the individual trash in various containers.

The recycling station is outdoors and, with its orderly cyclical track, I cannot escape being reminded of IKEA's labyrinthine consumer track at this sustainable graveyard for trash, including old IKEA furniture. It's where stuff goes to die and to be reborn as something else. Dad quickly chats to some other amateur recyclers — maybe they exchange tips — and I witness modern masculine Swedish bonding. Dad brings an extremely rigorous discipline to this, and prioritises this Saturday recycling run as a civic, neo-pastoral activity. My mum realised this in 2004 when she decided that an old Arla milk pack was a good place to hide a house key for a cousin who was coming to stay. Needless to say, the pack was recyclable and, without checking the content of the carefully washed out pack, Dad ensured it made its way to the recycling station. The result of his commitment to recycling was both my parents going bin-diving for the house keys at Skä Community's SITA recycling station. Recycling — despite the odd misadventures of my parents, you should try it some day as a practical form of meditation away from the digitally fast-paced consumer society. Some men go deer hunting; Swedish men go to the recycling station.

What about women? (And apologies for the gender stereotypes here.) Well, some knit. And not just Swedish women. Indeed, it seems not all millennials are constantly glued to their iPads. Many are instead glued to their old-school, analogue knitting kits. Peter Fitzgerald, a retail director at Google UK, says that while online searches for knitting-related terms have grown steadily since 2004, the knitting growth in 2011 has been really noticeable. 'Our data shows that searches for knitting have increased over 150 per cent just this year', he says. The term 'knitting for beginners' has increased by 250 per cent.[16] At John Lewis in the haberdashery department, sales figures for luxury Germany yarn brand Gedifra have risen by 126 per cent compared to 2010, while Rowan yarn is up 57 per cent. Worldwide, Rowan yarn sales have just about doubled in one year.[17] Penguin gave this analogue craft its stamp of approval when, in 2010, it released its first knitting book in the United Kingdom since the 1950s. Meanwhile, Stitchlinks, a community dedicated to promoting the use of therapeutic knitting and stitching through research, is drawing causal relationships between knitting and mental health.[18] And they have an increasing pool of data from knitting enthusiasts to draw from. This pastime has become retro chic, and enthusiasts are using it as an escape from digital consumerism, and as a way to create something using the digits on their hands, instead of their digital minds. Fascinating. And proof that my grandma Ingrid, who loved knitting, would have been supercool had she been alive today.

What is going on here? My contention is that we are seeking to balance the digital and the analogue in our own lives. While we are fascinated by the future, we get our grounding from the historic. Yes, recycling is not the same as collecting manure, processing it and then spreading it as organic fertiliser on Swedish strawberry fields, but it is a modern version of pastoral caring for the land. For those of us whose hands love the feeling of yarn, it is a welcome break from the ergonomic strain caused by Blackberry thumbs, iPad necks and Android eyes. It also connects us to something historic, something that used to be done in the absence of technology, in the light cast by a lonely oil lamp. Taking the time to reconnect to the small-scale farmers who make our food is a way to bring the humanity back into food, and a way to extend the expressive finger to hyper-efficiency and precision agriculture. Striking a lotus pose at a yoga retreat becomes a way to reconnect with ourselves and our inner longings, and our individual search for meaning. Coming together at our local mega church (or alternative religion, concert, sport or spiritual event) and feeling alive, spiritually connected and happy brings us something analogue that the digital cannot give us. All of these activities provide people with an inner analogue peace derived from a slower pace, which all humans crave. As Yogis say — Om.

We learn the analogue through the digital

The intriguing thing is that the digital world informs our analogue decisions and actions. Where do people learn to knit? YouTube. Where do they access new patterns? Via digital downloads. Where do tricks and tips get exchanged? Via online social networks. Yeah, sure, knitting enthusiasts still connect in the analogue world, but it dramatically differs from how I recall knitting being practised and taught.

When I was a child in Sweden, our family used to spend the summers an hour outside of Stockholm with my grandfather Per and grandmother Ingrid at their rented summerhouse, an old vicarage on Färingsö Island in the Lake Mälaren archipelago. Ingrid had grown up in West Gothland on the farm Riddarhagen near the city of Töreboda. Grandpa Per, yes the second generation owner–manager of Georg Sörman, had grown up in Stockholm but had spent his summers at the summerhouse Sörgarde nearby (thus the cupid acquaintance). They both had strong associations with country, and the summerhouse 'Prästgärden' (vicarage) in Stockholm became a manifestation of hobby farming.

Per had decided that the more than 120 apple trees in the property's orchard would be turned into a productive asset, and his three daughters, and their partners and children, became a handy workforce. Not my idea of fun. Pastoral, yes, but relaxing? Not in my books. Per was a technophobe and rather nostalgic about the virtues of spartan, tough living. (Having not grown up with the reality of cleaning up manure and milking cows by hand at 4 am, 365 days a year (like Ingrid), he might have had a romantic, yet misplaced, idealistic view of farming.) So he insisted on no radio, no TV and no technology — and even no shower at the house. This meant that evenings, from the perspective of a child, were pretty boring. Board games were limited to a 1960s version of Chinese chequers. When the adults sat up, the open fireplace was the only visual entertainment (ironically, during the days of DVDs a recording of an open fire became a digitised best-seller around the world). And grandma would knit. Sometimes she would show my mum or her sisters how to do a technical move, which would elicit an occasional 'Aha!'. The learning was analogue, and it was passed down inter-generationally. Sound romantic? Maybe. The point is there was nothing else to do. Today we have a choice, and the popularity of knitting illustrates that we choose to exercise it in favour of the old school and natural.

This longing for the analogue, natural, escapist old school creates business opportunities. The US outdoor sports and recreation industry is a $646 billion industry. Using our bodies, breathing fresh mountain air and connecting to nature is something we're prepared to pay a premium for. Not having mobile coverage at Burning Man or Injidup is seen as bliss (despite my initial misgivings). And the analogue enables us to reconnect with our own hearts, and turn off the chatter in our digital minds. This is something the digital world cannot replace in the short term. This creates a huge opportunity for business — to build community, to create attractive physical spaces, to sustainably use nature, to be more artisan and to create escapes where we can reconnect with ourselves and those emotionally near to us. Ask yourself, what is your business doing to facilitate connection for people?

Enabling reconnection by disconnection

Qantas allows me to reconnect by disconnecting. For that, I love them. And I know many other business travellers who feel the same about the Qantas First Class International Lounge at Sydney Airport. It's a veritable oasis on the way to somewhere else. As the website, complete with a digital tour, tells the uninitiated and aspiring,

Designed by Qantas Creative Director Marc Newson, the Lounges [at Sydney and Melbourne] are exceptional retreats offering unsurpassed levels of comfort, service and luxury, complete with day spa, restaurant, library, private work suites and concierge service.

It's perhaps unsurprising that one of my friends and close clients, let's call him Michael, once invited me to join him at the lounge four and a half hours prior to our flight's departure from Sydney to Dallas. For brunch, a glass of champagne and some midday desserts — in Australian celebrity chef Neil Perry's open kitchen. And, of course, being a metrosexual, I left Michael for 30 minutes and ducked off for a quick facial at the Payot day spa. All of this at zero expense. Okay, both Michael's company and mine give lots of business to Qantas, but this oasis is a great brand building exercise. It's the ultimate escape for weary globetrotters: a private room away from the noise and stress of an airport. Because of this analogue space, Qantas is able to provide a sticky experience that is an escape before the escape to somewhere else. It gave Michael and me a good chance to talk shop in a relaxed environment, away from the digital noise of our normal lives. And it always leaves me wondering why more companies aren't focusing on creating extraordinary analogue escapes for their clients. As Qantas frequent flyers will attest, this carrot is a truly sticky experience, meaning that Qantas knows who their 20 per cent are who provide 80 per cent of their revenue. Analogue escapism has dollar signs attached to it. And by the way — this is why my second carrier of choice, Virgin, has introduced a campaign of 'The Romance is Back', which speaks directly to our analogue hearts, and reconnects us to the virtues of this challenger brand. Meanwhile, frequent flyers are happily being courted by two suitors aiming to steal hearts.

Round 6. Analogue fights back. In this round, Analogue Anachronism whipped back with relaxed and centred energy, against the frantically defensive Digital Disruption. It became clear that our digitised minds crave balanced reconnect with analogue hearts. The more digitised our lives become, the more we seek to reconnect by disconnecting from the digital. Escaping into the analogue outdoors, recycling fast-moving consumer goods packaging, and knitting in front of an open fire are little luxuries we crave to centre and ground us, and get us away from our digital doldrums. We seek out reconnected meccas like Burning Man, spa retreats and branded spaces of relaxation to wind down instead of winding up. Analogue Anachronism made a heavy point in this round that being relaxed, disconnected and in the moment is a big commercial opportunity, and as a result keeps winning the hearts and minds of the audience. And in this round, the meticulous judging panel was also on its side.

Ask yourself...

Think about the following to get some analogue juices flowing:

1   Is there an opportunity for your business in providing a great analogue place where people can escape from the digital doldrums?

2   When was the last time you arranged a physical meeting or conference where your clients and prospects had a chance to tune out and slow down?

3   Have you recently re-balanced your digital, high-tech offerings with analogue, high-touch points?

4   How does your business provide clients with deep and meaningful experiences, and not just efficient, convenient service?

5   When you think of thanking your clients for their business, how do you thank them and with what analogue gifts or experiences?

> ## Apply it...
>
> Now create from analogue experiences:
>
> - Create a chilled-out, physical space where you can engage with clients in a low-tech, analogue way. This could be a temporary pop-up, conference exhibit or retail feature, or a permanent physical feature of your office, or bricks and mortar experience.
>
> - Reconnect with clients in nature. Take a walking meeting, or invite them for an experience outdoors—no smart phones allowed.
>
> - Take your next client-facing conference out of hotels and into an analogue space. Think ocean, bush or an island.

Endnotes

1   P. Alestig Blomqvist, 'O2 research', SvD, www.svd.se/naringsliv/karriar/ny-teknik-ger-nya-ergonomiproblem_7769078.svd, last accessed 9 April 2013.

2   P. Alestig Blomqvist, 'Ericsson report', SvD, www.svd.se/naringsliv/karriar/ny-teknik-ger-nya-ergonomiproblem_7769078.svd, last accessed 9 April 2013.

3   P. Alestig Blomqvist, 'Ericsson report', SvD, www.svd.se/naringsliv/karriar/ny-teknik-ger-nya-ergonomiproblem_7769078.svd, last accessed 9 April 2013.

4   P. Alestig Blomqvist, 'Ericsson report', SvD, www.svd.se/naringsliv/karriar/ny-teknik-ger-nya-ergonomiproblem_7769078.svd, last accessed 9 April 2013.

5   Unknown, 'Economic snapshot of the salon and spa industry', Professional Beauty Association (PBA) and National Cosmetology Association (NCA), probeauty.org/docs/blueprints/2012_Economic_Snapshot_Salon_Industry.pdf, last accessed 9 April 2013.

6   Unknown, 'The US spa industry—fast facts', International Spa Association, www.experienceispa.com/media/facts-stats/, last accessed 9 April 2013.

7   Unknown, 'Yoga Journal releases 2008 "Yoga in America" market study', Yoga Journal, www.yogajournal.com/advertise/press_releases/10, last accessed 9 April 2013.

8   S. Thumma & W. Bird, 'Changes in American megachurches: Tracing eight years of growth and innovation in the nation's largest-attendance congregations', Hartford Institute for Religion Research, hirr.hartsem.edu/megachurch/megastoday2008_summaryreport.html, last accessed 9 April 2013.

9   J. Tozzi, 'Entrepreneurs keep the local food movement hot', BusinessWeek, www.businessweek.com/smallbiz/content/dec2009/sb20091217_914398.htm, last accessed 9 April 2013.

10 T.P. DiNapoli, 'Farmers' markets double in New York', Office of the New York State Comptroller, www.osc.state.ny.us/press/releases/aug12/081112.htm, last accessed 9 April 2013.

11 M. Lindner, 'What people are still willing to pay for', Forbes, www.forbes.com/2009/01/15/self-help-industry-ent-sales-cx_ml_0115selfhelp.html, last accessed 9 April 2013.

12 Unknown, 'The outdoor recreation economy', Outdoor Industry Association, www.outdoorindustry.org/pdf/OIA_OutdoorRecEconomyReport2012.pdf, last accessed 9 April 2013.

13 Unknown, 'The outdoor recreation economy', Outdoor Industry Association, www.outdoorindustry.org/pdf/OIA_OutdoorRecEconomyReport2012.pdf, last accessed 9 April 2013.

14 Unknown, 'The outdoor recreation economy', Outdoor Industry Association, www.outdoorindustry.org/pdf/OIA_OutdoorRecEconomyReport2012.pdf, last accessed 9 April 2013.

15 S. Jones, 'Sweden wants your trash', NPR, www.npr.org/blogs/thetwo-way/2012/10/28/163823839/sweden-wants-your-trash, last accessed 9 April 2013.

16 P. Lewis, 'Pride in the wool: The rise of knitting', The Guardian, www.guardian.co.uk/lifeandstyle/2011/jul/06/wool-rise-knitting, last accessed 9 April 2013.

17 P. Lewis, 'Pride in the wool: The rise of knitting', The Guardian, www.guardian.co.uk/lifeandstyle/2011/jul/06/wool-rise-knitting, last accessed 9 April 2013.

18 Unknown, 'About us', Stitchlinks, www.stitchlinks.com/about_us.html, last accessed 9 April 2013.

Taking stock

The Digilogue Strategy Map and a banking case study

One of the most exciting industries with regard to digilogue opportunities is banking. Think of banking for a moment — most likely you think of your local branch office or perhaps an elaborate Art Nouveau building with high ceilings and lots of gold-emblazoned window bars. Banking is undergoing a massive shift as a result of digital disruption (as we discussed in chapter 1). In Australia, banking is dominated by the Big 4 banks — Commonwealth, Westpac, ANZ and NAB. Non-bank alternatives are available, however — for example, credit unions, which occupy an interesting space with regard to digilogue. Why? Well, credit unions are seen by Australian consumers as consumer advocates and an alternative to the 'big bad banks'. They are seen as locally engrained, community-driven and, because of their membership co-op structure, where profits are reinvested into the community of their members, they have been seen as a fair choice, where people are more important than profits. This should make for a digital media swan song, and massive branding opportunities. However, many credit unions have been slow to adopt digital into their strategies, and have instead invested in their core touch point, the branch, as the way to cement themselves into communities. But take a look at the following Digilogue Strategy Map (overleaf), which shows what our client is doing to provide value to digital minds and to connect with analogue hearts. The discussion following the map expands on the strategies.

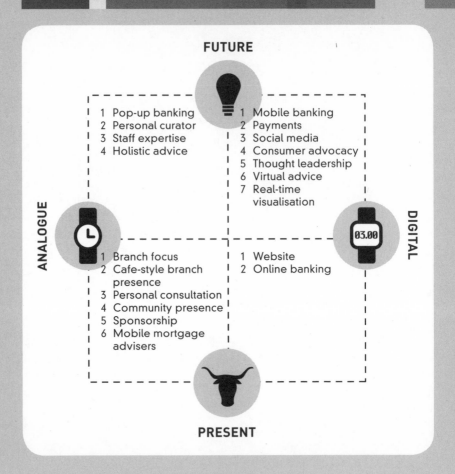

FUTURE

ANALOGUE			DIGITAL
1 Pop-up banking	1 Mobile banking		
2 Personal curator	2 Payments		
3 Staff expertise	3 Social media		
4 Holistic advice	4 Consumer advocacy		
	5 Thought leadership		
	6 Virtual advice		
	7 Real-time visualisation		
1 Branch focus	1 Website		
2 Cafe-style branch presence	2 Online banking		
3 Personal consultation			
4 Community presence			
5 Sponsorship			
6 Mobile mortgage advisers			

PRESENT

Analogue — present

This credit union has a strong and focused branch strategy. It wants to follow in Apple's retail footsteps and connect with its members in a face-to-face fashion. The branches are more like cafes, are on brand (in that they feel informal and are strongly community driven), and are located centrally in the communities where they play a part. Inside the branches, there are both private and public opportunities for engaging with staff. Members can access a mortgage specialist in a private room or more casually in a booth, for example. When we mystery shopped our client, we found the service extremely comforting, caring and personalised. This was in contrast to what we experienced in the Big 4 banks we visited. The credit union is also reaching out into communities via sports and charity sponsorship, which is amplifying its brand recognition and illustrating its strong commitment to community presence. Being 'out there' in the community is also backed by its mobile mortgage managers, who will consult regionally and in areas where the credit union doesn't yet have a physical branch presence.

Digital — present

The credit union has an engaging website, updates it and its blog regularly, and is socially integrated across the major digital networks like Facebook, Twitter and YouTube. It offers online banking (standard these days). Given that it, and its fellow credit unions, don't control the same extensive network of bricks and mortar real estate as its banking competitors, the digital world offers opportunity for the credit union to gain greater reach and brand recognition by providing further value to digital minds across Australia.

Analogue — future

The leadership of the credit union is committed to a high-touch approach to the future. It believes that in certain parts of banking, and in certain segments of retail customers, people will always look to a trusted banking expert for advice. Overlaying the demographic analysis shows that the average credit union member is 45 years old and ageing, which means retaining the working elements of the analogue strategy makes sense.

The question is what aspects of retail banking can be digitised, and which cannot. Given that day-to-day banking needs and real-time access to accounts and computerised analysis of mortgage repayments and net wealth can largely be done online, holistic, long-term, expert advice — in person — will still be a core element of the analogue strategy for the credit union. Excel can tell a savvy person a lot these days, but occasionally we need a curator or financial adviser to deconstruct what it all means and what the trends in our financial outlook actually say about our behaviours, and the types of financial improvements we can make. That advice may be delivered in a branch, or in the future via pop-up branches, which may look more like a hole-in-the-wall cafe than a bank of old.

Digital — future

To go digilogue, our credit union client needed to introduce more digital touch points. While the big banks and big corporates use social and digital media to seem small, nimble and personable, credit unions, by virtue of their membership focus, co-op structure and embedded culture, already are small, nimble and personable. But they need to tell that story in their communities, both on-and offline. Increasingly, banking is done online, on-the-go. And increasingly we want real-time access to information in visual form via our smart phones. Mobile banking via apps that illustrate key information and developments, combined with payments capability, will be a key for any banking player to retain client relationships, and our client is executing a strategy which is in tune with the fact that banking clients are increasingly taking financial matters into their own hands.

Having a stake in social media in a bigger way, and playing the consumer advocate and cheeky challenger card, is a big opportunity for the credit union — particularly in engaging a younger demographic who mistrust the banks. This plays out in its digitised community engagement, and its rolling out of advertisements highlighting the point that they are about members, not numbers. When you combine web-conference advice with face-to-face advice in-branch, and add it as a layer of customer touch points, you also increase the opportunity for the credit union to go national, and connect with hitherto untapped markets.

Summary

Our client is heading in the right direction. They are re-balancing their branch-centric strategy, by including more digital touch points. While they are expanding their network of branches, extending their digital branch and brand strategy is core to their future focus. The credit union is enjoying high single-digit growth, and is expanding in line with expectations. When the digital combines with the analogue, they will be in a position to truly become the consumer advocate that Australian banking clients deserve. This is in line with broader global trends and feeds directly into consumer behaviour, which indicates that clients and members are voting with their feet and that, increasingly, banking is a digital endeavour. In the future, striking the right digilogue balance will continue to be key.

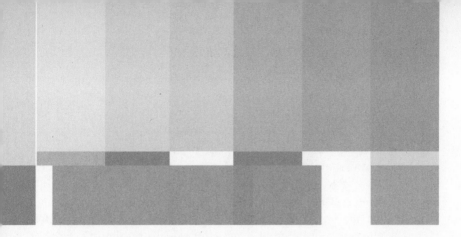

7: Digital diffusion of ideas

Slow and steady wins the race. Or...? The centred and well-grounded Analogue Champion's breathing is relaxed after a successful round. In this next round, things are about to speed up. We will see how digital speed can far surpass analogue focus, and why ideas can be diffused globally and locally at breathtaking speeds because of smart digital communications. We will case study how the Apple iPad became an idea virus that successfully diffused across society in a lightning fashion and created a new category of devices, which we now take for granted. And we will explore how this case study offers hope for both digital and analogue businesses selling services, solutions, ideas and products into tomorrow's marketplace. Fight.

Diffusing ideas in a digitally disrupted era

On 27 January 2010, Steve Jobs launched the digital media tablet — the iPad — in Cupertino, California. The crowds of geeks, tech writers and Apple observers received it with love, whoops and fandemonium. Of course. Steve's fanfare and hyperbole on stage hit new levels of amplification (and perhaps deservingly so). Steve prefaced the launch and demonstration of the iPad with the big hypothetical question — was there room for a 'third category' device? That is, a device that came somewhere between the smart phone and the laptop. He made Jobs-esque sweeping (yet impactful) statements, such as 'everybody uses a smart phone and a laptop', and invited us to consider that for there to be room for a third category device that didn't evolutionarily sit in the category of smart phone or laptop, the device had to meet a pretty high standard, and do a bunch of digital things better than both the smart phone and the laptop. It had to outperform these devices in the experience that it offers to people when they browse, email, photo share, watch videos, play games and read eBooks. Steve categorically dismissed the netbook that was starting to become popular at the time as filling this third category, saying that a netbook was not better at anything, and that it was just a cheap laptop. (There is nothing like dissing your competitors while the world is tuned in for your glorified sales pitch, is there?) Of course, this was Steve's chance to show us the real deal, the third category device of the future. We now call it the iPad. And it's very thin.

Steve went on. He told us that what this device does is 'extraordinary'. It's 'unbelievably great'. Arguing it lived up to the high standards he'd set for a third category device, Steve said the iPad was 'way better than a laptop', 'way better than a smart phone'. The iPad even enabled you to hold the 'internet in your hands'. It's 'phenomenal for mail', and it's a 'dream to type on'. He concluded the pre-demonstration hype by claiming that the iPad is 'more intimate than a laptop', and 'more capable than a smart phone'. Steve thus pulled off one of the greatest sales pitches in history. Why? Because he managed to sell us on a platypus-like concept — the digital tablet, which didn't neatly fit into any one evolutionary box. It wasn't a smart phone, it wasn't a laptop. Just like the platypus, which shares features with mammals (it lactates), birds (it lays eggs) and reptiles (it is venomous), the iPad became a curious, well-branded, unique 'thing' — it became a device we wanted for the future. It was a digital idea, a memetic virus, which took hold in the hearts and minds of innovators and early adopters around the world. The way its popularity spread was the epitome of the digital diffusion of an idea. And the iPad changed the media landscape — both in the way we consume media and in how companies need to think about the business that they are in. We didn't know what it was; we just knew we wanted one. Because Steve told us we needed it. Digital diffusion.

Anyone who is in marketing and sales is in the business of selling ideas. The iPad is a wonderful case study in how to sell a disruptive idea, and how to create demand for the adoption of services and products associated with that idea. The idea of the iPad quickly moved beyond the innovators and the early adopters, who are the first movers when it comes to new digital ideas, and into the mainstream. To explore this concept of digital diffusion of ideas, we will overlay sociologist Everett Rogers' idea of the Law of Diffusion of Innovations across the launch of the iPad (see figure 7.1, overleaf). When you read the following sections, think about how a new product or service launch in your business could be executed in a similar fashion and how both the digital and analogue aspects of the communications form an essential component in how you sell in ideas to your clientele and prospects. To really succeed in business, in a digitally disrupted era, you must learn how to diffuse ideas across society and ensure adoption — in, it turns out, a digilogue way.

Everett Rogers' idea in its essence is this: innovation, no matter how good it is, cannot be successfully adopted without social acceptance and behavioural change. This takes time, which is indicated on the X-axis (the horizontal line) in figure 7.1. Over time, as an innovation is adopted, market share and reach improve, which is indicated on the Y-axis (the vertical line). Everett Rogers studied the diffusion of innovations in different societal groups, and noticed a pattern in how ideas and innovations are adopted. Rogers noted different categories of adopters: innovators, early adopters, early majority, late majority and laggards. These different groups, and their indicative sizes in society, are visualised in figure 7.1, according to the wave-like curvature.

As you study figure 7.1, imagine that a digital idea takes root in the hearts and minds of the innovators on the left of the horizontal, and then in a wave-like fashion spreads to the laggards on the right. As it does, the S-like arrow climbs higher, indicating the diffusion of the idea in society at large, until it reaches critical mass. It is important to note here that adoption is an individual process, as a person goes through a series of stages from first hearing about an innovation to finally adopting it. The diffusion process, however, signifies a group phenomenon — one which suggests how an innovation spreads among consumers. Overall, the diffusion process essentially encompasses the adoption process of several individuals over time. The following sections cover how this played out in the case of the iPad, and how the diffusion process is relevant to how you market your ideas, services and products in a digilogue fashion in your business.

Figure 7.1: wave of selling a disruptive idea

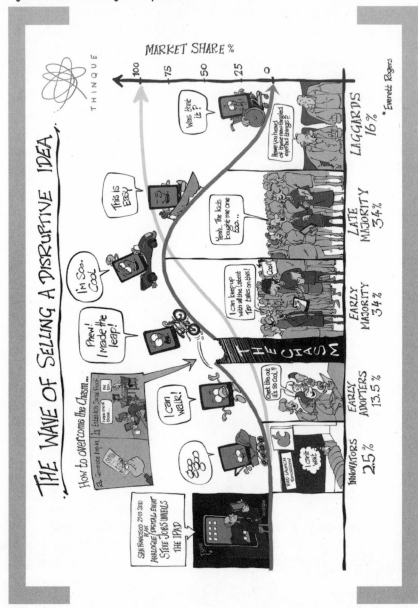

Innovators

When Steve Jobs launched the iPad, it was via a digilogue live event — both digital and analogue — (simulcast) around the globe, to a select group of influencers — the innovators. Innovators constitute around 2.5 per cent of the population at large, and they have a high risk tolerance. In fact, their risk tolerance sees them adopting new technologies that may ultimately fail, and many of them have a 'constant beta' mindset. They are of high social class, have financial resources, and tend to be well connected socially and professionally.

Apple didn't extend a democratic invite for its launch, but rather segmented and targeted a group of innovators to invite, and these innovators could blog about, rave about, fantasise about and geek out about the iPad. The innovators invited included investors, tech bloggers, and Apple fans, partners, channels and vendors (who were also interested in the success of the idea of the iPad and its wave of diffusion). For those who weren't invited to the actual event, some 'second-grade' innovators had to camp out outside of an Apple store for days, as the iPad made its way through to the retail channels around the world. (Perhaps the only time a retailer will be okay with people in sleeping bags on cardboard outside their entrance.)

If you're an innovator, there is merit in simply having the latest device, and being able to show it off, and (if you weren't invited to the Cupertino launch) to share the pictures of your sleep-out outside an Apple store. However, 2.5 per cent is not enough to cause a digital wave of diffusion. Apple was keen for more. And it knew it would come.

Early adopters

This is where the early adopters come in. They constitute 13.5 per cent of the population, and tend to be young in age. They have a high degree of opinion leadership, are influential and have more advanced education than the population at large. While they are slightly more cautious in their adoption of new technologies compared to the innovators, they are still among the first adopters, and their more judicious evaluation of new technologies provides them with a greater depth of thought leadership, persuasiveness and influence in society than the innovators. Their nuanced opinions matter, and they are more focused on making the 'right' adoption decision than making the 'first' adoption decision.

These were the people who waited until reviews, blogs and the social twittersphere had reacted to Steve Jobs's launch. They were the ones who compared the benefits and features of the iPad with the suspected benefits and features of future competitive devices. Combined with the innovators, the early adopters provide a bellwether indication of the likely success of digital diffusion, because they represent 16 per cent of the population combined. Without their support, the iPad wouldn't have gotten off the ground.

The chasm

Between this group of early adopters and innovators, and the next cohort of the market place — the early majority — there is a chasm that any idea needs to cross in order to create a crest of mass consumption. Rogers argues that there are five intrinsic characteristics of the innovation idea that are critical at this stage for diffusion. Importantly, they are a combination of digital as much as analogue factors — digilogue is critical for the diffusion of ideas, services and devices.

The five required characteristics are as follows:

- *Relative advantage* — how improved an innovation is over the previous generation. In the case of the iPad, this wasn't an issue because it was the first mainstream tablet that came onto the market, and that's why Steve pitched it as a third category device that had certain relative advantages compared to smart phones and notebooks.

- *Compatibility* — the level of compatibility that an innovation has that can enable it to be assimilated into an individual's life. The iPad was easily integrated into the lives of innovators and early adopters, particularly those who were already using Apple products and iTunes. The iPad easily plugged into this ecosystem and became compatible with existing behaviours. This compatibility made the leap across the chasm to the early majority more achievable.

- *Complexity or simplicity* — how easy an innovation is to use. If the innovation is perceived as complicated or difficult to use, an individual is unlikely to adopt it. Steve's demonstration of the iPad, and the subsequent viral videos of toddlers using the iPad and figuring it out intuitively, as well as the mobility of the iPad and its consequent public visibility in terms of ease of usage, became a powerful persuasive force.

- *Trialability* — how easily an innovation may be experimented with. If a user is able to test an innovation, the individual will be more likely to adopt it. Thus, it was critical for Apple to make the device available through its retail channels, so that people on both sides of the chasm could play around with the device and explore its benefits and ease of use.

DIGILOGUE

- *Observability*—the extent that an innovation is visible to others. An innovation that is more visible will drive communication among the individual's peers and personal networks and will in turn create positive or negative reactions more broadly. We know that the iPad was easily observable, and because of its mobility and novelty it quickly became a feature in cafes, on aeroplanes, at airports and in lecture halls.

Early majority

Because of the iPad's success over the five factors in the preceding section, the early majority became convinced — and the digital idea of the iPad made its way across the chasm. The early majority tend to be well educated, but are not as quick in their adoption of new technologies as the innovators and early adopters. They constitute 34 per cent of the population, so they are of critical importance in any mass adoption and the success of a digital idea. The early majority have above-average social status and are in contact with early adopters, but seldom hold positions of opinion leadership in the social system. In the context of iPads, I would rank reasonably tech-savvy baby boomers and generation Xers in this category. It may be that their kids or their early adopter or innovator friends and colleagues pestered them about the benefits of the iPad, until they eventually succumbed, and got onto the bandwagon of the idea. Once adopted, the early adopters realised the merit in a tablet, and what kind of media it gave them access to.

Late majority

The top of the crest of the wave of selling a disruptive idea is formed jointly by the early majority and the late majority. The late majority is represented by my parents. I gave them the first model of the iPad when I upgraded to the new iPad in 2012, and I received an email back from my mother telling me how some of her friends had sent her a 'how to use the iPad as a chopping-board' video. For a digital native like me, this sent shudders down my spine.

Six months after giving them the iPad, it had only been used once and I asked them to re-gift it to my brother, who would actually use it. And not as a chopping board. Like the early majority, the late majority constitute 34 per cent of the population, but these individuals approach an innovation with a high degree of scepticism and only approach the innovation (and then adopt it) after the majority of society has adopted it. Late majority members typically have below-average social status, and less financial lucidity than the previous categories. They are mostly in contact with others in the late majority and early majority, and very little with opinion leadership in society at large. (Don't tell my parents but, when it comes to technological adoption, they're on the verge between late majority and laggards.)

Laggards

The last category is the laggards. These guys constitute 16 per cent of the population, and may well think that iPads are newfangled eye-pads. They are the last to adopt an innovation. Unlike some of the previous categories, individuals in this category show little to no opinion leadership, typically have an aversion to change-agents and tend to be advanced in age. Laggards typically tend to be focused on 'traditions', and are likely to have the lowest social status and lowest financial fluidity, be the oldest of all other adopters, and are mostly in contact with only family and close friends.

In the case of the iPad, these individuals were not the key target market for Apple, but in a twist, the brand has been able to reach some of the laggards as well. In 2011, I was on my way back to Sydney from the United States, and I happened to be sitting next to an NBC executive who was travelling to the Australian Open golf event. We got to talking about digital diffusion and he started lamenting the decreasing contact he was having with his veteran-generation mother. Curious, I enquired about the reason. 'Well, you know Anders', he said, 'my mother lives in a retirement community in California, and we used to bring her home for family dinners on Sunday nights to spend time with us and the kids. However, the community has recently introduced an Apple user group on Sundays, which means Mum doesn't have time for us anymore, because she and her retiree buddies are being coached by an Apple genius and sharing tech advice among each other. However, we do hope that this means she might be able to join us on Skype in the future, Anders.' Digital diffusion of ideas never ceases to surprise.

Adoption numbers following the digital diffusion of the iPad idea

I should make clear here that the diffusion of innovations concept doesn't mean that every person in each group has adopted the idea and made a purchase. So let's take a look at the adoption numbers, and get a sense for the diffusion across the wave. Apple's tablet helped define a new market that is constantly growing. In the first days after its release in 2010, the original iPad sold barely 100000 units. A year later when the iPad 2 was released, analysts suggested that Apple sold around 800000 over the launch weekend. In 2012, the company proudly announced 3 million of the next release — 'The new iPad' — sold in the first days of availability.

Perhaps even more impressively, the iPad created the market for tablet devices, and this enabled competitors to ride on the wave Apple created. So, even though Apple's market share is predicted to fall from the highs of 83.9 per cent in 2010 (when it first sold the disruptive idea) to 47.1 per cent in 2015 as competition ramps up,[1] that is within the context of the growing third category device market that it created.

According to Intel, the tablet market will continue to grow between 73 per cent and 88 per cent till 2014,[2] which buoys the market for media companies producing content, and for application developers, software providers and hardware companies. In 2012, there were 54.8 million tablet users in the United States alone (more than 16 per cent of the total population), with predictions this will reach 89.5 million users by 2014.[3] These statistics show that a large percentage across several of the adoption groups has accepted the diffused idea, and that many have also backed the idea with their hard-earned cash. The tablet has now gone well and truly from its initial adoption phase among innovators and early majority, and made significant inroads into the early majority, late majority and laggards groups. The wave of diffusion keeps rolling.

So what can we learn from this case study of the iPad's rise to digital fame, and from the model of the digital diffusion of ideas? Here are some key takeaways points:

- The model is a great illustration of how ideas and concepts spread in society. Whether you are selling an idea or a product, ask yourself who you should be targeting the product or service for. Is it crucial that innovators and early adopters take up the product or service, or should you be targeting one of the other groups?

- If you're selling soft services such as consulting and thought-leadership, who are the innovators and early adopters you need to influence? These guys shape opinions and are likely to give you the kudos required to launch your ideas into the marketplace. So make sure you write a list of the associations, conferences and blogs these guys hang out at so that you can connect with them, and get your services in front of them.

- When you're thinking about digital or analogue ideas that you want to spread in society, ask yourself whether your ideas have relative advantage, compatibility, simplicity, trialability and observability. If not, how do you tweak their packaging to make it easier for groups to adopt your innovation?

- When selling an idea or concept, or your thought leadership, firstly consider which stakeholders you need to communicate with and persuade of its merits, and which communication channels can best influence them. Then consider how much time might be required to influence the influencers, and how they might in turn influence the masses. Lastly, ask yourself which social system you're selling the idea into and what cultural barriers might exist against adopting the idea.

For a quick visual refresh of the model, refer back to figure 7.1 on page 140.

We're all in media now

The uptake of the iPad is a great example of how a disruptive idea became mainstream, but the device has also influenced behaviour and business practice. Because of the iPad we are all in media. Its very nature as a media tablet gives all of us a chance to sell our ideas, services and products into an increasingly digital marketplace. And diffusion happens faster as a result of the mobile smart phone and tablet world. A tablet, whether an iPad or a Samsung Galaxy, enables businesses to both broadcast and narrowcast their messages, and see those ideas go mobile. Whether you are a media company or not, you need to start thinking like media companies.

What does this mean? Media companies are in the industry of sending out messages, and ensuring their messages are diffused to the masses. Distribution of the ideas used to be controlled by large analogue players like major newspapers or TV stations. Today, that is no longer the case, and a small player — a small business or a small brand — can all of a sudden play a big game. Equally, non-traditional media players like large retailers, food and beverage companies, professional services firms and management consultancies need to adopt a media mindset, where distribution and diffusion of their ideas and messages is critical. This asks a lot of leaders, but with Cisco predicting that 90 per cent of all data by 2014 will be comprised of video messages,[4] it also serves as an opportunity for leaders to ask themselves how they can represent themselves and their organisations as active media players. Tablet media devices are there in the hands of the target market, waiting for a strong signal in the noise — waiting for a good idea to be diffused.

And the social nature of the tablet amplifies the social diffusion of ideas. Let us compare and contrast some statistics. It took:

- the analogue phone 75 years to reach 50 million users
- the analogue radio 38 years to reach 50 million users

- the analogue TV 13 years to reach 50 million users

- the digital Facebook 3.5 years to reach 50 million users

- the digitally mobilised Angry Birds 35 days to reach 50 million users.[5]

In other words, the rate of adoption of new technology is speeding up, and has reached dizzying speed. And the question is: what does this mean for your business?

Mobile media

We have reached a tipping point. In 2011, more smart phones and tablets were shipped than PCs and notebooks. And this trend of people taking the web into their own hands is likely to continue, as you can see in figure 7.2. This trend is also buoyed by the decreasing prices of tablets and smart phones, which means these devices will penetrate emerging markets and become the preferred way to access the web.

Figure 7.2: global mobile smart phone and tablet sales tipping point

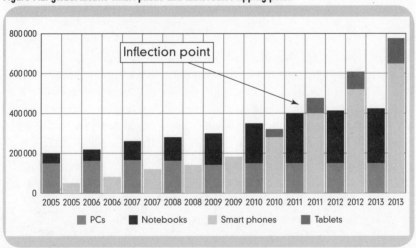

Source: Based on data from KPCB '2012 Internet Trends' report.

As Steve Jobs foresaw, we now have the internet 'right there in our hands'. The disruptive idea of the iPad tablet is a wonderful case study in driving the diffusion of an idea, and seeing the mass adoption of a new technology concomitant with that idea. Contemporaneously, it has also forced companies to play a more front-of-mind role in filling the media tablet with content. This is a huge opportunity in the context of digilogue.

Imagine what your brand can do in terms of storytelling, in terms of inviting the customer or client into your world, in providing them with a human analogue experience of what your brand is like in real life, not just in Second Life. This digital diffusion of innovation is a critical opportunity to level the playing field and allow slower players and bricks and mortar businesses to access a new world of information, and to provide the mobile world of the internet with great media about your brand.

High-speed adoption rates and the high usage of tablet devices mean you need to think seriously about mobile media. And they mean that you need to create engaging content, just like a media company — whether you're a farmer, a fertiliser company, a law firm, a thought leader, a consultant, a retailer or a grocer. Blog to your heart's content, record HD video, tweet weird yet interesting content, Instagram enticing photos relevant to your clients, fans and prospects. Invite people into your world, and use mobile media to both capture fascinating content and distribute it to those millions of devices that sit in the hands of a captive audience waiting for a strong signal among the noise.

For your ideas to resonate you need to invest in your own media platform. This is also known as *earned media*, and it's the kind of media that people opt in to hear about. It's the kind of media that those same people will come back for — again and again. And tell their friends about. It is estimated that management consultancies spend up to 5 per cent of gross revenue on what the consultancy industry call 'thought leadership', which in many ways is a glorified term for free educational information.[6] Online, content is king, and increasingly we access engaging content on the go, socially and locally. Think of each of these screens as a way to digitally diffuse your ideas, and position your brand.

I cannot help but think about the opportunities this provides for my mum's business, Georg Sörman. As I venture and trendspot around the world, I see bricks and mortar retailers increasingly using digital tablets to their advantage. It may be my tailor in Sydney, Patrick Johnson, who captures my measurements via an iPad on Evernote, monitors the good and bad shifts in my seasonally adjusted size, and keeps my details up to date via his customer relationship management system. It may be Haberdash in Chicago, which keeps me up to date with their events and new product launches via their blog, which is conveniently optimised for browsing on the go. It may be Grandpa in Stockholm, which delivers an interesting, thought-leading digital newsletter to my inbox each month and gives fashion tips. It may also be Best Buy's mobile app, which enables me to shop with them digitally while I am browsing their store physically.

Digital diffusion of ideas shouldn't scare analogue businesses, because it enables analogue ideas, brands and stories to be digitally distributed and drive offline business traffic. We are all in media today, and all of a sudden my mum (like all business leaders) needs to start thinking like media companies. The customer is waiting for your thought-leading ideas.

Round 7. Digital. 5–4. With some strong counter-punches by Analogue at the end of the round. Digital Disruption displayed that diffusion of ideas can happen quickly and globally. Analogue Anachronism threw a couple of lazy punches — radio, television and phone. Digital countered with the exponential growth of Facebook and Angry Birds. We saw how the digital idea of the iPad made its way through society across the innovators, early adopters, early majority, late majority and laggards, and what lessons brands can learn from this in terms of winning the hearts and minds of customers. But meanwhile, we also saw the diffusion of innovation doesn't happen in a digital vacuum. The key convincers are relative advantage, compatibility and complexity versus simplicity, trialability and observability — all of which tend toward the analogue hearts. Importantly, Digital Disruption threw out a clarity punch — because of media tablets, we all need to think like media companies. Analogue kept counter-punching and we saw that the existence of digital tablets opens up great opportunity for digital hors d'oeuvres and thus analogue attraction, and previews of analogue experiences in physical reality.

Ask yourself...

Here are some ways to start thinking about disruption and your business:

- What is a disruptive idea that you believe is critical for the market to adopt, which is aligned with a new product or service you're launching? (For example, what is your equivalent of a third category device that we all need?)

- What would the market have to believe in order for that idea to be adopted by individuals, and digitally diffused amongst innovators, early adopters, early majority, late majority and laggards? List them. (For example, the early adopters could believe 'This will make me more productive', or the laggards could believe 'This will enable me to Skype with the grandchildren in Sydney'.)

(continued)

Ask yourself... *(cont'd)*

- Which media platforms do you believe are most aligned with your brand? Is it YouTube, iTunes, Twitter, Instagram, a corporate blog and/or Pinterest? Pick the ones that you think and your research shows can be most easily integrated with your communication behaviours, and which has the most chance of reaching your clients and prospects.

- If you're an analogue business, what parts of your analogue, high-touch brand personality can most easily come alive via digital media tablets?

- If you're a digital business, which parts of your digital, high-tech brand personality can most easily come alive via digital media tablets?

Apply it...

Now put your thoughts and research into practice with the following:

- Set up a content creation schedule. My rule of thumb is 4–2–1. That is, four digital blogs per month, two digital newsletters and one new video. You may be more ambitious, but the 4–2–1 publishing schedule keeps our content fresh, and amps Thinque's digital engagement.

- Automate your content creation—prepare once, use often. In other words, speak at an industry conference, record it on video and publish it on YouTube, export the presentation as a PDF for your blog or Slideshare, and tweet for engagement.

- Get inspired. Read great blogs about your industry and become an avid consumer of awesome content. Link to great content, and redistribute it.

Endnotes

1 Unknown, 'Media tablet OS share estimates', Gartner, telebeem.
   com/blog/wp-content/uploads/2011/04/tablet-share-chart.jpg, last
   accessed 9 April 2013.

2 J. Paczkowski, 'Intel: Tablet market will grow between 73 per cent
   and 88 per cent by 2014', AllThingsD, http://allthingsd.com/20100511/
   intel-tablet-market-will-grow-between-73-percent-and-88-percent-
   by-2014, last accessed 9 April 2013.

3 E. Schonfeld, 'Estimate: 90 Million US tablet users by 2014; iPads
   drop to 68 per cent share', TechCrunch, techcrunch.com/2011/11/21/
   estimate-90-million-u-s-tablet-users-by-2014-ipads-drop-to-68-
   share, last accessed 9 April 2013.

4 T. Kelly, 'Cisco predicts 90 per cent of network data traffic will be video
   by 2013', Logicalis, www.logicalis.com/news-and-events/analyst-
   relations/cisco-predicts-90-of-network.aspx, last accessed 9 April 2013.

5 G. Kofi Annan, 'Radio took 38 years to get 50 million users; Angry Birds
   Space took 35 days', Trickle-Up, innovation.gkofiannan.com/radio-
   took-38-yrs-to-reach-50-million-users-o, last accessed 9 April 2013.

6 Unknown, 'Free thinking: Why expensive consultancy firms are
   giving away more research', *The Economist*, www.economist.com/
   node/16994439, last accessed 9 April 2013.

Taking stock

The Digilogue Strategy Map and a pharmaceutical case study

Since 2009, Thinque's been working closely with one of the major pharmaceutical companies in the world. The world of pharma is being upended, and digital disruption is one of many threats and opportunities for the industry. Patients are now sharing advice, data and stories on digital platforms like PatientsLikeMe, starting Facebook groups, and creating digital, unofficial conversations about originator brands like my client's. Curiously, doctors, the traditional gateway to the prescription pad, and a key communication channel for pharmaceutical companies, are also becoming more digitally savvy. While this is an interesting case of patient and partner empowerment, it also can place the pharmaceutical companies in stressful situations with the US FDA (Food and Drug Administration), and it changes the nature of the game for Pharma's traditionally oversized marketing and sales functions.

We have had a chance of working horizontally across functions and vertically across levels within this US-based organisation — including within BioMedicines, Diabetes, Oncology, Data Science and Solutions, HR, the Development Centre of Excellence and IT. In common is a relentless focus on innovation, including adapting to the digital world, in both the company's internal and external communications. The company is a firm believer in the customer service value chain, which is the idea that great leadership (internally) leads to great engagement by staff, which in turn leads to awesome customer experiences, which in turn drives business results. The following Digilogue Strategy Map shows some of the digilogue initiatives the company is driving in the United States at the time of writing. The discussion following the map expands on the strategies.

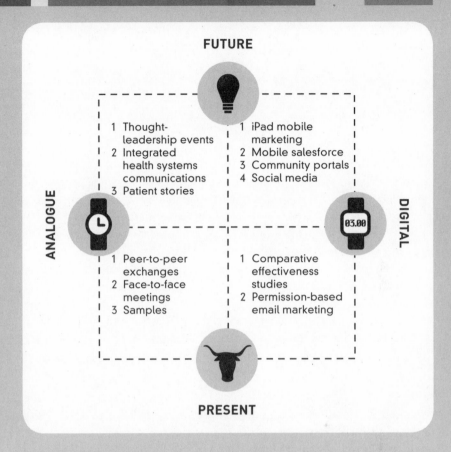

FUTURE

ANALOGUE

1 Thought-
  leadership events
2 Integrated
  health systems
  communications
3 Patient stories

1 iPad mobile
  marketing
2 Mobile salesforce
3 Community portals
4 Social media

DIGITAL

1 Peer-to-peer
  exchanges
2 Face-to-face
  meetings
3 Samples

1 Comparative
  effectiveness
  studies
2 Permission-based
  email marketing

PRESENT

Analogue—present

If you've seen the movie *Love and Other Drugs*, let me confirm it's a reasonably accurate reflection of how marketing and sales was done in the pharmaceutical world of the 1990s. Large salesforces aggressively targeted one group of the three main stakeholders in the pharma game—providers of professional medical care (the other two stakeholders being patients and payers, or insurance companies). The game has changed since then, largely due to heavy regulation, and the touch points between originator brands and doctors are less intimate, and thus less influential. Gone are the days when a pharmaceutical company (any pharmaceutical company) could fly its highest prescribing doctor clients to Hawaii for a conference. But most pharmaceutical companies still market in a highly analogue manner to this segment of stakeholders—via face-to-face meetings in busy medical clinics, handing out samples, and occasionally getting medical thought leaders to engage in peer-to-peer exchanges with fellow MDs talking about the health benefits of a particular drug. While the effectiveness and consequent future longevity of some of these strategies is in doubt, some form of personal rapport building is likely to continue, as sales reps are also seen as a doctor's partner in some circumstances.

Digital—present

Because of FDA regulations, pharmaceutical companies have traditionally been restricted to communicating with patients via their medical doctors (depending on the jurisdiction). While direct-to-consumer advertisement is allowed in some jurisdictions, this has created policy issues, as doctors' medical advice can be seen to be compromised by pushy patients who believe the drug they saw during a Super Bowl ad break is the only one that will fix them. However, many pharmaceutical companies have engaged in digital marketing focused on education of doctors on an opt-in basis. This meant that doctors could access medical data and comparative effectiveness studies, and decide whether the originator brand's drug was truly better than the generic competition's or other originator brand's pills.

Pharmaceutical companies have also had a certain web presence but, depending on jurisdiction and regulation, it's been challenging for them to figure out how to engage in local markets digitally. Often, the websites have been a repository of corporate and legal speak, aimed more at patients reporting adverse events than at the company building a branded and empathic connection with patients, providers or payers.

Analogue—future

The powers that be in the American market are slowly shifting, and so our client is shifting its focus. While historically there has been a strong focus on doctors as the gatekeeper to the prescription pad, payers (insurance companies) and integrated health services are more and more being seen as the key stakeholder groups to be influenced. This presents opportunities for smart thought leadership, analogue events (like co-branded thought leadership conferences) and identifying new touch points with payers—within clearly defined and regulated boundaries. Being seen as an innovator, and thus meriting a premium value in an increasingly cost-conscious market, will be key, and so thought leadership internally and externally will be key in pharma.

One of the ways in which the comparative effectiveness of drugs can be communicated is by visualising a drug's positive impact by connecting both the hearts and minds of the payers. Increasingly, pharmaceutical companies are deploying patient stories and big data visualisation to make their cases, and to establish both an emotive and heavily dollar-driven conversation with those that have the deep pockets, and who, ultimately, decide what doctors can recommend and what patients can ask for. For example, the company has combined patient education and visual cartographies of diabetes care in its influencing of Asian governments to show how its diabetes drugs and devices, combined with education, can lead to longer adherence and thus better patient outcomes. Combined with a visually appealing re-design of medical data, which used to be only targeted at regulatory approval, clinical trial data is also being turned into more humane visualisations that insurers and payers can easily understand — thus converting statistical significance into commercial significance.

Digital — future

Pharmaceutical companies have needed to play within the boundaries of existing legislation, and to err on the side of caution in exploring new territories such as digital and social media. Little government leadership or direction in this space has meant that most pharma brands, including our client, have been forced to engage in social media in a somewhat stunted fashion. They have been allowed to listen, but not to speak. In other words, they have been allowed to monitor peer-to-peer conversations about their brands, but not to actively participate in those conversations.

This may be shifting, because big data, patient stories and medical information abounds online. As mentioned in chapter 4, Nike has made inroads into the preventative healthcare space by combining music, data monitoring and GPS capabilities in its Nike+ suite of digital curation. Slowly, pharma companies are noticing this disruption, and mobile glucometers and health monitoring are now springing up as patients are taking healthcare matters into their own hands. The web is where patients, doctors, and the decision-makers and researchers at payer companies access information, and thus pharma companies need to play in this space. There is an opportunity to participate in these conversations, via both public and private portals, and increasingly we see soft branding initiatives on Facebook and elsewhere across the web, where originator brands, or more aptly their parent companies, have a presence in either curating the conversation or engaging with clients or prospects. Our client is helping curate the diabetes conversation by digitising a diabetes patient's journey across a cartography that educates the patient, the provider and the payer about healthy management of the disease.

Our client is also determined to shift how its marketing and salesforce interact out there in the field. Thus, it is rolling out a replacement of laptops with iPads across the whole salesforce, enabling its salesforce to connect face to face with doctors, demonstrating the efficacy of their drugs via iPad screens, while working more efficiently and providing real-time data via the customer relationship management system on the road.

Summary

Our client company is making some smart inroads into the digilogue world of the future. Many initiatives from the unregulated days of Big Pharma are now defunct, and pharmaceutical brands have had to rethink who they should be communicating with, and how. This is changing what they can say and where. Face-to-face interactions will not disappear, but with an increasing transparency and treasure-trove of information online for payers, patients and providers, pharma companies like our clients will increasingly play a role in this digital future.

8: Analogue location drives digital innovation

Round 7 homed in on how innovative ideas, products and services can capture the imaginations of the marketplace, and analysed the diffusion process across societal cohorts. This round examines where those innovative ideas come from and the insight that analogue location drives digital innovation. Through this chapter, you'll observe the critical importance of place in innovation and business model applicability, and why analogue, personal connections are a key in the digital world we are increasingly spending time in. We will take a look at world-class MBAs and where the true value in a business degree lies, before examining how digital natives may affect collaboration and digital innovation in the future. But first, join me at Sprout Cafe in Palo Alto, Silicon Valley. Digital Disruption is about to have its digital bubble burst by an analogue needle.

The importance of local connections

Silicon Valley. This name conjures up word and brand associations like high-tech, digital, Facebook, Google, venture capitalism, Stanford, Palo Alto, Cupertino, Infinite Loop, LinkedIn, Cisco, cloud, big data, Xerox PARC. But think about the name in two parts. Firstly, *silicon* — the raw material for most commercial semiconductors; the backbone of the digital world. Next, *valley* — a physical description of a geological depression with predominant extent in one direction. Put the two words together and we have the metonym for the US high-tech industry. This physical, analogue place has been driving digital innovation and the creation of cyberspace for decades. There is a certain irony in this. One of the promises of digital innovation is that we can increasingly interact in digital cyberspace via globally diversified teams, that we can collaborate across time zones, and that we can video conference with partners and clients. Yet the history of digital innovation is one that has been consistently emerging from a specific geographical location — Silicon Valley. This ecosystem has been the central playground for digital ideas, and the mecca for innovators and venture capitalists since the name made its way into popular jargon in the 1980s.

This is why I find myself at Sprout Cafe on University Avenue in Palo Alto. Sprout, a salad cafe, represents the essence of California — everything about it oozes local, vegetarianism, artisan, farm-to-table, fresh, organic and sustainable. I sit there awaiting the arrival of Mark Zuckerberg, Sheryl Sandberg, Sergey and Larry, or Tim Cook. Hmm, not today, apparently. I keep munching on my mung beans and line-caught tuna. But the digital rock stars are here (albeit in the ether), and they are here for a specific reason. Silicon Valley is *the* digital talent cluster. Despite the best efforts of the United Kingdom, Russia, South Korea and India, digital pioneers still migrate to Silicon Valley. While the Valley reminds me of a large business park, its cafes, campuses and educational institutions speak to the history of innovation, and provide the habitat where digital ideas can flourish. It's immersive, it's an eco-chamber, it's incestuous, and it's highly analogue. Why? Trust, introductions, physical demos, handshakes, morning bike rides, children in the same school, serendipity, hype, constant pitching, caffeinated idea exchanges, Stanford alumni, MBAs. This analogue stuff matters — particularly when it comes to digital innovation.

Analogue spikes

Coincidentally, I am in Silicon Valley to advise one of the largest internet security firms on big data and its impact on privacy. In person. While cyberspace is awash with information, research, reports, case studies, and data visualisations on this topic, being in Silicon Valley somehow feels more authentic. It's the equivalent of running a political PR agency in Washington DC, a design studio in Milan, a fashion label in Paris or a hedge fund in London. I mean there is a reason we go to industry conferences, isn't there? As a friend from Cisco, vendor of the famed

Telepresence Webconferencing facility, says, there are still no virtual beers and virtual golf. We still crave the human connectivity, the mingling and the networking. Even if you're an introvert and don't usually like that kind of thing, there is a sense of belonging that goes with these types of analogue places. A sense of being in the know, a sense of knowing who's who, a sense of being in tune with the latest thinking in the industry.

Thus it is helpful for us to connect with Stanford faculty, to hang out with VCs, to have dinner with budding entrepreneurs and spend time at the Google campus. Often the good stuff — the best information and the insider tips — can be found in the physical gaps and the informal conversations. We have all been to the conferences where we realise that the speakers (and I might be shooting myself in the foot here) merely provide the social lubricant that enables delegates to talk real shop in the hallways. And make deals. While having coffee. The same is true in Silicon Valley. Here the analogue enables the digital. And it does so because it's a spiky region.

Let me explain. In his 2005 book, *The World Is Flat*, author Thomas Friedman suggested that because of globalisation, the world was becoming flatter, borders were being removed, and distance and place were becoming irrelevant. Richard Florida countered this argument in a way that supports the contention in this chapter — that analogue place and location matters. In his book *Who's Your City?*, Florida shows maps of population growth, economic activity, innovation (as demonstrated by patent registration) and scientific discovery (as demonstrated by residence of the most heavily cited scientists). Silicon Valley, or the San Francisco Bay Area, as Florida calls it, is one of the top talent clusters, or 'spiky' regions, in the world — a world which, according to Florida, is far from 'flat'. Instead, geographical spikes are created by talented individuals who tend to cluster with one another, creating a (non-linear) multiplier effect that attracts additional talented individuals to that geographical area. This is certainly true of Silicon Valley.

Let's explore this analogue place by numbers for a moment:

- In 2006, the *Wall Street Journal* found that 12 of the 20 most inventive towns in America were in California, and 10 of those were in Silicon Valley. San Jose led the list with 3867 utility patents filed in 2005, and number two was Sunnyvale, at 1881 utility patents.

- Silicon Valley has the highest concentration of high-tech workers of any metropolitan area, with 285.9 out of every 1000 private-sector workers.

- Largely a result of the high-technology sector, the San Jose–Sunnyvale–Santa Clara, CA metropolitan statistical area has the most millionaires and the most billionaires in the United States per capita.

- Over 34 per cent of Silicon Valley residents belong to the Creative Class — a key driving force for economic development in post-industrial societies — and the professional purpose of the class is 'problem solving, [and] their work may entail problem finding' as well as drawing on 'complex bodies of knowledge to solve specific problems using higher degrees of education to do so'.[1]

- Real annualised house price growth in the area between 1950 and 2000 was the highest in the United States at 3.53 per cent, where the national US average was 1.7 per cent.

- Over 40 per cent of people in Silicon Valley have a bachelor's degree or higher academic degree.

Yes, Silicon Valley and its surroundings is a nice place to hang out. There are Napa vineyards nearby and great cafes. It's near the Pacific Ocean, skiing is a few hours away and culture abounds in San Francisco. And the people in the Bay Area rank highly on Richard Florida's Gay-Bohemian Index.[2] And, ironically, Silicon Valley flies in the face of globalisation pundits, who argue that analogue place is becoming irrelevant. Technology has been sold as holding the promise of a boundless world, yet seemingly we still crave the analogue place. We are free to live wherever we want, and the digital world has supposedly made analogue location and geography extraneous. It's a wonderfully democratising idea, yet in the case of Silicon Valley, the epicentre of digital transformation, it's wrong. Talent, capital, creativity, innovation happens 'here', and 'here' still connotes a physical location. And, in the case of Silicon Valley, analogue location has been driving digital innovation for decades.

Digital foot in the analogue door

One of those innovations is LinkedIn. And shortly after my uneventful star-spotting outing to Sprout Cafe, I had a chance to hang out at the LinkedIn HQ in Mountain View, Silicon Valley. What is fascinating at LinkedIn is how much they are drinking their own digital Kool-Aid, without a recognition that this digital network still relies on analogue connections. I listened curiously to a senior member of staff who described the LinkedIn network. While I am a big fan of this particular social network and its services, a certain irony exists in the story that he told me.

I'll call this person Jack, and he is a young 30-something executive with an Ivy League MBA, and a background at one of the top global management consultancies. He was very excited about how he got his job at LinkedIn — via LinkedIn. He shows me a data visualisation of the connection he had via a management consultancy to one of the leaders at LinkedIn, and how this LinkedIn connection led to employment — with LinkedIn. Here, I couldn't help but be a bit cynical and point to the fact that, at the end of the day, this digital connection was built on analogue foundations. Jack's merits (Ivy League MBA, top management consultancy) and analogue network, digitally duplicated via LinkedIn, led to a new connection to a person, who, yes, could digitally do his due diligence via LinkedIn about Jack. This is an example of the digital and the analogue working together, but while I have admiration for Jack's intellect and his accomplishments, we cannot forget about the human need for local, analogue connections, which underpins the digital element of social networking. Analogue trust and merit feature heavily in this digital story of LinkedIn's success.

Nonetheless, this Silicon Valley story is impressive, as shown through the following:

- LinkedIn has more than 200 million members in over 200 countries and territories.

- It is significantly ahead of its competitors Viadeo and XING.

- The membership grows by approximately two new members every second.

LinkedIn is the white-collar professional's dream. I am in no way dissing it. What I am arguing, though, is that we too easily forget about its analogue underpinnings. Local connections are key, and these enabled Jack's move to Silicon Valley.

This digital innovation has disrupted the old notion of the CV, the paper testament to analogue, personal accomplishments. If the CV sucked as an executive summary of your life's work, LinkedIn provides a much more dynamic version, which enables analogue connection, via the digital space. Recruiters, potential partners and future employees can quietly browse your profile and conduct digital due diligence, accessing your recommendations, references, areas of expertise, client list, portfolio, blogs, case studies, publications, 500-plus list of connections and academic articles. Your reputation has become your webutation, and it's available for all to see on LinkedIn. This has now become your digital foot in the analogue door. But deep trust is created elsewhere.

Analogue trust, analogue connections

Deep trust ensues in the here and now, in the analogue place of location. That is why immersive experiences can be so powerful. I think of some of the most enduring friendships and professional relationships I have formed, and it's not always the case that my most rewarding rapport has been formed with people I've known for the longest period of time. I would argue that many strong rapports and relationships ensue from deep, immersive and meaningful experiences.

Let me give you an example. I have a friend, Joel, who I spent time with at university. We were living in the same university hall for one year, at opposite ends of the corridor. We studied law together, and hung out in the same circle, but it wasn't until a joint road trip that we became close. At the end of that year, Joel and I travelled from Canberra to Brisbane — a trip of 1203 kms, or 13 hours — in his Nissan 300ZX sports-car. We stopped at strategic locations like the Hawkesbury River for a weekend hangout with university friends, Port Stephens for a necessary overnight motel visit, and Byron Bay with new friends in a funky hostel. During that youthful trip we met ladies, had a couple of encounters with speed cameras, and suffered from boredom listening to the English indie band Travis on repeat on the Central Coast, north of Sydney. But it was an immersive experience and, even though Joel now lives in Melbourne and I do not get to see him as often as I'd like, every time we catch up it's like we only saw each other yesterday, and we strike up a great rapport. The same goes for the friends I grew up with in Sweden, like CP, Fredrik, Therese, Lars and Julia. Going through formative experiences, whether the experience be a road trip, puberty, your first teenage hangover (it happens, folks) or love, creates glue and trust. It is something a computer interface cannot yet replicate. Deep trust ensues in the here and now, and business, despite the best intent of digital disruptors, hasn't been able to cross this rapport chasm yet.

I was reminded of this by a client, and dear friend of mine, recently. Incidentally, we were spending time at Stanford doing executive education, and my client and friend — let's call him Rob — told me of the cost his company incurs each year for their annual conference. My friend is very numbers-focused and has used Thinque's keynote speaking and executive consulting services on a few occasions. As a tough negotiator, he gladly comes to me with commercial 'feedback'. This time he told me that domestic flights, accommodation, transfers and food at the one-day airport hotel conference costs his company at least $250 000, even before productivity costs were taken into account. In his next breath, he told me that his CFO had told him that my services were extremely good value. (I wasn't sure whether to take this as a compliment or to raise my fees.) The bigger point for you, though, is that his company, of which he is the managing director in Australia, still chooses to fly in people from across a large nation continent so they can spend face-to-face time with each other.

Location matters. And it matters because this is how generation after generation has worked. Face to face. Admittedly, this may be about to shift with digital natives playing an increasingly important role in organisations. Meanwhile, though, face to face still drives team-building activities, creative brainstorms, ideation sessions and future strategy rollouts. Seems that golf and beers matter more than we might have thought.

Analogue alumni

Another example of the importance of analogue location and human connections is the famous MBA — the Master of Business Administration. This degree comes under a lot of scrutiny, with *BusinessWeek*, *The Financial Times* and *The Economist* all providing international rankings, and universities, alumni and future students pay close attention to which educational institution is edging ahead of the pack and whether their particular institution of choice is going strongly.

One of the business schools that consistently ranks in the top 10 internationally across the board is the Graduate School of Business at Stanford. In 2012, I had a chat with my Australian friend — I'll call him Matt — who is doing his two-year MBA at Stanford. I asked Matt about the main features and benefits of this exclusive program, which costs him around US$100 000 per year in tuition, books, medical insurance and a study trip. Matt responded, frankly, that it was an insurance policy. 'What?' I reacted. 'Tell me more, Matt.' 'Well, you see, Anders', he said, 'everyone here is already successful in their careers, and will be even more so moving forward. Because of the glue and rapport we build, and the cognitive drills we have been through together, we will always help each other out. This means that if I ever lose a job or fail in my next start-up, I will simply call on one of my cohort members and, at worst, they will give me a job — if not connect me with someone from their network who can help me out.' Aha. Got it. Matt is right. And these connections translate into innovation and cash — both on a micro and macro level. Stanford is a \$4.4 billion enterprise, but it sits on an endowment fund of \$17 billion, built by alumni and the Silicon Valley network, which contributes 21 per cent of the operating budget for the university.[3] This analogue glue that Matt, his mates, their predecessors and their successors have built up over generations of MBA graduates cements rapport, connection and reinvestment in local, digital ideas.

This is what I loved about my own MBA. In an age of digital learning and online MBAs, the oldest university in Australia, the University of Sydney, had to innovate to get on the front foot. So they launched their Global Executive MBA. While the course delivers digital content and the MBA Director, Nick Wailes, is a fellow digirati, it is the analogue immersion in location that sets this degree apart. Part of the course (40 per cent) is delivered to a select cohort group of 20 at the University of Sydney Business School, while the remaining 60 per cent is delivered at other top-end universities around the world. The focus for the executives who participate in the program is immersion and experiential learning through local application.

We thus spent 20 per cent of our time at the Indian Institute of Management, Bangalore, 20 per cent at Stanford University, Silicon Valley, and 20 per cent at the London School of Economics, England. The design thinking behind the program is to deliver global lessons applied locally in four different continents, across different stages of the business life cycle — start-up, emerging, mature and late-stage markets. What really transpires, though, are real relationships, friendships and experiences with your cohort colleagues, cemented during these global road trips. Your emotional intelligence gets tested, your strategic thinking stretched, and your cultural sensitivities shifted as you throw yourself at digital and analogue problems around the world. What you leave with is a rich tapestry of experiences but also, and even more importantly, local, analogue connections that drive your organisation's future innovations.

Digital natives challenging analogue location

So against this 'local' context, my futurist question remains. The question is whether the (re)entry and cementation of digital natives, like Matt, in the workforce will change the importance of local connections. Some believe it will. And the reason for this is digital collaboration. Millions of people are now online, playing and collaborating on multi-player platforms like World of Warcraft, and they are creating new types of behaviours, leadership skills and communication efficiencies. World of Warcraft has 10 million–plus subscribers, who are collaborating in virtual teams across the world and co-creating a virtual reality. For them, analogue place is less important than cyberspace when they're in in-game mode. According to John Seely Brown and Douglas Thomas, the in-game mode, and the gamer's disposition, develops digital skills[4] that should translate into analogue behaviours that modern, globally diversified organisations ought to seek. Seely Brown and Thomas describe gamers as:

- *Bottom-line oriented* — they seek to be evaluated and are focused on results. The goal is meritocracy and improvement.

- *Understanding of the power of diversity* — the strongest teams in virtual worlds are teams with diverse talents, and by definition each team member is incomplete.

- *Thriving on change* — nothing is constant in a game. Gamers create change and expect flux.

- *Seeing learning as fun* — learning in-game provides knowledge on how to overcome obstacles and adds to a gamer's resourcefulness.

- *Marinating on the 'edge'* — gamers innovate and create on the edge of the possible, and solving a task is about the challenge, but it's also about the enjoyment of finding a novel way to do so.

These five attributes, so goes the argument, make for employees who are flexible, resourceful, improvisational, eager for a quest, believers in meritocracy and foes of bureaucracy. Many of our *Fortune* 500 clients would love to see these types of attributes in their employees.

Interestingly, even cyberspace has a sense of place, though, because a world like World of Warcraft has geography and distinctness to it, and is created in a virtual physicality called Azeroth that is not dissimilar to the 'real worlds' we have built in locations like Silicon Valley.

So does this translate into the digital tipping point — where digital cyberspace becomes more important than analogue location? Not yet. While we have seen online collaboration and innovation work well in the case of open source software creations like MySQL and Linux, the corporate world at large has been slow to break their nexus with analogue location. It is certainly foreseeable, and in fact largely inevitable, that the digital gamer attributes outlined in this section will become more mainstream. As generation Zers (born 1994 to 2010) start graduating, and shaping the workforce with new demands and contributions, digital collaboration will increase. If you were born after 1994, you don't know a world without the internet, and if you have grown up on a diet of MySpace, Facebook, Justin Bieber, Twitter and *Big Brother*, your analogue skills are likely to differ from those of a baby boomer, generation Xer or millennial. On the flipside, your technical and digital acuity could well be out of this world. Literally. This holds promise for the future of digital.

Business is still local

But in the meantime, business is still local. Since the 1980s we have been told by management consultants to 'think global, act local'. *Glocalisation* refers to the practice of conducting business according to both local and global considerations. Two of the flag bearers that you might associate with this global–local trend are HSBC — with their mantra of 'The World's Local Bank' — and IBM — with their 'Solutions for a Small Planet'. Both tune into the analogue, the micro, the local.

Equally, business strategists have known for a while that local adaptation is key in their go-to-market strategies, and thus multinational corporations are encouraged to build local roots. Coca-Cola's locally adapted World Cup 2010 vocal advertorials and McDonald's Chicken Maharaja Mac are examples of localising products for their analogue locations. It's kind of the reverse of the provenance marketing I covered in chapter 3. The latter speaks highly of where a product or service emanates (iPhone — 'Designed by Apple in California. Assembled in China') while the former, like the Maharaja Mac, takes local considerations and diet into account in the way the product is pitched in a particular market. The key in local adaptation is turning a global brand into one that is locally relevant. According to John Quelch and Katherine Jocz, the authors of *All Business Is Local*,[5] the argument is that customers just want the best products and the best experiences in the area they know best — their own neighbourhoods. For all the talk of thinking global, having our heads in the cloud and being digitally connected, it seems customers still have their feet firmly planted on the ground — locally.

The analogue local affects the most digitally disruptive of them all — like Silicon Valley firms. In 2013, Marissa Meyer, CEO of Yahoo (ex-Googler and first-time mum), made a very analogue move and, in doing so, ended telecommuting and the days of free remote collaboration at Yahoo — by declaring that staff must now start showing up in its analogue offices. In its strategy to re-establish its digital innovation merits, this analogue move is an interesting one to observe. Google. This Silicon Valley firm is not beyond geography and physical boundaries. Instead, geography and analogue location is affecting its digital reach and its innovation strategies — including China. In 2010, Google closed down the analogue and digital presence of Google.cn and, because of consistent privacy breaches and censorship regulations, Google moved its offices and servers to Hong Kong, which annoys me (and anyone else with a dissident slant) each time I work in China and have my Gmail, YouTube and Google Maps messed with. At the same time, perhaps ironically, Google faced invasion of privacy lawsuits in Germany after its StreetView product was found to be photographing private homes.[6] This digital company, with no physical product in the marketplace in 2010, suddenly became very aware of analogue geography and location-specific eccentricities. Meanwhile in China, the local search engine Baidu got to cement its market leadership. Digital disruptors are not immune to analogue location, and brands more broadly need to create innovations that are localised.

Analogue location is eccentric. It's specific, rooted in geology and shaped by geography. It's about people and is shaped by local customs and regulations. It's about glue, subtle signs, sweaty palms in a handshake. It's about serendipity and passion. And talent is attracted by passionate, analogue spikes in creativity. Like Silicon Valley. It's human to want to be immersed in location, in the present moment. Namaste. And in the near term it will not be replaced by the digital. In fact, we have seen in this chapter that analogue location drives digital innovation, and that the most disruptive of digital players still get affected by analogue location.

Round 8. Analogue. Technical Knockout. Digital slowly gets back on its feet.

Wrapping up the fight

One of Georg Sörman's most famous clients was world heavyweight champion Ingemar 'Ingo' Johansson (1932–2009). In 1959 he spellbound Swedes with his upset victory in Madison Square Garden against title holder Floyd Patterson, after flooring Patterson seven times in Round 3. Radio Luxembourg's commentator Lars-Henrik Ottoson's excited voice on the crackly analogue radio shouted out the TKO in Round 3 to Swedes around the nation glued to their radio sets. One New York magazine's front cover had the catchy title 'Ingo — it's Bingo', and in interviews, the pronunciation-challenged Ingo referred to his right fist as 'toonder and lightning'. The commentators at the time jumped on this and called his right paw 'The Hammer of Thor'.

Ingo's professional career spanned 26 wins, 2 losses and no draws, and he was inducted in the World Boxing Hall of Fame in 1998, and into the International Boxing Hall of Fame in 2002. He was popularly referred to as the Cary Grant of boxing (check out Ingo's love of fine cloths in figure 8.1), and eventually featured in Hollywood movies like *All the Young Men* next to Sidney Poitier. He is the only Swedish man in history featured in *Sports Illustrated*'s 'Sportsman of the Year' (in 1959). And as a young boxer in my early teenage years, I remember my excitement when my grandpa Per brought me home a Georg Sörman business card autographed by Ingo, as he claimed with pride that we looked after the premiere gentlemen of Sweden in our store. Back then, like most kids, I collected these autographs, and it instilled a sense of pride.

Ingo Johansson never fought in a single draw and, indeed, draws are relatively rare in boxing. The scoring system is geared to discourage draws — three referees (in most instances) award points, and the scoring system is designed toward decisive determination of a winner. In the aggregate, the total points awarded can on occasion result in a draw, based on the mathematics of the points awarded by the two judges or, alternatively, if one judge awards the match to one contestant, the second to the other contestant, and the third judge holds it to be a draw.

In 2004, Manny Pacquiao — the PacMan — from the Philippines and Juan Manual Marquez — the Dinamita — fought it out in one of the most famous drawn bouts in boxing history. These two boxers are often considered the top two pound-for-pound boxers in the world, and in December 2004 they connected at the MGM Grand Garden Arena in Las Vegas for the first match in a long rivalry, with the referee eventually raising two paws in the air. No-one had expected this match to go all the way to the finish line, and no-one had predicted this result — especially when the PacMan came out swinging and floored the disrupted Dinamita three times in the first round. Normally, this would be considered a technical knockout (TKO), but the Dinamita kept coming back before the countdown was up and, despite a gushing nose, showed extreme resilience in a match that scored 115–110, 110–115 and 113–113.

Figure 8.1: Ingemar Johansson at Grossinger's, New York (1960)

Source: © Bettmann/CORBIS

Both sides felt like they should have been awarded the victory, but the aggregate view saw it differently. The rivalry between Pacquiao and Marquez has since continued over three more fan-friendly and memorable matches, and now spans nearly a decade of sparring. In our version of the PacMan and Dinamita bout, PacMan (or Digital Disruption) came out swinging early on and nearly TKOd Analogue Anachronism. But the latter fighter came back strongly, and evened the fight. And the further the fight went on, the more the players realised how equally matched they were, and a grudging respect formed. In the aggregate, the judges are forced to award a draw and raise two paws from widely different champions. Instead of a continuing rivalry, perhaps these players could learn to meet in the middle?

Ask yourself...

Use the following to think about your analogue roots:

- What can you learn from an organisational design perspective from Silicon Valley's success? What makes it unique? Which bits can be replicated to boost your digital innovation?

- How are you currently feeding the analogue souls and creativity of your staff? Look at the Silicon Valley ecosystem and its environment and ask how you can bring some of this flavour in-house.

- Are people attracted by the location-specific passion of your teams? How can you better communicate the vibe of your workplace and make it attractive for both clients and future employees?

- How are you nurturing analogue networks to add value to your team members and partners (like Stanford or Sydney University Business Schools)?

- What is the right positioning for your brand between the global and the local? How do you utilise the tension inherent in glocalization and between digitally distributed teams to boost innovation output?

Apply it...

Now use your thoughts to create strategies:

- Immerse yourself in a learning experience focused on innovation. Travel to an innovation hot spot. Explore customer needs from a new angle. Take your executive team on an out-of-the-box road trip to glue them together.

- Combine analogue and digital elements at your next conference. Run Open Space or World Cafes (creative facilitation methodologies), and combine them with digital capturing of ideas and collaboration with off-siders around the world. Track the output and reach of the analogue ideas generated in the digital world.

- Audit your global and local employer brand. Will you attract innovators or laggards (refer to chapter 7)? If your company was a city, would it be more like Silicon Valley or more like Florida? Create a 2020 vision of the analogue environment you want to create. Write down what the virtual environment would look like next to the analogue environment.

Endnotes

1 R. Florida and S. Johnson, 'Class-divided cities: San Francisco edition', The Atlantic Cities, www.theatlanticcities.com/neighborhoods/2013/04/class-divided-cities-san-francisco-edition/4832, last accessed 9 April 2013.

2 R. Florida and S. Johnson, 'Class-divided cities: San Francisco edition', The Atlantic Cities, www.theatlanticcities.com/neighborhoods/2013/04/class-divided-cities-san-francisco-edition/4832, last accessed 9 April 2013.

3 Unknown, 'Stanford facts: Administration', Stanford, facts.stanford.edu/administration/finances, last accessed 9 April 2013.

4 J. Seely Brown and D. Thomas, 'The gamer disposition', *Harvard Business Review*, blogs.hbr.org/cs/2008/02/the_gamer_disposition.html, last accessed 9 April 2013.

5 G. McDonough-Taub, '"All business is local" — new book offers survival tips for going and growing global businesses', CNBC, www.cnbc.com/id/46103942, last accessed 9 April 2013.

6 G. McDonough-Taub, '"All business is local" — new book offers survival tips for going and growing global businesses', CNBC, www.cnbc.com/id/46103942, last accessed 9 April 2013.

PART III

Hanging up the gloves and meeting in the middle

The tension along the X-axis of the Digilogue Strategy Map may have eased off in the epic draw between the two champions. On the left in the red corner we had Analogue Anachronism, and on the right in the blue corner we had Digital Disruption. They stretched out as two extremes on the horizontal X-axis. In this part, it's time to zoom out from this plane and introduce the further dimension of the Y-axis, which is temporal in nature. At the bottom of the Y-axis is the present, and at the top of the Y-axis is the future.

We will look at how firms can combine the best initiatives from their past—brand equity, customer touch points, story, communication tactics and events, which may still operate with some success in the present moment—with innovative ideas that can reinvent, rejuvenate, rebirth and reposition the business, in the future. The next two chapters look at the temporal plane of past, present and future, and examine how businesses can survive and thrive through time, without losing their identity.

9: Standing the test of time

Few businesses can stand the test of time — and it seems that time is speeding up. Even fast seems too slow in today's business environment. In this chapter, we will look at how three companies have faced the temporal flow, and how they have strategically decided on unique directions in their ambition to survive and thrive through time. We will investigate how Kodak, Georg Sörman and LEGO have reacted and proactively responded to unique changes in their environment, and look at how the business life cycle stages shed further light on their strategic moves — both the wise and unwise moves. We will explore the metaphorical application of a life cycle to the idea of business and disruption, and look at how digital disruption in particular has affected these three players in different ways. Doing so will highlight that responding successfully to digital disruption is possible — if it's done in a way that combines timeless wisdom with rejuvenation, reinvention and rebirth, and in a timely fashion.

The business life cycle

The only certain things in life are death and taxes. (*How inspiring*, you may be thinking. *I really want to get out there and enjoy the moment now, Anders.*) Unfortunately, there is some truth to this statement. Now that all your zest for life has been punched out of you, consider this. Businesses have a life cycle too — complete with birth, death and taxes (see figure 9.1, overleaf). And that life cycle seems to be shrinking in length as a result of digital disruption. By 2009, the *Fortune* 500 had to say goodbye to 238 peers over the 10 years since 1999, nearly a 50 per cent turnover. Only 71 companies from the original *Fortune* 500 list of 1955 are still on the list. Digital Darwinism is speeding up the rate of change, and turnover of these companies is picking up as a result. This means that leaders need to get on the front foot and consider how to constantly ensure their relevance through revival, reinvention and rebirth. Whether your business is a start-up, in the growth phase, in the maturity stage or staring out over the cliff of decline, change and reinvention will be key to your future strategy.

Figure 9.1: business life cycle

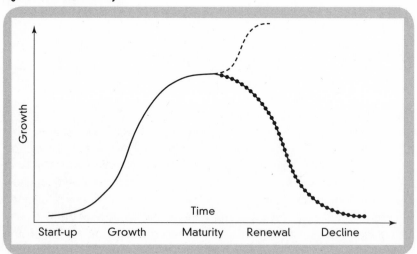

Let's explore how this business life cycle model is relevant to digital disruption and to this book's primary case study, Georg Sörman. The idea behind the business life cycle is metaphorical. It holds that a business, just like an organism, goes through different stages in life — birth, growth, maturity and death. There is a twist toward the end though — businesses can have the opportunity of renewal or rebirth.

You can see from the curve in figure 9.1 that growth is depicted along the Y-axis and that time is depicted along the X-axis. As a business moves through time, it follows a certain pattern. This is not applicable to all businesses, of course, because many collapse before even being born or die in a metaphorical early infant stage. (Apologies for the emotional metaphors here — I didn't make them up. The model, however, is useful.) Different challenges or areas of focus attach to the different gradients and stages of the curve, depending on the growth created and the time that has lapsed since inception (see figure 9.2). While at the start-up phase, businesses are largely focused on concept creation. Concept development and market development become the focus during the growth phase, while business optimisation is the key in the maturity phase. At the critical juncture between renewal and decline, a leader must decide whether to reinvent the business in the hope of a future upswing or to harvest during the downswing, leading to the eventual death of the company. Take a moment and plot your own organisation along this curve. Where do you sit, and does the model apply to your company's market strategies?

Figure 9.2: business life cycle — focus areas

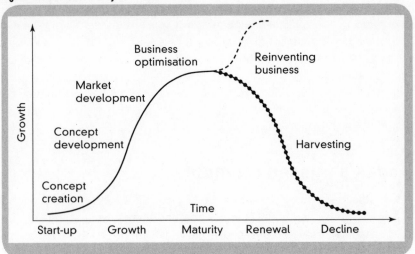

Throughout this book, we have seen countless examples of companies that have failed at the critical juncture between renewal and decline. Think of Kodak. Borders. Blockbuster's US business. Numerous record companies. Multiple newspapers. Several publishing companies. As discussed, disruption is nothing new and, in the past, there were massive disruptions. However, digital disruption makes the impetus for revival, reinvention and rebirth even stronger. Imagine that you're a mature technology company. You have gone through the start-up phase, you survived the growing pains of the growth phases and you went through your IPO at the end of the 1990s. Your strategy for the past 10 years has been driven by mergers and acquisitions, as you've focused on gaining an extra few per cent market share in your mature market. A lot of the entrepreneurial spirit has died through the increasing corporatisation of your firm. You're missing the good old days when things were exciting. Now, your M&A strategy of incorporating start-ups and funky geeks with cool new technology into your suited-up culture is one of applying engineering thinking to a structural problem, instead of applying cultural thinking to an anthropological issue. You argue about assimilation versus integration.

All of a sudden, a management consultant whips out the business life cycle at a strategic review and asks 'How innovative are you?' You cannot answer the question, but you remember a time line of the company's history and point out that you were conceived in a garage in the Bay Area. So you must be innovative, right? At least once upon a time. Now, you've probably grown a bit wiser, and you're not as unruly as you once were. Yes, you and your colleagues miss the crazy nights of cracking out code and flying by the seat of your pants, but you all have mortgages now. The management consultant points out that the solutions and software that your corporate behemoth charges for are now being provided by nimble start-ups — who offer it for free. Open sourced, digital players, who have innovative business models, and no public shareholders to report to. The management consultant tells you that you need to think about reinventing yourselves.

Kodak's missed moment

A business reinventing itself is a tough task, right? Yet, this is the kind of situation our clients at Thinque often find themselves in. Indeed, the preceding scenario is a real scenario playing out in the strategy alignment of a company we have been working with. Digital disruption at the top of the curve. Just when you're the most comfortable and your focus has narrowed to a few core, stable, known competitors, each of you carving out a nice percentage in the pie chart of the market, a disruptor enters and disturbs your mature slumber. All of a sudden, there is impetus for change. But is it too late? In the case of Kodak it was.

In 2001, Kodak was ranked 27th on Interbrand's Top 100 Most Valuable Brands in the World survey. But by then the company had hit the peak and started the slide of the downwards curve. It tried to reinvent. It launched digital solutions. It even launched a massive campaign in 2006 called

'Winds of Change'. But, ironically, it failed to reinvent the company, and the video is now a somewhat cynical YouTube totem to a failed reinvention. It's worthwhile checking it out.[1] As a warning sign.

Kodak's big mistake was its tunnel vision at the apex of the business life cycle, where it assumed that people would take photos that they would later develop and print into analogue memories. This led them to a failure to commercialise the opportunity of digital. Check out this time line of Kodak's demise, and check if you can track its decline with your index finger (also shown in figure 9.3, overleaf) from the apex to the eventual bankruptcy:

- 1975 — invents the digital camera
- 1985 — based on its analysis of Kodak, the *Wall Street Journal* argues that it was 'Hard to find anything with margins like colour photography which is legal'[2]
- 1997 — CEO Fisher says 'Electronic imaging will not cannibalize film!'
- 1999 — cuts 19 900 jobs
- 1999 — has 27 per cent market share in digital, but is losing up to $60 on each sale
- 2004 — stops selling film cameras in Europe; 15 000 job losses
- 2006 — sells digital camera interests to FlexTronics
- 2007 — sells Light Management Film Business
- 2009 — records a fourth-quarter loss of $137 million
- 2010 — is removed from the S&P 500 index
- 2011 — stock drops to all-time low of $0.54, and drops 80 per cent over year
- 2012 — receives warning from NY Stock Exchange notifying it that its average closing price was below $1.00 for 30 consecutive days, and that it's running the risk of being delisted
- 2012 — files for Chapter 11 bankruptcy (with stock at $0.36).

Figure 9.3: Kodak and the business life cycle

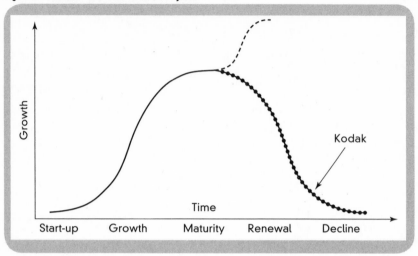

Georg Sörman

Honestly, I get disturbed by the business life cycle model. Why? Because I believe it has merit. And that through its impact, curvature and warning signals we can derive lessons. Its shape is illustrative. And I believe my mum's company is precipitously close to the edge. It's a mature company that was spoilt for many years in a mature market. The start-up went well from 1917 when, under the tutelage of Georg Sörman, it started off with concept creation. The company grew and was able to add the concept of tailored men's fashion, and it developed a market in Stockholm and beyond. Per Sörman, my grandfather, was able to optimise the company and, while some of his management style and processes were old school, they worked up until the turn of the millennium. But times had sped up, and my mum took over a company staring out over the cliff of renewal versus decline — perhaps, like Kodak, digitally disrupted in a market that was beyond recognition. Retail had changed. Consumers had changed. And all of a sudden my mum was told she had to change — and to change everything. To reconsider what business she was in. To go digital or die. Even though I wish it didn't, the business life cycle, for a business at age 96, makes sense. It would be an affront to call the company archaic, geriatric or blind. Yet, some analysts have. And it's a stark relief. It's a mirror on the reality facing my mother. Reinvent or die. Face digital disruption. But do not pretend it's not there. In this instance, she cannot have her change back. This is the crucial juncture in her company's existence (see figure 9.4).

Figure 9.4: Georg Sörman on the business life cycle

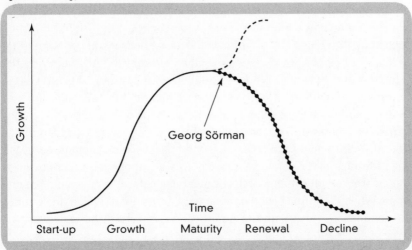

It's been said that the definition of insanity is doing the same thing over and over and each time expecting a different result. This may be a bit trite, but when it comes to mature businesses sitting at the vertex of the life cycle, reinvention, innovation and renewal are key. And it's the hardest thing to do. You're being asked to question your identity, and made to feel like your identity is worth less in a context of sexier alternatives. You're being told to change your diet, to undergo an extreme makeover, to hit the gym and to buy some cellulite cream. Your self-esteem suffers. You feel displaced. You might even become depressed. This goes to the core of who you are. And it goes to the core of a business's identity, too. Culture, history, processes, thought processes, behaviours, locations, business models, colleagues — all potentially being thrown out. This is existential stuff. It's a matter of survival, of life and death. And it requires serious consideration. Do you stay in or do you get out? If you stay in, how are you going to 'turn up'? Are you going to do something differently, or simply wait for your retirement cheque before the company goes belly-up? Or harvest what can be saved by selling off the company's assets and opting out. Throw in the towel. My mum owns the company, but she doesn't have the retirement option — six years before Swedish retirement age, she has to make this work. This wave-like curve needs to turn into an S-curve for her. Revival and renewal is her only option. My mum's a fighter. The towel is not an option. At age 59 she is picking a battle with digital disruption, and she knows that more of the same old same old will mean bankruptcy. The chips are on the table. And I reckon she's in with a chance — on certain terms.

Those terms are reinvention, rejuvenation, rebirth. At the risk of engaging in metaphorical overkill, Georg Sörman needs to arise out of the ashes like the mythological Phoenix. You remember the story, right? In Greek mythology the Phoenix is a bird that lives for hundreds of years and that cyclically practises its own regeneration or rebirth. According to the myth, the bird goes into the forest, builds a funeral pyre, burns itself in ceremony to rise from its ashes, young again, and live through another cycle of years. The Phoenix is associated with the sun, and this ceremony symbolises renewal. This concept of rebirth is timely given that zenith has passed. Yes, I know I am stretching the logical underpinnings here, but I like the idea that businesses, my mum's or otherwise, can rejuvenate and be reborn. It's happened successfully before, and it can happen again. What's required? Hindsight, insight, foresight. Passion and energy. A willingness to let go and to adapt. Curiosity, brutal honesty with the state of affairs and a clear definition of what you want to create. A 20/20 vision that stretches beyond the horizon. Sweat, and equity. Rejuvenation, reinvention, rebirth — as anyone who has set out on a successful personal fitness strategy will know — requires discipline and commitment. A willingness to change habits.

Waves of change

Waves of change will keep rolling in, as you'll remember from chapter 1. And the key for business leaders, like my mum, is figuring out how to re-position your businesses in a way that takes advantage of these trends. Remember, we need to:

- Spot what's going on around us. Identify the trends and where you're at in the life cycle. Take a brutally honest view of your internal and external environment.

- Get a feel for the underlying currents of change. Ask yourself what they mean, where they stem from and how the change in customer behaviour affects your business.

- Position (or re-position) your business to ride the waves of change. Work out which of your core capabilities are in tune with the wave of change, and figure out how to communicate those in an engaging way to the market. If you need to repackage aspects of your story or evolve your business model, this repackaging needs to form part of your re-positioning strategy.

In the following sections, I take you through ways Georg Sörman can best navigate the crest of the business life cycle, and ensure the sun doesn't set on the business.

Spotting major trends

The major trends relating to Georg Sörman are the following:

- *Online retail* — continuing to grow. For example, in 2012 in the United States (digital) Cyber Monday online sales reached $1.47 billion, up 17 per cent from 2011, representing the heaviest online spending day in history, according to digital tracking firm comScore.[3] This firm also reported that (analogue) Black Friday (the day following Thanksgiving) sales reached $1.04 billion, besting 2011's amount by 28 per cent. Online spending for the month of November 2012 reached $20.4 billion, a 15 per cent increase over the corresponding period in 2011, according to comScore research.

- *Mobile retail* — growing at exponential rates. Mobile sales overall (during Cyber Monday), which include smart phones and tablets, had a huge boost — up more than 96 per cent from 2011.[4]

- *Showrooming* — the customer behaviour of doing research online, visiting a store for fit, and then ordering online.

- *Factory outlets* — low-cost suburban shopping destinations selling directly from the manufacturer to the consumer, without a retailer facilitating or curating the transaction.

- *Digital era* — we are living in a digital and price-sensitive era where shopping habits are increasingly digitised, and goods and services are becoming commoditised as a result.

- *Retro and luxury revival* — on the flip side, segments of the market are willing to pay extra for great service and a high-touch, analogue experience. Worldwide spending in luxury products rose by 13 per cent in 2010 and 10 per cent in 2011, exceeding the previous results recorded before financial markets collapsed.[5] Importantly, some consumers derive status from being seen to be shopping in certain venues.

Getting a feel for the underlying change

The underlying change in retail includes the following:

- *Information asymmetry is no longer* — retailers used to be able to get away with charging an extra percentage point or two to boost their margins. Now the web has empowered customers to make informed decisions based on real-time data. This means that, increasingly, customers are opting for the most cost-efficient way (at face value) of receiving the goods.

- *Thrifty cool* — being thrifty is seen as smart since the global financial meltdown. And justly so. Spending lots of time aimlessly walking around the streets comparing stores is less efficient than a comparison website.

- *Offline versus online* — while more and more customers are becoming channel agnostic, they are also seemingly comfortable with the idea of receiving a service for free, only to then transact somewhere else.

- *Economic reality* — since the global financial meltdown, we hold on to our disposable income. When we make a decision to spend instead of to save, we want to know that we are getting value for our money.

- *Timeless sustainability* — luxury brands would argue that the ultimate in sustainability is to buy quality rather than fast fashion. Certain consumer segments agree, and buying classic fashion with a timeless quality has a certain long-term sustainable allure to it.

Riding the waves of change

The factors and strategies Georg Sörman could use to position (or re-position) itself to ride the waves of change include the following:

- *Timeless fashion* — Georg Sörman sells timeless, high-quality pieces for which there is stable demand. It's not a fast fashion retailer, and it doesn't compete on price.

- *Analogue core customer* — the store's core clientele are an older demographic, which is not as prone to digital retailing. Engaging this demographic in new ways, while also communicating with a younger demographic that enjoys analogue experiences will be key moving forward.

- *Digital storytelling of analogue brand* — in re-positioning and rejuvenating the brand, telling the analogue story of the family brand via digital media will also be key.

- *Creating a great in-store experience* — given the necessity for reinventing and rebirthing the brand based on its stage in the business life cycle, an increased focus on creating unparalleled in-store experiences will be key. This can drive foot traffic and amplify the analogue story via digital channels.

- *Amplifying digital and analogue word-of-mouth* — curating an analogue experience and making it easy to share that story via analogue word-of-mouth and digital word-of-mouse can become a way to boost referrals and access new psychographics.

- *Moving upmarket* — creating in-store experiences and boosting customer service to even higher levels means that a certain clientele with a higher disposable income may demand even more upscale brands. While creating even more value for clients, the store also needs to ensure that it captures that value.

- *Online–offline integration* — an opportunity exists to boost digital retail, and use digital e-commerce to drive in-store visits by giving clients the option of picking ready packages up in store, and amplifying cross- and up-selling.

Georg Sörman, analogue business models, and bricks and mortar businesses more generally can be turned around, and they can be saved from demise. But the only way forward is reinvention, renewal and rebirth. In chapter 10, you'll see a more detailed Digilogue Strategy Map for how we are going about reinventing the business Georg Sörman in the face of digital disruption.

Toy Story reinvented: LEGO

You probably remember the famous plot. It's one we can all relate to (if we still remember childhood). In Disney/Pixar's animated movie *Toy Story*, the toys pretend to be inanimate when humans are around but come alive when we look away. Woody is a pull-string cowboy toy, and the leader of all the toys belonging to a fellow called Andy. Andy's family is about to move house, so his mum hosts an early birthday party for Andy with his friends, and the toys, worried about the new, innovative, cool toys Andy is about to receive, stage a reconnaissance mission. To their shock and horror, they find a teched-up space ranger — Buzz Lightyear — whose impressive features intimidate the more old-school Woody. The inevitable happens, as it often does with kids, and Andy's affections turn away from Woody and are instead concentrated on Buzz Lightyear. Kids' tastes can change quickly, and this movie tunes into the emotional side of toys and the feelings of being discarded in favour of something new. Yes, the toys in the movie are anthropomorphised, but the movie became a hit also because it is easy to relate to this story as a human being. Anyone who's been in a romantic relationship that ended, or experienced a friendship where tastes changed with time, knows what I mean.

The movie reminds me of my own toy story, one centred on a temporal romance with LEGO. These toys are in serious business. And, although they are a bit of a timeless act, they have moved with the times in a timely fashion. I remember how distinctly LEGO featured in my upbringing in Sweden in the early 1980s. I also remember being somewhat of a pest each weekend. There was a toy store called 'Stor & Liten' (Big and Small) in Stockholm that my father dreaded going to because of the crowds, and the way they amped up the heating. LEGO was ingrained into my mind, and their brand had brandwashed me, to the great delight of my parents — particularly my dad who looked after my brother, Gustaf, and me on Saturdays because my retail mum was working at Georg Sörman. I must have been an extreme pest, or just delightfully charming, because my dad would indulge Gustaf and me by going to 'Stor & Liten' and occasionally, when we had saved up enough 'veckopeng' (weekly allowance), we could invest in LEGO.

One particular Christmas, just before the annual re-run on TV Channel 2 of the 1982 epic *Ivanhoe*, based on Sir Walter Scott's novel of the famous knight, I was given the LEGO Castle set. At the time, I was really into history (or rather knights, as my dad would point out). And I remember spending a whole winter's day with Dad in the kitchen, engineering this castle to perfection. I still remember the feeling and today, when I Google image searched the LEGO set, I had physical 'LEGO goosebumps' — childhood memories came flashing back.

To me LEGO was so cool. So much so, that I had dreams of going on a family holiday to LEGOLAND in Denmark, the provenance of the brand. As a child does, and slightly envious of my friend Fredrik who had travelled there with his parents and brother, I once wrote a fictitious tale at the age of six of going to this theme park. This was really embarrassing, because at kindergarten I had apparently claimed that it was a non-fictitious, biographical account of our family holiday. It became even more embarrassing when my dad presented me with a binder of my kindergarten musings in 2011. In his diplomatic way, he said something to the effect of 'Well, you've always had a rich imagination'. I blame LEGO.

My love affair with LEGO lasted throughout the 1980s, and formed a rich part of my childhood. These little bricks had something magical to them. It was like IKEA, but for kids (and Danish, rather than Swedish). My love affair with LEGO was eventually discontinued by my own version of Buzz Lightyear–technology. In 1988 I was introduced to the eight-bit Nintendo at my friend Tim's place. Digital video games started coming between me and my stationary, analogue connection with LEGO's bricks.

Timeliness and timelessness

But this is not a story about my fascination with LEGO, and later rejection of it in favour of Nintendo. It's a story about how a little brick managed to sustain relevance and emotional connection through time, despite the macro threats of digital disruption and the entry of nimble start-ups in the toy market. And what I love about LEGO is how it has combined timeless wisdom with timeliness in its reinvention moves.

LEGO is an old family company and is still privately held. While the bricks were introduced in 1958, the company dates back to 1934, when the founder, Ole Kirk Christiansen, decided to call his wooden toys company LEGO, after the Danish expression 'leg godt' (play well). The modern LEGO brick was patented on 28 January 1958, and bricks from that year are still today compatible with current bricks — something that speaks to the timeless quality of the bricks. Since inception, LEGO has had an obsession with quality, and its original mantra that 'det bedste er ikke for godt' ('the best is never too good') is still part of the company's quality framework. The brick has withstood the test of time as a result of this quality focus — but also because of the company's willingness to continuously innovate.

In 2012, the BBC asked Open University's engineering students the quality question 'How many LEGO bricks, stacked one on top of the other, would it take to destroy the bottom brick?' According to their calculations it would take a stack of 375 000 bricks to cause the bottom brick to collapse, which represents a stack 3591 metres (11 781 feet) in height. The brick system in and of itself adds further dimensions to the enduring quality and interest in the bricks, as six eight-stud bricks can be arranged in 915 103 765 different ways. This enduring quality, marinated in continuous innovation, is one of the reasons LEGO was chosen by *Fortune* magazine as the toy of the 20th century, ahead of Barbie and the teddy bear.[6] Not bad for a collection of bricks.

This toy brand is not only serious play but also serious business. Annual production of LEGO bricks now averages approximately 36 billion, or about 1140 elements per second. If all the LEGO bricks ever produced were to be divided equally among a world population of six billion, each person would have 62 LEGO bricks (as you can tell, production has picked up more recently). According to an article in *BusinessWeek* in 2006, LEGO could also be considered the world's number one tire manufacturer; the factory produces about 306 million tiny rubber tires a year.

However, at the beginning of the 2000s, the business was in serious trouble. Digitisation of play, changing tastes among kids, expired patents, piracy, increasing buyer consolidation and power, and a fat supply chain had led the company into the red. Annual sales collapsed by 25 per cent in value to 8.5 billion Danish kroner in 2003. Net losses doubled between 2003 and 2004. Between 1998 and 2004, four of the seven years resulted in losses. It laid off 12 per cent of its payroll in 2004. At this time, LEGO executives estimated that the company was destroying €250 000 in value — every day. LEGO was a strong brand going through a deep crisis. The family who had founded and run the company knew they had to change direction. As Kjeld Kirk Kristiansen, grandson of the firm's carpenter founder, said, 'When the map no longer corresponds to the field, one ought to change the map'.[7] Wise move, it would turn out.

LEGO brought in new thinking — reinvention, rejuvenation, rebirth — at the last hour. The company promoted Jorgen Vig Knudstorp to the position of CEO. His mantra — 'there are no sacred cows' — symbolises the coming reinvention of the company. Leadership zoomed out and canvassed the supply chain and the external environment, focused the company on their core capabilities, and re-positioned it to be on trend. Over the last couple of decades the company had diversified into theme parks (like the one I imagined I had been to), clothing lines and movies. Simplicity now replaced complexity, and its classic strength in old-school non-electronic toys became its lifeline again. Just 30 products generated 80 per cent of sales, while two-thirds of the company's 1500-plus stockkeeping units were items it no longer manufactured. Three quarters of the company's sales each year were from non-electronic games, and the company's brand strength among its key consumer group (1 to 11 year olds) was undisputed, and became a key building block in the company's rejuvenation efforts. (Excuse the pun.) Transparency and brutal honesty with reality became culturally engrained. Knudstorp admitted 'that the focus on electronic competition was really a blame game'.[8] Digital and analogue — digilogue — was the way forward.

So what did they do? LEGO's product line was reinvented to focus on its historic core, and on profitability and simplicity, which meant that the previously complex supply chain became extremely lean. Importantly, the products LEGO chose to focus on were both timeless and timely. Some LEGO lines were inspired (and on licence) by blockbuster movies like *Avatar*, *Harry Potter* and *Batman*, and complemented on occasion by branded digital video games as in the cases of LEGO *Star Wars* and LEGO *Pirates of the Caribbean* games. The company also won digital minds in the digital space by enabling LEGO Digital Design, a software program for elder LEGO fanatics who want to co-create their own models with LEGO, for customised assembly.

By 2005, stock turnover had increased by 12 per cent, and in 2005 LEGO had recorded its first profits — €61 million — since 2002.[9] LEGO kept on this path and since 2005 has tripled in size, while keeping profitable, and ensuring that 50 per cent of its revenues come from its classic lines. In 2011, it posted revenues of €2.510 billion, with an operating income of €759 million and net income of €558 million. Knudstorp said in a 2011 interview that 'the best way to protect ourselves against the crisis in Europe and North America is to continue to reinvent the product and the brand'.[10] He matches this timeliness with timeless wisdom by saying that

> It's very important that LEGO never changes its heritage. We are about putting bricks together, being able to play with them, and taking them apart again, and that is the same whether we are in London or Beijing. That will never change.

Timelessness and timeliness combined. And what has ensued is that instead of entering the harvesting period of decline, the brand has kicked into a new paradigm — rather than riding a wave-like curve, it is instead creating an S-curve (see figure 9.5).

Figure 9.5: LEGO's reinvention at the top of the crest

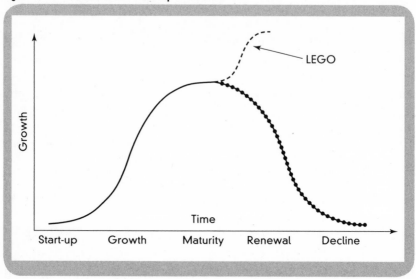

LEGO's reinvention strategy hit the nail on the head, through the following:

- LEGO spotted what was going on around them. They identified the trends affecting them and where they were at in the life cycle. Consequentially, they took a brutally honest view of their internal and external environment.

- They got a feel for the underlying currents of change. They asked themselves what these currents meant, where they stemmed from and how any (perceived) change in customer behaviour affected their business.

- They re-positioned their business to ride the waves of change. They worked out which of LEGO's core capabilities were in tune with the waves of change, and they figured out how to communicate those capabilities in an engaging way to the market.

This doesn't mean LEGO's leadership can rest on their laurels, but it does mean that they navigated an extremely sensitive time in the business life cycle. Their rejuvenation, and the results stemming from that renewal, are testament to their ability to withstand the test of time. Not by relying on what they have always done, but rather by focusing on both timeless wisdom and timely, strategic action.

Digital Darwinism is putting businesses out of business at an unprecedented pace. The velocity of the business life cycle is speeding up. Constant renewal, reinvention and rebirth is required to stay relevant, afloat and profitable. Brutal honesty with the state of affairs is required, and as business leaders we must constantly spot, feel and reposition to ensure our coordinates are in tune with the changing currents around us. According to a popular paraphrase of Charles Darwin

> It is not the most intellectual of the species that survives; it is not the strongest that survives; but the species that survives is the one that is able best to adapt and adjust to the changing environment in which it finds itself.[11]

Sage advice. Adaptation is the only way to withstand the test of time — reinvention, rejuvenation, rebirth. Some fail — like Kodak. Some are in the middle of the journey — like Georg Sörman. And some have managed to make a paradigm shift — like LEGO. Because of LEGO's timelessness and timeliness, I may one day, despite whatever digital disruption may be on the horizon, find myself sitting with my kids, building LEGO castles. Brick by brick. Timeless and timely is the answer to standing the test of time. This analysis should soothe your analogue hearts, and inspire your digital minds, confirming that ongoing connection to your clients' hearts and minds is possible. Through time, the territory does change, and when it does change, you need to change the map. Adapt or die.

Ask yourself...

Go through the following questions to start thinking about your business and its life cycle:

* Where in the life cycle are you? Can you track your older and younger competitors on the curve? Plot them out on the curve, and think about how this differentiates your challenges. What opportunities can you spot in this?

(continued)

Ask yourself . . . *(cont'd)*

- How do you combine the timeless wisdom inside your organisation and your core capabilities in a timely fashion to beat the competition?

- If you were brutally honest, what waves of change do you spot around you that will affect you, and maybe even speed up the business life cycle and shorten the amount of response time you have? List them.

- What underlying currents of change are driving internal and external threats, and how are digitised customer expectations disrupting your old game plan? Does your strategy map to reality? If not, change it.

- What is your vision for 2020, and does it align with where you are at and where you want to be on the business life cycle? What portions of the business model and your positioning do you need to tweak, throw out or totally reposition to stay relevant and profitable? Book in your executive committee now for a serious chat.

Apply it . . .

And here are some powerful life cycle strategies:

- At your next strategy review, use the business life cycle to communicate the necessity for real innovation, reinvention and rebirth. Get your executive committee and the most promising gen X/gen Y/gen Z talent (with communication abilities) to brainstorm how to position the company for 2020.

- Use the spot, feel and (re)position framework to do a thorough scan of the external environment. Digitise the framework and use it on your in-house social network to communicate to the whole organisation what changes are disrupting your business and how reliant on global forces you are for success. Identify which factors you can affect and which you cannot. Empower your team to launch innovations that are in tune with larger macro trends.

- Hire Thinque to facilitate your next 2020 Scenario Plan. We'd love to help.

Endnotes

1 Unknown, 'Winds of change: Kodak is back!', Kodak, www.youtube.com/watch?v=GtYXGY4wB-0, last accessed 4 April 2013.

2 Unknown, 'What's Wrong with This Picture: Kodak's 30-year Slide into Bankruptcy', Knowledge@Wharton knowledge.wharton.upenn. edu/article.cfm?articleid=2935, last accessed 11 April 2013.

3 M. Kaplan, 'Analyzing Black Friday, Cyber Monday, and overall retail sales', *Practical eCommerce*, www.practicalecommerce.com/ articles/3832-Analyzing-Black-Friday-Cyber-Monday-and-Overall-Retail-Sales, last accessed 4 April 2013.

4 J. O'Donnell and H. Malcolm, 'Cyber Monday sales finish strong', *USA Today*, www.usatoday.com/story/money/personalfinance/2012/11/26/ cyber-monday-fast-bumpy-start/1727005, last accessed 4 April 2013.

5 N. Anzivino and M. Lazzaro, 'Market vision luxury', PWC, www.pwc.com/it/it/publications/assets/docs/marketvision-luxury-2012.pdf, last accessed 4 April 2013.

6 J. Pisani, 'The making of ... a LEGO', *BloombergBusinessweek*, www.businessweek.com/stories/2006-11-29/the-making-of-a-legobusinessweek-business-news-stock-market-and-financial-advice, last accessed 4 April 2013.

7 B. Lévy, 'Following record losses, Lego is going to lay off close to one thousand employees', *Le Monde*, reproduced on lapasserelle, www.lapasserelle.com/courses/corporate_finance/Lesson5/ miscellaneous/lego_eng.htm, last accessed 4 April 2013.

8 K. Oliver, E. Samakh, and P. Heckmann, 'Rebuilding Lego, brick by brick', Booz & Company, www.booz.com/media/uploads/ RebuildingLego.pdf, last accessed 4 April 2013.

9 K. Oliver, E. Samakh, and P. Heckmann, 'Rebuilding Lego, brick by brick', Booz & Company, www.booz.com/media/uploads/ RebuildingLego.pdf, last accessed 4 April 2013.

10 J. Vig Knudstorp, CEO, 'LEGO Group announces annual results 2011', LEGO Company, www.dailymotion.com/video/xp5csp_lego-group-announces-annual-results-2011_news#.UV1suBz-HUk, last accessed 4 April 2013.

11 L. Megginson, 'Lessons from Europe for American business', *Southwestern Social Science Quarterly* (1963) 44(1): 3–13, p. 4.

Taking stock

The Digilogue Strategy Map and a printing and digital solutions case study

In 2010, I was the keynote futurist during a series of events for a major player in the printing and digital solutions industry (incidentally, a competitor of Kodak's in some markets). The printing industry has seen major upheaval as a result of digital disruption, and forward-thinking printing businesses are adopting digital print technologies instead of analogue, offset print techniques in their work. The industry, which for many years was really in the manufacturing (of print) industry, was facing a paradigm shift into digital print and services, with many printers heavily invested in old equipment and processes, and resisting the coming digital revolution in print. The client's brief was to pitch the merit of digital print technology and look at how the digital and the analogue could merge in their business models and services. The events were held in diverse cultural and economic markets like Taiwan, Beijing, Kuala Lumpur, Sydney and Auckland, and the series was a thought-leadership initiative and a soft branding B2B exercise. Our clients' clients and sales executives had a chance to upskill and update their insights on the latest trends, disruptions and technologies affecting the industry. From the event series, our client made $12.3 million in sales directly attributable to the event, and achieved a return on investment above 500 per cent. Take a quick look at the following Digilogue Strategy Map, which shows how the client applied digilogue strategies to make this the most successful thought-leadership event in its history. The discussion following the map expands on the strategies.

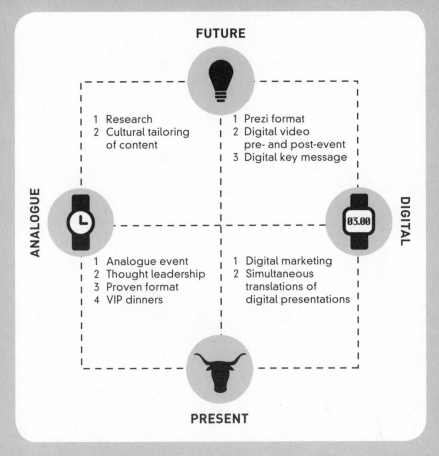

FUTURE

ANALOGUE

1 Research
2 Cultural tailoring
  of content

1 Prezi format
2 Digital video
  pre- and post-event
3 Digital key message

DIGITAL

1 Analogue event
2 Thought leadership
3 Proven format
4 VIP dinners

1 Digital marketing
2 Simultaneous
  translations of
  digital presentations

PRESENT

Analogue — present

The client had found a workable and profitable model in the thought-leadership conference series in Asia–Pacific. It was keen to ensure the continued success of a highly analogue, face-to-face event that would ensure it got its thought-leadership message out to key markets in the fast-growth Asia–Pacific marketplace. The event also enabled it to stage an invite-only VIP event for its top clients, and give its sales executives direct access to analogue time with those accounts. This face-to-face time with VIP client is invaluable, and equals hundreds of hours in phone calls, travel times and meetings.

Digital — present

Given that our client was selling digital and analogue (offset) print equipment, it had been surprisingly inactive in the digital communications space. Its digital efforts had been restricted to electronic direct marketing efforts, and the simultaneous translation via projection screens of the keynote speakers' content into the key languages represented in the various locations.

Analogue — future

One of the key new initiatives that we agreed on with the client was to amplify the research that the speakers (including my own team) were required to do in preparation for each of the events. This was deemed critical to ensure relevance and cultural application in each location. Thus, we visited with the client's clients and headquarters in Sydney, Singapore, Adelaide and Stockholm, and Thinque's researchers developed location-specific materials that would resonate with the audiences.

Digital — future

Our client was keen to ensure that the digital communications in this event series would align with their key message that digital technology is the way forward. While this client is an innovator in both digital and analogue (offset) printers, the growth in the industry is from digital print, which necessitates using digital printers. So, together we co-created digital video messages, both pre- and post-event, to engage our client's key clients. This was integrated with the client's web channels, and showcased that the client is serious about all forms of digital communications, including digital video. We also wanted to make the thought-leadership content even more engaging for the non-native English speakers, some of whom were solely reliant on simultaneous translations to access the nuggets of wisdom shared at the event. Consequently, we used the Prezi visual presentation format, and created five separate presentations to ensure maximum understanding and value creation — which was in line with our client's wish to align itself with the message that 'digital enables tailoring'.

Summary

This thought-leadership event was a success. Across the touch points in Taiwan, Beijing, Kuala Lumpur, Sydney and Auckland, the feedback was phenomenal. Most importantly this digilogue strategy translated into directly attributable dollars for our clients, a continuation of the program, and a sustained collaboration between Thinque and this *Fortune* 500 company in their strategy and digilogue marketing work.

10: Going digilogue

We sometimes reminisce about the good old days. Sometimes we pretend that things will go back to normal. Occasionally, our nostalgia blurs how hard things used to be. We remember margins, resources and the seasons in which we reaped the rewards. But when I look at the history of my mum's, Per's and Georg's business, it has always been tough. And adaptation, innovation and creativity have always been the answers to external change. The store was started during the First World War, survived the Great Depression, the Second World War, two inter-generational buy-outs, and new market entrants. It's been standing the test of time until now. Where to from here?

This chapter looks at how we are applying digilogue at Georg Sörman to help it reinvent, rejuvenate and rebirth, and prove that bricks and mortar retail can survive in a world of digital disruption. And in the process, we're hoping to ensure that another generation of Swedish men and foreign clients can continue to transition from boyhood to adulthood, to manhood and, ultimately, to gentlemanhood — in style.

Strategising for future success

I speak digital. My mum speaks analogue. During that fateful conversation in Stockholm in May 2012, we couldn't seem to understand each other. The last time we had spoken about the business, the conversation had ended with me saying, 'So your clients only know how *not* to buy from you online'. Hardly a constructive conversation. But with time, perhaps, we get wiser. What was the translational sweet spot between the high-touch, which she represents, and the high-tech, which I represent? Well, I think you know the answer by now. Digilogue (for a refresh, see figure 10.1).

Figure 10.1: digilogue convergence

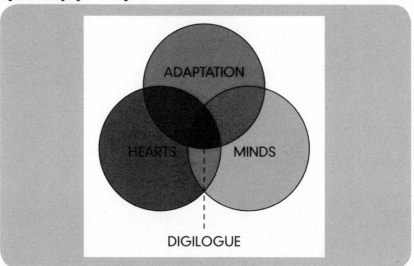

Like many of my clients, my mum likes the idea of staying true to history (like the icon shown in figure 10.2), respecting the past and making historical brand equity connect with hearts and minds in today's quickly changing business landscape. Yet, like so many traditional business folks, she hates the idea that she is singing in an analogue shower, with an analogue echo. In a way that is failing to connect her brand more broadly with customers who should be interested in what she has to say. This is her, and many of our clients', pain point.

Figure 10.2: the Georg Sörman icon today

Digital has the capacity to amplify a brand, and to make the telling of an analogue story easier. So couldn't we use the digital and the analogue together? Well, why the heck not. We decided to give it a crack. As with all things new, we needed to pay credence to things old. We wanted to retain the legacy and history, while making sure the legacy and history didn't end today. And while you think about this and the other case studies through the book, bear in mind how the Digilogue Strategy Map can translate into your own business. You'll be pleased to know that the solutions that I share with you err on the simple side. I hope that's okay — at Thinque, we think there is something beautiful about simple.

First, a quick preamble of the operating principles we needed to establish as a family before beginning a project like this. As you know, all families are psychotic, and this was a way of preventatively dealing with emotional angst:

- First, Birgitta (my mum) and Lars-Olof (my dad) established an operating framework. Everything we did had to have analogue and digital elements. This meant that we would all need to upskill in each other's native tongues. I would need to learn more about the analogue story, and they would need to learn more about digital amplification.

- We decided to get close to the customer. We asked ourselves the two essential questions. How do we better provide value to our customers' digital minds? And how do we better connect with our customers' analogue hearts?

- We agreed that history and future would need to co-exist. Thus, we started mapping the communication touch points, and areas of the business that really worked and we wanted to keep. Remember, don't throw the analogue baby out with the bathwater. But, equally, we had to ask ourselves the tough question of what new communication touch points and digitisation areas we'd need to introduce.

- We needed to get a better understanding not only of the story, but also of the numbers that made up the story of the store's history. What became clear through a financial analysis of the business was that there was latent potential for improvements and cost efficiencies through digitisation.

- We asked ourselves the questions: what has been the core capacity of the store Georg Sörman, including its internals (see figure 10.3)? And what brand associations do the logo and name bring alive for our customers, including any past advertising associations (see figure 10.4, overleaf)?

- Equally, what did we want the brand to be when the store turns 100 years in 2017? This enabled us to think more long term, and to start crafting small tactical actions that would feed into the future digilogue strategy.

- Finally, we asked ourselves the futuristic question: which touch points in our business environment can be digitised, and which ones can never be? Bricks and mortar is still key to our strategy, but knowing that the customer increasingly does their digital due diligence, we wanted to curate an analogue experience that couldn't be matched in the digital world.

Figure 10.3: the Georg Sörman store internals

Figure 10.4: Georg Sörman's *Mad Men* history

As I have through this book, we used the Digilogue Strategy Map to think about digital disruption, and our future business model. The following sections cover what we kicked off with at Georg Sörman.

Analogue—present

How do the dimensions and dynamics of the different quadrants play out in the case of Georg Sörman? We begin with the bottom left quadrant, focusing on present—analogue. The question is what analogue present activities or brand features are we executing that are reaping results, and that we should continue or amplify? In other words, this question comes under the mantra of don't throw the baby out with the bathwater. For Georg Sörman, we listed six activities and brand features that are key (as shown in figure 10.5).

Figure 10.5: present—analogue at Georg Sörman

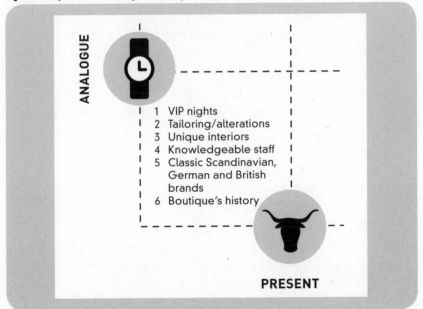

The six activities and features are as follows:

- The store organises a biannual VIP event in store. It's an analogue event aimed at loyal customers, and key clients are connected with sales professionals and representatives from the brands that the store stocks. It's invite-only and the event has an air of exclusivity about it. Clients of Georg Sörman appreciate the opportunity to be in store and spend quality time with the sales professionals, and have a chance to meet the faces behind the classic Scandinavian, British and German brands that the store stocks. From the store's perspective, these are two of the highest grossing days of the year, and they are profitable. In other words, a win–win for the store and for the clients.

- The store employs two tailors, with decades of experience between them and skill in ensuring that your clothes fit you the way they are meant to. While this service exists, it's not something that is widely communicated outwards. We identified this as an attribute we needed to continue, but also needed to amplify.

- The store is an antique lover's dream. The interiors are originally from the fashionable Stockholm haute couture store Nordiska Kompaniet's (NK) 1930s outfit, and they have been complemented with unique antique design pieces over the decades with a focus on classic, timeless items. While a store visit is a journey through time, we identified that the story about the store and its interiors, and the male Swedish legends who had come through its doors, like world-champion boxers and ice hockey players, could be told and amplified even more.

- The store has a good mix of older and younger staff, all of whom are dedicated to great service. The combined men's wear retail experience of the 11 staff at Georg Sörman is 183 years.

- Georg Sörman stocks classic Scandinavian, British and German brands. The brands are aligned to the interior feel of the store in the sense that there is a focus on timeless fashion and classic pieces like business shirts, suits, pocket squares and gentlemen accessories.

- The store's 96 years of history and cross-generational family background are a major strength, as are its logo and brand, which were immortalised in its 10-year feature as the best neon sign in Stockholm, and subsequent shaping into the Oscar of logos in the Swedish capital (see chapter 1).

These six factors were major analogue strengths of the company's history and present activities, and we agreed these were aspects we wanted to continually highlight, and in some cases amplify and emphasise in the company's communications and client touch points.

Digital—present

Of course, speaking of the analogue past and present were nostalgic moments for my mum. This is also where the brand had a lot of strength. We thus had to turn our attention to the bottom right corner of the Digilogue Strategy Map: digital — present. The question here is what present digital activities or brand features are we executing that are reaping results, and that we should continue or amplify? In other words, which parts of our current digital strategy are working and could be amplified to get even better results? We identified four factors (see figure 10.6). Not surprisingly, Georg Sörman was not as strong in this space.

Figure 10.6: present — digital at Georg Sörman

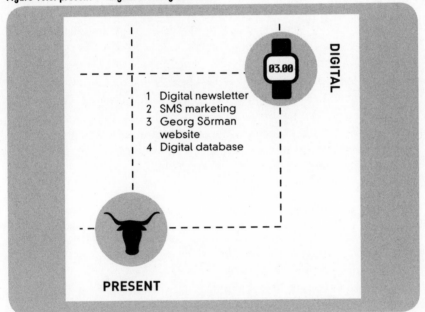

Here are the four digital factors we identified:

- For clients who had opted into its permission-based digital marketing, Georg Sörman sends weekly newsletters. In the past these had been promotion and price driven, and they were delivering some results for the firm.

- Georg Sörman also communicates via SMS to its clients, and once again these SMS had been driving some traffic to the store. This strategy was by no means a slam-dunk, but was seen as a part in a multichannel communication strategy moving forward.

- The store's website www.georgsorman.se had been redesigned and given a refresh in 2010, but when we started our digilogue strategy, the website was receiving only 800 unique visits per month. Nonetheless, the website would continue to exist, with the right tweaks and content updates.

- Georg Sörman has a rich databank of information about its loyal clients. Sizes, purchases, email and physical addresses. This customer relationship databank had been built over 95 years, and been digitised and updated from its old analogue A5 cards of the past. This digital information would need to be optimised and analysed, in order to segment and tailor client touch points in the future.

Now, you might be thinking that the analogue and digital strengths are mismatched and indeed you'd be correct. In many ways, the digital touch points were minimal and the present strategies are imbalanced in favour of the analogue. This needed to be rectified. The good news is that there was lot of potential in the future digital space that Georg Sörman could start developing. But before we turn our attention in that direction, I first visit what we decided to do in the top left corner: future — analogue. What analogue features should we introduce in the future to ensure positive word-of-mouth and impressive experiences for our clients?

Analogue—future

This future analogue quadrant is critical. We asked ourselves what touch points can be digitised and which ones can never be digitised. We agreed that there are pertinent aspects to the face-to-face interaction that can never be digitised, and that clients would travel physically to have an in-store experience with analogue wow-factor. But we were also frank with ourselves and realised that the St Eriksgatan venue could do with a refresh and some new concepts. So the future analogue strategy (shown in figure 10.7) would focus on creating a physical in-store experience with pizzazz, and on creating positive word-of-mouth and word-of-mouse reviews to amplify the brand story.

Figure 10.7: future — analogue at Georg Sörman

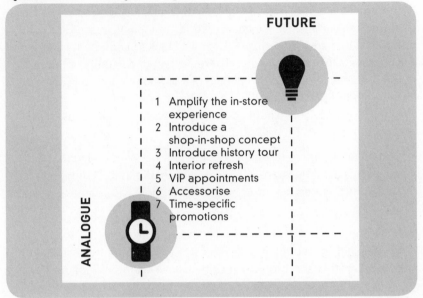

The strategies were as follows:

- The overall objective became to amplify the in-store experience. We needed to communicate the elements of professionalism and personalised service. For example, the tailors were hidden away in the back right corner of the store, so many clients didn't know about this high-touch, analogue service. Instead, the tailors and my mum — as the brand ambassador and third-generation owner — needed to be front and centre, so that clientele got a sense for the human beings who were reliant on their custom, and who could add value to their classic fashion choices.

- By looking at the demographic segmentation of the store, it became clear that the mean client was a 52-year-old man. Incidentally, the foot traffic demographic in the neighbourhood was a 38-year-old professional. Georg Sörman wanted to retain the custom of its core clientele, but complement it with a younger demographic. Thus, we decided that we would create three stores-in-store, each catering to the unique needs of three psychographic segments — the young professional, the man and the gentleman. This way, our future adaptation strategy didn't need to exclude our core clientele, but we would also be able to service the needs of a younger psychographic, one that incidentally is interested in timeless fashion.

- Georg Sörman has enjoyed the custom of famous Swedish gentlemen through its lifetime. For example, legends like former world-champion boxer Ingo Johansson, ice hockey world champions Lasse Bjorn and Rolle Stoltz, and soccer great Lennart Nacka Skoglund had all been customers through the ages. There was an opportunity to tap into the timeless class and Swedish sense of fashion by showcasing their clothing choices and linking them closely with the Georg Sörman brand. This would connect with both younger and older Swedish men.

- We can all do with a facelift from time to time. We decided that certain areas of the store needed a refresh and that the lighting should be updated in line with the look and feel we wanted to create.

- One of the areas we were seeing being digitised was client service. More and more, prospects would come into the business, try on clothes, get measured for size (and some would shamelessly write down their measurements, which they have every right to), and then order the item on line. But you'd probably agree it wasn't time well spent by a tailor with over 30 years' experience. Thus, this analogue service should be valued according to the expertise it was imbued with, and we decided it was time to make the tailor's time accessible by appointment. That is, to get the full service of the tailor, customers needed to make an appointment.

- Men are efficient by nature. Particularly when they shop. They may want to be able to buy an eau de toilette, have a haircut, freshen up their razors or have their shoes polished while they are in store. Offering services like this by a shopping concierge, and being able to cross-sell the concept of a man's cave and Georg Sörman as a destination became key.

- There are certain times in the retail calendar that are slower than others. Overall, between 10 am and midday is generally slow, but the store still incurs overheads such as staff wages and electricity. To spread activity, we decided that an early-bird special would apply to incentivise Georg Sörman's customers to come in and do their shopping, and to be able to spend quality analogue time with them.

Analogue is not a thing of the past but, as you can see, there were opportunities to introduce some new analogue touch points and modify some older ones, too. We believed that it was essential to bring out the authentic, analogue story in every client interaction, and to respect the time the clients invested in coming into the analogue space. Georg Sörman realised that clients have a choice how and where to spend their time. Time is perhaps the most precious commodity we have, and bricks and mortar retailers need to respect it, by creating experiences that can never be digitised.

Digital—future

The question remained, though — how would we amplify the story of the analogue brand and drive new traffic digitally? Now we needed to focus on the top right corner: future — digital. As you'll recall, the present digital quadrant was imbalanced compared with present analogue, and showed areas for massive improvement. What new areas of activity would we need to introduce in digital channels in order to stop singing in the shower, and instead enable clients and new prospects to digitally window-shop, and consequently shop from the store both digitally and analogue? (Strategies shown in figure 10.8.)

Figure 10.8: future — digital at Georg Sörman

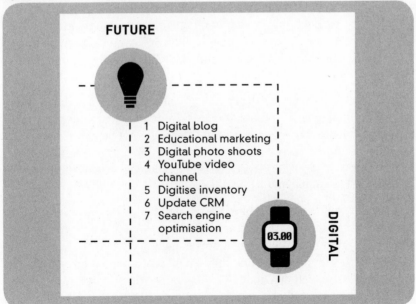

FUTURE

1 Digital blog
2 Educational marketing
3 Digital photo shoots
4 YouTube video
   channel
5 Digitise inventory
6 Update CRM
7 Search engine
   optimisation

DIGITAL

Here's where we decided to focus our future digital efforts:

- Blogging is one of the best ways of driving traffic and optimising search engine results. Georg Sörman didn't have a blog, so this became one of the first digital initiatives at www.georgsorman.tumblr. com. The blog would focus on education, fashion and timeless tips relevant to the core psychographics whose custom the store relied on, and the models on the blog are members of the fourth generation of Sörman (-Nilssons), my brother (with more talent as a model) and me.

- We decided that the electronic direct marketing newsletter needed to move beyond price and promotions, and instead should focus on digital education and adding value. At specific times, we could opt to alert the client base about special offers, but to boost the brand and add value to our clients' lives, the digital marketing needed to shift.

- The store is filled with antiques, interesting furniture and old-school gadgets like cash registers and typewriters. We decided that we needed to capture this analogue story, digitally, and use it on the website and in the electronic direct marketing newsletters.

- YouTube is the world's second largest search engine. Cisco predicts that 90 per cent of digital data in 2014 will be video. As such, video presents a great educational and marketing platform for retail. Video tips on such topics as grooming, shoe polishing and how to be a gentleman all became opportunities for capturing timeless wisdom, and sharing educational tips, while engaging a younger demographic.

- Inventory was managed by old-school, manual methods. This was inefficient and slow, and resulted in sub-optimal customer service. Digital inventory management and sales were not being practised, which resulted in overstocking of supplies, and less than ideal turnover of sales. Additionally, clients didn't have a choice to buy online. This needed to be addressed.

- Georg Sörman sat on 96 years' worth of data. Yes, some of it was a mess, but there was also some gold there. Purchase patterns, sizes, colour preferences and contact details galore existed, but hadn't been mined to the extent that they could. When combined with digital promotions, we all of a sudden had an opportunity to optimise both our analogue and digital communications with clients.

- The website was producing little traffic. Online, content is king, and there was little content on the page, which would be rectified by the presence of an updated blog. Equally, Georg Sörman had overspent on inefficient and hard-to-measure newspaper advertisements, and could shift part of its advertisement spend to result-driven Google Ads and social media engagement. With mobile e-commerce taking off, and an increasingly web-savvy audience, Georg Sörman needed to get onto the front foot and ensure digital visibility.

We had a lot of work ahead of us. The stakes were, and still are, high. We agreed on a staged plan from 2012 to 2017 (summarised in the complete Digilogue Strategy Map in figure 10.9), which we are executing. The initial results are promising, with margins healthier, increased web traffic and more efficient management of inventory.

Figure 10.9: the complete Digilogue Strategy Map for Georg Sörman

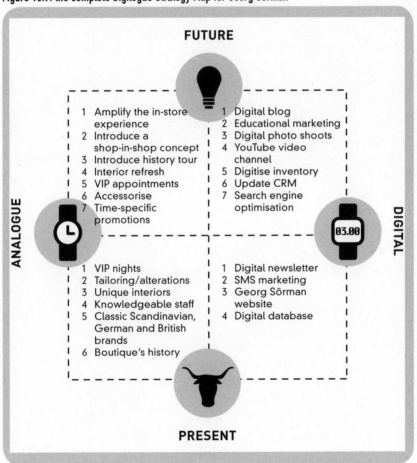

FUTURE

| ANALOGUE | | DIGITAL |

1. Amplify the in-store experience
2. Introduce a shop-in-shop concept
3. Introduce history tour
4. Interior refresh
5. VIP appointments
6. Accessorise
7. Time-specific promotions

1. Digital blog
2. Educational marketing
3. Digital photo shoots
4. YouTube video channel
5. Digitise inventory
6. Update CRM
7. Search engine optimisation

1. VIP nights
2. Tailoring/alterations
3. Unique interiors
4. Knowledgeable staff
5. Classic Scandinavian, German and British brands
6. Boutique's history

1. Digital newsletter
2. SMS marketing
3. Georg Sörman website
4. Digital database

PRESENT

One digilogue activity captures the spirit of the Digilogue Strategy Map particularly well. My father in many ways is the quintessential gentleman and nowadays sits squarely in the gentlemen psychographic of the store's client profile. He also happens to have some oratory skill, and featured in a digital video clip for the store. The scene is shot in store, and in it my father shows the audience in an educational manner how to do a four-in-hand (or universal) tie knot. The great thing about this education is that it's a value add and it's something every man needs to know. Even more impressively, the video featured in the digital edition of Sweden's main Swedish newspaper, *Svenska Dagbladet*. Within a day, the video had been tweeted and Facebook-liked across the digital world, and the analogue story of Georg Sörman had been digitally amplified. This in turn drove sales, and was an early reinforcement for my mum, Birgitta, that digital wasn't just a foe but was also an opportunity to engage in digilogue strategies that would ensure the continued survival and revival of her family's heritage and her legacy.

Digilogue strategies ensure reinvention. They ensure rejuvenation. And they support rebirth. Through time, and over time — over decades, and even over centuries. And our trust, belief and conviction is that digilogue will ensure the sustainable survival and thriving of Georg Sörman until and beyond its centenary birthday in 2017, just as we have shown that digilogue strategies are helping our heritage retail banker and two of our centenarian *Fortune* 500 clients adapt to a world where waves of change will keep smashing into our shores. It's time for you to go digilogue.

Ask yourself...

Here I've repeated the questions I asked at the end of chapter 4, to give you another chance to really get yourself thinking about your digilogue strategies:

- Which parts of your customer interactions can be digitised, and which ones can never be?

- What are you currently doing to provide value to digital minds, and to connect with analogue hearts?

- Which bits of timeless wisdom and analogue interactions are unique to your business, and ought to be continued?

- What digital initiatives are already reaping results, and could be amplified?

- What new analogue touch points and digital connections can you make to better position your business in the hearts and minds of your current clients, and tomorrow's prospects?

Taking stock

The Digilogue Strategy Map
and you (part two)

You also have a chance now to revisit your own Digilogue Strategy Map (if you need to, refer to chapter 4 to refresh your memory about what's required). Remember — you can download and print a strategy map from www.thinque.com.au/digiloguemap.

Again write down at least three things in each quadrant — either things you are doing at present, or things you will be introducing in the future to provide value to digital minds and connect with analogue hearts. This time around, try to come up with three new strategies across each of the future quadrants.

CONCLUSION

In 2006, my dad, LO, was invited to a centenary birthday celebration for a man named Sven — a rather famous man in Swedish commerce. This man made a big impression on my dad. Sven was born in 1906, 11 years before Georg Sörman opened its doors. My dad was invited because Sven was a reservist in the Swedish Army, and my dad and he had come into contact through the armed forces and my dad's career in the army. Sven had specifically asked for my dad's audience, which made Dad curious, since they didn't know each other that intimately. And when a centenarian asks for you to come to his birthday celebration, you don't say no. My dad represented his troops and, as always, turned up as the gentleman he is. He was struck by how lucid the 100-year-old Sven was, and how large his social network was. My dad enjoyed the occasion, and left somewhat inspired by the youthful energy of the 100 year old.

To his surprise, an email popped up a week after this celebration. It was Sven, inviting Dad for a private lunch to discuss the future of military technology. My dad did a double take and checked the sender email. He scratched his head. He paused and thought to himself. Here was a 100 year old who wanted to talk about the future of digital drones, cyber warfare and counter-terrorism. But my dad obliged, and travelled to Djursholm north of Stockholm where the senior gentleman resided. After exchanging pleasantries, the gentlemen sat down to discuss real business — the future of military technology. But just as Dad was about to engage in a conversation, Sven's mobile phone rang. 'Excuse me, LO', he said. 'It's my great-grandchild in Brazil. I'll just be a moment.' My dad was stunned. But, Sven was a gentleman and asked his great-grandchild whether he could return the call to Rio de Janeiro later. Curiously, my dad quizzed Sven about his use of this consumer technology. Sven responded, 'You don't get to live to a 100 if you don't use technology. You don't get to connect with your great-grandchildren if you don't use technology, if you ostracise yourself by disconnecting from the world'. Sven continued, 'You cannot be afraid of change, you have to be open to change. You have to take advantage of change, Lars-Olof!'

Sven's message is simple. And his timeless wisdom is extremely timely. His willingness to adapt, to rejuvenate and to reinvent is inspiring on an individual level, and is heartening on a commercial level. Because it gives hope to Georg Sörman, the enterprise, to live to a 100, and to experience its centenary birthday as a profitable and healthy business that combines the best of the old school and the new school, that synthesises high-touch and high-tech, and that blends the analogue with the digital. That is the way forward. That is the way to win the digital minds and analogue hearts of tomorrow's customer. Adaptation. Digilogue.

INDEX

Anders Sörman-Nilsson

Futurist. Strategist. Scenario Planner.

Anders Sörman-Nilsson is a futurist and innovation strategist who helps people and organisations decode trends and plan for what lies ahead.

THINQUE

Ask yourself...

- Is your organisation being digitally disrupted?

- Does your future strategy lack clarity and buy-in?

- Is your analogue business model ready for the digital future?

- Do you currently balance the timeless wisdom of your organisation with timely and future-compatible strategies and action?

- Are your teams inspired by your brand's 2020 Vision?

- Do your leaders and staff suffer from change fatigue?

Anders Sörman-Nilsson (LLB MBA) is the founder of Thinque — a strategy think tank that helps executives and leaders convert these disruptive questions into proactive, future strategies. As an Australian-Swedish futurist and innovation strategist he has helped executives and leaders on four continents map, prepare for, and strategise for foreseeable and unpredictable futures. Since founding Thinque in 2005, he has worked with and spoken to clients like Apple, Johnson & Johnson, Cisco, Eli Lilly, SAP, IBM, Xerox, ABN Amro Bank, Commonwealth Bank, McCann Erickson and BAE Systems, across diverse cultural and geographic contexts.

Anders is an active member of TEDGlobal (Oxford 2009/Edinburgh 2011), has keynoted at TEDx, guest lectured at the University of Sydney and University of Technology Sydney Business School, and Anders is a Million Dollar Round Table speaker. He has completed executive education at the Indian Institute of Management Bangalore, Stanford University Graduate School of Business, and London School of Economics.

Anders' two most popular keynote topics globally are 'Digilogue: how to win the digital minds and analogue hearts of tomorrow's customer', and 'Waves of Change: global trends that will disrupt your slumber'.

Anders is represented by Ode Management globally, and you can enquire about his availability to speak at your next conference or strategy review by contacting odemanagement.com today. He splits his time between Stockholm, New York and Sydney.

www.anderssorman-nilsson.com

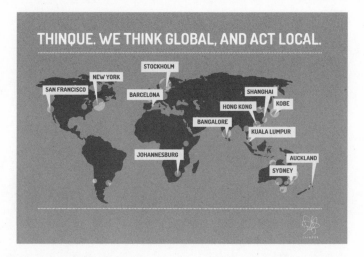

THINQUE. WE THINK GLOBAL, AND ACT LOCAL.

Thinque is a strategy think tank that helps executives globally plan for foreseeable and unpredictable futures. Since 2005 we have been scenario planning and executing Digilogue strategies across the globe with *Fortune* 500 companies.

We would love to talk to you about:

- Digilogue Communication Strategies
- 2020 Vision and Future Planning
- Disruptive Trend Research Reports
- Executing Your Future Strategy
- Providing You with Trusted Futurists-in-Residence
- Thought Leadership Strategies and Management Consulting

For more information on Thinque including client case studies and references please visit us on www.thinquetank.com.

THINQUE

Learn more with practical advice from our experts

microDomination
Trevor Young

First Be Nimble
Graham Winter

The One Thing
Creel Price

Dealing with the Tough Stuff
Darren Hill, Alison Hill and Dr Sean Richardson

The Art of Deliberate Success
David Keane

The New Rules of Management
Peter Cook

Outlaw
Trent Leyshan

Power Stories
Valerie Khoo

The Ultimate Book of Influence
Chris Helder

Available in print and e-book formats WILEY